Lecture Notes in Computer Science 4862

Commenced Publication in 1973
Founding and Former Series Editors:
Gerhard Goos, Juris Hartmanis, and Jan van Leeuwen

Norbert Fuhr   Jaap Kamps   Mounia Lalmas
Andrew Trotman (Eds.)

Focused Access
to XML Documents

6th International Workshop of the Initiative
for the Evaluation of XML Retrieval, INEX 2007
Dagstuhl Castle, Germany, December 17-19, 2007
Revised and Selected Papers

 Springer

Volume Editors

Norbert Fuhr
University of Duisburg-Essen, Department of Computational
and Cognitive Sciences, 47048 Duisburg, Germany
E-mail: norbert.fuhr@uni-due.de

Jaap Kamps
University of Amsterdam, Archives and Information Studies/Humanities
1012 Amsterdam, The Netherlands
E-mail: kamps@science.uva.nl

Mounia Lalmas
University of London, Department of Computer Science
Queen Mary, E1 4NS, London, United Kingdom
E-mail: mounia@dcs.qmul.ac.uk

Andrew Trotman
University of Otago, Department of Computer Science
9015 Dunedin, New Zealand
E-mail: andrew@cs.otago.ac.nz

Library of Congress Control Number: 2008934300

CR Subject Classification (1998): H.3, H.4, H.2

LNCS Sublibrary: SL 3 – Information Systems and Application,
incl. Internet/Web and HCI

ISSN   0302-9743
ISBN-10   3-540-85901-2 Springer Berlin Heidelberg New York
ISBN-13   978-3-540-85901-7 Springer Berlin Heidelberg New York

Springer is a part of Springer Science+Business Media

springer.com

© Springer-Verlag Berlin Heidelberg 2008
Printed in Germany

Typesetting: Camera-ready by author, data conversion by Scientific Publishing Services, Chennai, India
Printed on acid-free paper   SPIN: 12522002   06/3180   5 4 3 2 1 0

Preface

Welcome to the proceedings of the 6th workshop of the Initiative for the Evaluation of XML Retrieval (INEX)! Now in its sixth year, INEX has become an established evaluation forum for XML information retrieval (IR), with over 100 participating organizations worldwide. Its aim is to provide an infrastructure, in the form of a large XML test collection and appropriate scoring methods, for the evaluation of XML IR systems.

XML IR is playing an increasingly important role in many information access systems (e.g., digital libraries, web, intranet) where content is becoming more and more a mixture of text, multimedia, and metadata, formatted according to the adopted W3C standard for information repositories, the so-called eXtensible Markup Language (XML). The ultimate goal of such systems is to provide the right content to their end-users. However, while many of today's information access systems still treat documents as single large (text) blocks, XML offers the opportunity to exploit the internal structure of documents in order to allow for more precise access, thus providing more specific answers to user requests. Providing effective access to XML-based content is therefore a key issue for the success of these systems.

The aim of the INEX 2007 workshop was to bring together researchers in the field of XML IR who participated in the INEX 2007 campaign. During the past year participating organizations contributed to the building of a large-scale XML test collection by creating topics, performing retrieval runs and providing relevance assessments. The workshop brought together the results of this large-scale effort, summarized and addressed the issues encountered, and devised a work plan for the future evaluation of XML retrieval systems.

In total seven research tracks were included in INEX 2007. These studied different aspects of XML information access: ad-hoc, document mining, multimedia, heterogeneous, entity ranking, book search, and link-the-wiki. The consolidation of the existing tracks, and the expansion to new areas offered by the new tracks has enabled INEX to extend its scope. This volume contains 37 papers selected from 50 submitted ones (74% acceptance rate). Each paper was peer-reviewed.

INEX is funded by the DELOS Network of Excellence on Digital Libraries, to which we are very thankful. We thank Schloss Dagstuhl (Leibniz Center for Informatics) for housing the INEX workshop in its unique and inspiring atmosphere, and Springer for publishing the results of INEX 2007 in these final proceedings. We gratefully thank the organizers of the various tasks and tracks who did a superb job. Finally, special thanks go to the participating organizations and people for their contribution.

May 2008

Norbert Fuhr
Jaap Kamps
Mounia Lalmas
Andrew Trotman

Organization

Project Leaders

Norbert Fuhr	University of Duisburg-Essen, Germany
Mounia Lalmas	Queen Mary, University of London, UK
Andrew Trotman	University of Otago, New Zealand

Contact People

Saadia Malik	University of Duisburg-Essen, Germany
Zoltán Szlávik	Queen Mary, University of London, UK

Wikipedia Document Collection

Ludovic Denoyer	University Paris 6, France

Document Exploration

Ralf Schenkel	Max-Planck-Institut für Informatik, Germany
Martin Theobald	Stanford University, USA

Topic Format Specification

Birger Larsen	Royal School of Library and Information Science, Denmark
Andrew Trotman	University of Otago, New Zealand

Task Description

Jaap Kamps	University of Amsterdam, The Netherlands
Charles Clarke	University of Waterloo, Canada

Online Relevance Assessment Tool

Benjamin Piwowarski	Yahoo! Research Latin America, Chile

Effectiveness Measures

Gabriella Kazai	Microsoft Research Cambridge, UK
Benjamin Piwowarski	Yahoo! Research Latin America, Chile

Jaap Kamps	University of Amsterdam, The Netherlands
Jovan Pehcevski	INRIA Rocquencourt, France
Stephen Robertson	Microsoft Research Cambridge, UK

Document Mining Track

| Ludovic Denoyer | University Paris 6, France |
| Patrick Gallinari | University Paris 6, France |

Multimedia Track

| Thijs Westerveld | CWI, The Netherlands |
| Theodora Tsikrika | CWI, The Netherlands |

Entity Ranking Track

Arjen de Vries	CWI, The Netherlands
Nick Craswell	Microsoft Research Cambridge, UK
Mounia Lalmas	Queen Mary, University of London, UK
James A. Thom	RMIT University, Australia
Anne-Marie Vercoustre	INRIA Rocquencourt, France

Link the Wiki Track

| Shlomo Geva | Queensland University of Technology, Australia |
| Andrew Trotman | University of Otago, New Zealand |

Book Search

| Gabriella Kazai | Microsoft Research Cambridge, UK |
| Antoine Doucet | University of Caen, France |

Table of Contents

Ad Hoc Track

Book Search Track

XML-Mining Track

Entity Ranking Track

Interactive Track

Link-the-Wiki Track

Multimedia Track

Overview of the INEX 2007 Ad Hoc Track

Norbert Fuhr[1], Jaap Kamps[2], Mounia Lalmas[3], Saadia Malik[1],
and Andrew Trotman[4]

[1] University of Duisburg-Essen, Duisburg, Germany
{norbert.fuhr,saadia.malik}@uni-due.de
[2] University of Amsterdam, Amsterdam, The Netherlands
kamps@science.uva.nl
[3] Queen Mary, University of London, London, UK
lalmas@dcs.qmul.uk.ac
[4] University of Otago, Dunedin, New Zealand
andrew@cs.otago.ac.nz

Abstract. This paper gives an overview of the INEX 2007 Ad Hoc
Track. The main purpose of the Ad Hoc Track was to investigate the
value of the internal document structure (as provided by the XML markup) for retrieving relevant information. For this reason, the retrieval results were liberalized to arbitrary passages and measures were chosen to
fairly compare systems retrieving elements, ranges of elements, and arbitrary passages. The INEX 2007 Ad Hoc Track featured three tasks: For
the *Focused Task* a ranked-list of non-overlapping results (elements or
passages) was needed. For the *Relevant in Context Task* non-overlapping
results (elements or passages) were returned grouped by the article from
which they came. For the *Best in Context Task* a single starting point
(element start tag or passage start) for each article was needed. We discuss the results for the three tasks, examine the relative effectiveness of
element and passage retrieval. This is examined in the context of content
only (CO, or Keyword) search as well as content and structure (CAS, or
structured) search.

1 Introduction

This paper gives an overview of the INEX 2007 Ad Hoc Track. The main research question underlying the Ad Hoc Track is that of the value of the internal
document structure (mark-up) for retrieving relevant information. That is, does
the document structure help in identify where the relevant information is within
a document? This question has recently attracted a lot of attention. Trotman
and Geva [13] argued that, since INEX relevance assessments are not bound to
XML element boundaries, retrieval systems should also not be bound to XML element boundaries. Their implicit assumption is that a system returning passages
is at least as effective as a system returning XML elements. This assumption is
based on the observation that elements are of a lower granularity than passages
and so all elements can be described as passages. The reverse, however is not

N. Fuhr et al. (Eds.): INEX 2007, LNCS 4862, pp. 1–23, 2008.

true and only some passages can be described as elements. Huang et al. [5] implement a fixed window passage retrieval system and show that a comparable element retrieval ranking can be derived. In a similar study, Itakura and Clarke [6] show that although ranking elements based on passage-evidence is comparable, a direct estimation of the relevance of elements is superior. Finally, Kamps and Koolen [7] study the relation between the passages highlighted by the assessors and the XML structure of the collection directly, showing reasonable correspondence between the document structure and the relevant information.

Up to now, element and passage retrieval approaches could only be compared when mapping passages to elements. This may significantly affect the comparison, since the mapping is non-trivial and, of course, turns the passage retrieval approaches effectively into element retrieval approaches. To study the value of the document structure through direct comparison of element and passage retrieval approaches, the retrieval results for INEX 2007 were liberalized to arbitrary passages. Every XML element is, of course, also a passage of text.

The evaluation measures are now based directly on the highlighted passages, or arbitrary best-entry points, as identified by the assessors. As a result it is now possible to fairly compare systems retrieving elements, ranges of elements, or arbitrary passages. These changes address earlier requests to liberalize the retrieval format to ranges of elements [1] and later requests to liberalize to arbitrary passages of text [13].

The INEX 2007 Ad Hoc Track featured three tasks:

1. For the *Focused Task* a ranked-list of non-overlapping results (elements or passages) must be returned. It is evaluated at early precision relative to the highlighted (or believed relevant) text retrieved.
2. For the *Relevant in Context Task* non-overlapping results (elements or passages) must be returned, these are grouped by document. It is evaluated by mean average generalized precision where the generalized score per article is based on the retrieved highlighted text.
3. For the *Best in Context Task* a single starting point (element's starting tag or passage offset) per article must be returned. It is also evaluated by mean average generalized precision but with the generalized score (per article) based on the distance to the assessor's best-entry point.

The *Thorough Task* as defined in earlier INEX rounds is discontinued. We discuss the results for the three tasks, giving results for the top 10 participating groups and discussing the best scoring approaches in detail. We also examine the relative effectiveness of element and passage runs, and with content only (CO) queries and content and structure (CAS) queries.

The rest of the paper is organized as follows. First, Section 2 describes the INEX 2007 Ad Hoc retrieval tasks and measures. Section 3 details the collection, topics, and assessments of the INEX 2007 Ad Hoc Track. In Section 4, we report the results for the Focused Task (Section 4.2); the Relevant in Context Task (Section 4.3); and the Best in Context Task (Section 4.4). Section 5 details particular types of runs (such as CO versus CAS, and element versus passage), and on particular subsets of the topics (such as topics with a non-trivial CAS query). Finally, in Section 6, we discuss our findings and draw some conclusions.

2   Ad Hoc Retrieval Track

In this section, we briefly summarize the ad hoc retrieval tasks and the submission format (especially how elements and passages are identified). We also summarize the metrics used for evaluation. For more detail the reader is referred to the formal specification documents [2] and [9].

2.1   Tasks

Focused Task. The scenario underlying the Focused Task is the return, to the user, of a ranked list of elements or passages for their topic of request. The Focused Task requires systems to find the most focused results that satisfy an information need, without returning "overlapping" elements (shorter is preferred in the case of equally relevant elements). Since ancestors elements and longer passages are always relevant (to a greater or lesser extent) it is a challenge to chose the correct granularity.

The task has a number of assumptions:

Display the results are presented to the user as a ranked-list of results.
Users view the results top-down, one-by-one.

Relevant in Context Task. The scenario underlying the Relevant in Context Task is the return of a ranked list of articles and within those articles the relevant information (captured by a set of non-overlapping elements or passages). A relevant article will likely contain relevant information that could be spread across different elements. The task requires systems to find a set of results that corresponds well to all relevant information in each relevant article. The task has a number of assumptions:

Display results will be grouped per article, in their original document order, access will be provided through further navigational means, such as a document heat-map or table of contents.
Users consider the article to be the most natural retrieval unit, and prefer an overview of relevance within this context.

Best in Context Task. The scenario underlying the Best in Context Task is the return of a ranked list of articles and the identification of a best-entry-point from which a user should start reading each article in order to satisfy the information need. Even an article completely devoted to the topic of request will only have one best starting point from which to read (even if that is the beginning of the article). The task has a number of assumptions:

Display a single result per article.
Users consider articles to be natural unit of retrieval, but prefer to be guided to the best point from which to start reading the most relevant content.

2.2 Submission Format

Since XML retrieval approaches may return arbitrary results from within documents, a way to identify these nodes is needed.

XML element results are identified by means of a file name and an element (node) path specification. File names in the Wikipedia collection are unique so that (with the .xml extension removed), for example:

```
<file>9996</file>
```

identifies 9996.xml as the target document from the Wikipedia collection. Element paths are given in XPath, but only fully specified paths are allowed. For example:

```
<path>/article[1]/body[1]/section[1]/p[1]</path>
```

identifies the first "article" element, then within that, the first "body" element, then the first "section" element, and finally within that the first "p" element. Importantly, XPath counts elements from 1 and counts element types. For example if a section had a title and two paragraphs then their paths would be: title[1], p[1] and p[2].

A result element, then, is identified unambiguously using the combination of file name and element path, for example:

```
<result>
  <file>9996</file>
  <path>/article[1]/body[1]/section[1]/p[1]</path>
  <rsv>0.9999</rsv>
</result>
```

Passages are given in the same format, but extended for optional character-offsets. As a passage need not start and end in the same element, each is given separately. The following example is equivalent to the element result example above since it starts and ends on an element boundary.

```
<result>
  <file>9996</file>
  <passage start="/article[1]/body[1]/section[1]/p[1]"
      end="/article[1]/body[1]/section[1]/p[1]"/>
  <rsv>0.9999</rsv>
</result>
```

In the next passage example the result starts 85 characters after the start of the paragraph and continues until 106 characters after a list item in list. The end location is, of course, after the start location.

```
<result>
  <file>9996</file>
  <passage start="/article[1]/body[1]/section[1]/p[1]/text()[1].85"
      end="/article[1]/body[1]/section[1]/normallist[1]/item[2]/text()[2].106"/>
  <rsv>0.6666</rsv>
</result>
```

The result can start anywhere in any text node. Character positions count from 0 (before the first character) to the *node-length* (after the last character). A detailed example is provided in [2].

2.3   Measures

We briefly summarize the main measures used for the Ad Hoc Track (see Kamps et al. [9] for details). The main change at INEX 2007 is the inclusion of arbitrary passages of text. Unfortunately this simple change has necessitated the deprecation of element-based metrics used in prior INEX campaigns because the "natural" retrieval unit is no longer an element, so elements cannot be used as the basis of measure. We note that properly evaluating the effectiveness in XML-IR remains an ongoing research question at INEX.

The INEX 2007 measures are solely based on the retrieval of highlighted text. We simplify all INEX tasks to highlighted text retrieval and assume that systems return all, and only, highlighted text. We then compare the characters of text retrieved by a search engine to the number and location of characters of text identified as relevant by the assessor. For best in context we use the distance between the best entry point in the run to that identified by an assessor.

Focused Task. Recall is measured as the fraction of all highlighted text that has been retrieved. Precision is measured as the fraction of retrieved text that was highlighted. The notion of rank is relatively fluid for passages so we use an interpolated precision measure which calculates interpolated precision scores at selected recall levels. Since we are most interested in what happens in the first retrieved results, the INEX 2007 official measure is interpolated precision at 1% recall (iP[0.01]). We also present interpolated precision at other early recall points, and (mean average) interpolated precision over 101 standard recall points (0.00, 0.01, 0.02, ..., 1.00) as an overall measure.

Relevant in Context Task. The evaluation of the Relevant in Context Task is based on the measures of generalized precision and recall [10], where the per document score reflects how well the retrieved text matches the relevant text in the document. Specifically, the per document score is the harmonic mean of precision and recall in terms of the fractions of retrieved and highlighted text in the document. We are most interested in overall performances so the main measure is mean average generalized precision (MAgP). We also present the generalized precision scores at early ranks (5, 10, 25, 50).

Best in Context Task. The evaluation of the Best in Context Task is based on the measures of generalized precision and recall where the per document score reflects how well the retrieved entry point matches the best entry point in the document. Specifically, the per document score is a linear discounting function of the distance d (measured in characters)

$$\frac{n - d(x, b)}{n}$$

for $d < n$ and 0 otherwise. We use $n = 1,000$ which is roughly the number of characters corresponding to the visible part of the document on a screen. We are most interested in overall performance, and the main measure is mean average generalized precision (MAgP). We also show the generalized precision scores at early ranks (5, 10, 25, 50).

3   Ad Hoc Test Collection

In this section, we discuss the corpus, topics, and relevance assessments used in the Ad Hoc Track.

3.1   Corpus

The document collection was the Wikipedia XML Corpus based on the English Wikipedia in early 2006 [3]. The Wikipedia collection contains 659,338 Wikipedia articles. On average an article contains 161 XML nodes, where the average depth of a node in the XML tree of the document is 6.72.

The original Wiki syntax has been converted into XML, using both general tags of the layout structure (like *article, section, paragraph, title, list* and *item*), typographical tags (like *bold, emphatic*), and frequently occurring link-tags. For details see Denoyer and Gallinari [3].

3.2   Topics

The ad hoc topics were created by participants following precise instructions given elsewhere [14]. Candidate topics contained a short CO (keyword) query, an optional structured CAS query, a one line description of the search request, and narrative with a details of the topic of request and the task context in which the information need arose. Figure 1 presents an example of an Ad Hoc topic. Based on the submitted candidate topics, 130 topics were selected for use in the INEX 2007 Ad Hoc track as topic numbers 414–543.

The INEX 2007 Multimedia track also had an ad hoc search task and 19 topics were used for both the Ad Hoc track and the Multimedia track. They were designated topics 525–543. Table 1 presents the topics shared between the Ad Hoc and Multimedia tracks. Six of these topics (527, 528, 530, 532, 535, 540) have an additional ⟨mmtitle⟩ field, a multimedia query.

The 12 INEX 2006 iTrack topics were also inserted into the topic set (as topics 512-514, and 516-524) as these topics were not assessed in 2006. Table 2 presents the 12 INEX 2006 iTrack topics, and their corresponding Ad Hoc track topic numbers.

```
<inex_topic topic_id="414" ct_no="3">
  <title>hip hop beat</title>
  <castitle>//*[about(., hip hop beat)]</castitle>
  <description>what is a hip hop beat?</description>
  <narrative>
    To solve an argument with a friend about hip hop music and beats, I
    want to learn all there is to know about hip hop beats. I want to know
    what is meant by hip hop beats, what is considered a hip hop beat,
    what distinguishes a hip hop beat from other beats, when it was
    introduced and by whom. I consider elements relevant if they
    specifically mention beats or rythm. Any element mentioning hip hop
    music or style but doesn't discuss abything about beats or rythm is
    considered not relevant. Also, elements discussing beats and rythm,
    but not hip hop music in particular, are considered not relevant.
  </narrative>
</inex_topic>
```

Fig. 1. INEX Ad Hoc Track topic 414

Table 1. Topics shared with the INEX 2007 Multimedia track

Topic	Title-field
525	potatoes in paintings
526	pyramids of egypt
527	walt disney land world
528	skyscraper building tall towers
529	paint works museum picasso
530	Hurricane satellite image
531	oil refinery or platform photographs
532	motor car
533	Images of phones
534	Van Gogh paintings
535	japanese garden old building -chapel
536	Ecuador volcano climbing quito
537	pictures of Mont Blanc
538	photographer photo
539	self-portrait
540	war map place
541	classic furniture design chairs
542	Images of tsunami
543	Tux

3.3   Judgments

Topics were assessed by participants following precise instructions [11]. The assessors used Piwowarski's X-RAI assessment system that assists assessors in highlight relevant text. Topic assessors were asked to mark all, and only,

Table 2. iTrack 2006 topics

iTrack	Ad hoc	Title-field	Type	Structure
1	519	types of bridges vehicles water ice	Decision making	Hierarchical
2	512	french impressionism degas monet renoir impressionist movement	Decision making	Hierarchical
3	520	Chartres Versailles history architecture travelling	Decision making	Parallel
4	516	environmental effects mining logging	Decision making	Parallel
5	521	red ants USA bites treatment	Fact finding	Hierarchical
6	513	chanterelle mushroom poisonous deadly species	Fact finding	Hierarchical
7	522	April 19th revolution peaceful revolution velvet revolution quiet revolution	Fact finding	Parallel
8	517	difference fortress castle	Fact finding	Parallel
9	523	fuel efficient cars	Info gathering	Hierarchical
10	514	food additives physical health risk grocery store labels	Info gathering	Hierarchical
11	524	home heating solar panels	Info gathering	Parallel
12	518	tidal power wind power	Info gathering	Parallel

relevant text in a pool of documents. The granularity of assessment was roughly a sentence. After assessing each article a separate best entry point decision was made by the assessor. The Focused and Relevant in Context Tasks were evaluated against the text highlighted by the assessors, whereas the Best in Context Task was evaluated against the best-entry-points.

The relevance judgments were frozen in January 2008. At this time 107 topics had been fully assessed. Moreover, 13 topics were judged by two separate assessors, each without the knowledge of the other. All results in this paper refer to the 107 topics with the judgments of the first assigned assessor.

- The 107 assessed topics were: 414-431, 433-436, 439-441, 444-450, 453, 454, 458, 459, 461-463, 465, 467, 468-475, 476-491, 495-500, 502, 503, 505-509, 511, 515-523, and 525-543.
- All 19 Multimedia topics, 525-543, were assessed.
- Only 8 of the 12 iTrack 2006 topics, 516-523, were assessed.

Table 3. Statistics over judged and relevant articles per topic

	total		# per topic				
	topics	number	min	max	median	mean	st.dev
judged articles	107	65,503	600	703	610	612	13.55
articles with relevance	107	6,491	2	479	36	61	70.91
highlighted passages	107	11,482	2	832	62	107	150.20

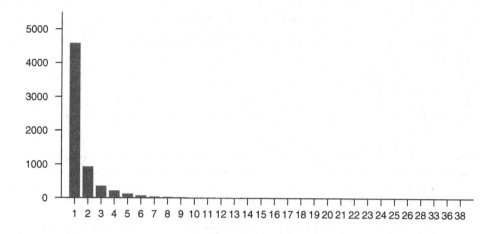

Fig. 2. Distribution of passages over articles

Table 3 presents statistics of the number of judged and relevant articles, and passages. In total 65,503 articles were judged. Relevant passages were found in 6,491 articles. The mean number of relevant articles per topic is 61, but the distribution is skewed with a median of 36. There were 11,482 highlighted passages. The mean was 107 passages and the median was 62 passages per topic.[1]

Figure 2 presents the number of articles with the given number of passages. The vast majority of relevant articles (4,581 out of 6,491) had only a single highlighted passage, and the number of passages quickly tapers off.

3.4   Questionnaires

At INEX 2007, all candidate topic authors and assessors were asked to complete a questionnaire designed to capture the context of the topic author and the topic of request. The candidate topic questionnaire (shown in Table 4) featured 20 questions capturing contextual data on the search request. The post-assessment questionnaire (shown in Table 5) featured 14 questions capturing further contextual data on the search request, and the way the topic has been judged.

The responses to the questionnaires show a considerable variation over topics and topic authors in terms of topic familiarity; the type of information requested; the expected results; the interpretation of structural information in the search request; the meaning of a highlighted passage; and the meaning of best entry points. There is a need for further analysis of the contextual data of the topics in relation to the results of the INEX 2007 Ad Hoc Track.

---

[1] Recall from above that for the Focused Task the main effectiveness measures is precision at 1% recall. Given that the average topic has 107 relevant passages in 61 articles, the 1% recall rougly corresponds to a relevant passage retrieved—for many systems this will be accomplished by the first or first few results.

Table 4. Candidate Topic Questionnaire

B1	How familiar are you with the subject matter of the topic?
B2	Would you search for this topic in real-life?
B3	Does your query differ from what you would type in a web search engine?
B4	Are you looking for very specific information?
B5	Are you interested in reading a lot of relevant information on the topic?
B6	Could the topic be satisfied by combining the information in different (parts of) documents?
B7	Is the topic based on a seen relevant (part of a) document?
B8	Can information of equal relevance to the topic be found in several documents?
B9	Approximately how many articles in the whole collection do you expect to contain relevant information?
B10	Approximately how many relevant document parts do you expect in the whole collection?
B11	Could a relevant result be (check all that apply): a single sentence; a single paragraph; a single (sub)section; a whole article
B12	Can the topic be completely satisfied by a single relevant result?
B13	Is there additional value in reading several relevant results?
B14	Is there additional value in knowing all relevant results?
B15	Would you prefer seeing: only the best results; all relevant results; don't know
B16	Would you prefer seeing: isolated document parts; the article's context; don't know
B17	Do you assume perfect knowledge of the DTD?
B18	Do you assume that the structure of at least one relevant result is known?
B19	Do you assume that references to the document structure are vague and imprecise?
B20	Comments or suggestions on any of the above (optional)

Table 5. Post Assessment Questionnaire

C1	Did you submit this topic to INEX?
C2	How familiar were you with the subject matter of the topic?
C3	How hard was it to decide whether information was relevant?
C4	Is Wikipedia an obvious source to look for information on the topic?
C5	Can a highlighted passage be (check all that apply): a single sentence; a single paragraph; a single (sub)section; a whole article
C6	Is a single highlighted passage enough to answer the topic?
C7	Are highlighted passages still informative when presented out of context?
C8	How often does relevant information occur in an article about something else?
C9	How well does the total length of highlighted text correspond to the usefulness of an article?
C10	Which of the following two strategies is closer to your actual highlighting: (I) I located useful articles and highlighted the best passages and nothing more, (II) I highlighted all text relevant according to narrative, even if this meant highlighting an entire article.
C11	Can a best entry point be (check all that apply): the start of a highlighted passage; the sectioning structure containing the highlighted text; the start of the article
C12	Does the best entry point correspond to the best passage?
C13	Does the best entry point correspond to the first passage?
C14	Comments or suggestions on any of the above (optional)

4   Ad Hoc Retrieval Results

In this section, we discuss, for the three ad hoc tasks, the participants and their results.

4.1   Participation

216 runs were submitted by 27 participating groups. Table 6 lists the participants and the number of runs they submitted, also broken down over the tasks (Focused, Relevant in Context, or Best in Context); the used query (Content-Only or Content-And-Structure); and the used result type (Element or Passage). Participants were allowed to submit up to three CO-runs per task and three CAS-runs per task (for all three tasks). This totaled to 18 runs per participant.[2] The submissions are spread well over the ad hoc retrieval tasks with 79 submissions for Focused, 66 submissions for Relevant in Context, and 71 submissions for Best in Context.

4.2   Focused Task

We now discuss the results of the Focused Task in which a ranked-list of non-overlapping results (elements or passages) was required. The official measure for the task was (mean) interpolated precision at 1% recall (iP[0.01]). Table 7 shows the best run of the top 10 participating groups. The first column gives the participant, see Table 6 for the full name of group, and see Appendix 6 for the precise run label. The second to fifth column give the interpolated precision at 0%, 1%, 5%, and 10% recall. The sixth column gives mean average interpolated precision over 101 standard recall levels (0%, 1%, ..., 100%).

Here we briefly summarize what is currently known about the experiments conducted by the top five groups (based on official measure for the task, iP[0.01]).

Dalian University of Technology. Using the CAS query. Only index the content contained by the tags often occur or retrieved by users. Use the BM25 retrieval model and pseudo-relevance feedback. Both document retrieval and document parts retrieval, and then combine the document score and document parts score. Further special handlings on the category of topics finding images, by removing the returned elements whose structural paths contained "image" or "figure" tags to the top one by one. Overlap was removed in the order of the resulting run.

Ecoles des Mines de Saint-Etienne. Using the CO query. Runs are based on the use of the proximity of the query terms in the documents. The proximity of an XML element to a query is based on the summation of the

---

[2] As it turns out, three groups submitted more runs than allowed: *mines* submitted 1 extra CO-run, and both *lip6* and *qutau* submitted 6 extra CO-runs each. At this moment, we have not decided on any repercussions other than mentioning them in this footnote.

Table 6. Participants in the Ad Hoc Track

Participant	Full name	Foc	RiC	BiC	CO	CAS	Ele	Pas	Total
cmu	Language Technologies Institute, School of Computer Science, Carnegie Mellon University	1	0	0	1	0	1	0	1
eurise	Laboratoire Hubert Curien - Université de Saint-Etienne	2	0	0	2	0	2	0	2
indstainst	Indian Statistical Institute	2	0	0	2	0	2	0	2
inria	INRIA-Rocquencourt- Axis	3	3	3	9	0	9	0	9
irit	IRIT	0	0	2	1	1	2	0	2
justsystem	JustSystems Corporation	6	6	6	9	9	18	0	18
labcsiro	Information Engineering lab, ICT Centre, CSIRO	1	0	0	1	0	1	0	1
lip6	LIP6	5	5	5	15	0	15	0	15
maxplanck	Max-Planck-Institut fuer Informatik	4	4	4	6	6	12	0	12
mines	Ecoles des Mines de Saint-Etienne, France	3	4	3	10	0	10	0	10
qutau	Queensland University of Technology	7	7	7	15	6	21	0	21
rmit	RMIT University	1	1	1	3	0	3	0	3
uamsterdam	University of Amsterdam	6	6	6	9	9	18	0	18
udalian	Dalian University of Technology	6	6	6	9	9	18	0	18
udoshisha	Doshisha University	2	0	0	1	1	2	0	2
ugrenoble	CLIPS-IMAG	3	3	3	9	0	9	0	9
uhelsinki	University of Helsinki	2	0	0	2	0	2	0	2
uminnesota	University of Minnesota Duluth	1	2	2	5	0	5	0	5
uniKaislau	University of Kaiserslautern, AG DBIS	3	3	0	6	0	6	0	6
unigordon	Information Retrieval and Interaction Group, The Robert Gordon University	3	3	3	9	0	9	0	9
unigranada	University of Granada	3	3	5	8	3	11	0	11
unitoronto	University of Toronto	2	0	0	0	2	2	0	2
uotago	University of Otago	3	3	3	9	0	0	9	9
utampere	University of Tampere	3	3	3	9	0	9	0	9
utwente	Cirquid Project (CWI and University of Twente)	3	2	1	6	0	6	0	6
uwaterloo	University of Waterloo	2	0	4	6	0	6	0	6
uwuhan	Center for Studies of Information Resources, School of Information Management, Wuhan University, China	2	2	4	8	0	8	0	8
Total	runs	79	66	71	170	46	207	9	216

normalized proximity score of each term position in the XML element. The proximity model is extended to take into account the document structure. The most simple and most used structure in document is the hierarchical one with sections, subsections, etc. where each instance at each level has got

Table 7. Top 10 Participants in the Ad Hoc Track Focused Task

Participant	iP[0.00]	iP[0.01]	iP[0.05]	iP[0.10]	MAiP
udalian-15	0.5633	0.5271	0.4697	0.4041	0.1689
mines-2	0.6056	0.5164	0.3677	0.2984	0.1221
uwaterloo-0	0.5335	0.5108	0.4284	0.3916	0.1765
cmu-0	0.5924	0.5083	0.4000	0.3435	0.1351
maxplanck-3	0.5780	0.5066	0.4006	0.3430	0.1307
utampere-5	0.5460	0.4998	0.3915	0.3007	0.0981
udoshisha-0	0.5262	0.4975	0.3970	0.3360	0.1460
qutau-8	0.5120	0.4924	0.4493	0.4234	0.2025
inria-2	0.4986	0.4835	0.4540	0.4118	0.2132
rmit-0	0.4995	0.4834	0.4545	0.4172	0.2238

a title. With this kind of structure, we define the proximity to a position in a title as 1 (maximum value) over all the positions in the corresponding section.

University of Waterloo. Using the CO query. Query terms were formed by transforming each topic title into a disjunctive form, less negative query terms. Wumpus [15] was used to obtain positions of query terms and XML elements. The most frequently occurring XML elements in the corpus were listed and ranked using Okapi BM25. Nested results were removed for the Focused task.

Carnegie Mellon University. Using the CO query. XML elements are ranked using a hierarchical language model that estimates the probability of generating the query from an element. The hierarchical language models incorporate evidence from the document, its parent, and its children, using a linear combination of several language models [12].

Max-Planck-Institut für Informatik. Using the CAS query: the basis for this run is an ad hoc CAS run were the target tag was evaluated strictly, i.e., a result was required to have the tag specified as target in the query and match at least one of the content conditions, whereas support conditions were optional; phrases and negations in the query were ignored. To produce the focused run, elements were removed in case they overlap with a higher scoring element for the same topic.

Based on the information from these and other participants:

- Both the best scoring team and the fifth rank team used the CAS query. Hence using the structural hints, even strict adherence to the target tag, seemed to promote early precision
- More generally, limiting the retrieved types of elements, either at indexing time (by selecting elements based on tag type or length) or at retrieval time (by enforcing CAS target elements, or using length-priors), seems to promote early precision.
- The systems at rank nine, *inria-2*, and at rank ten, *rmit-0*, are retrieving only full articles.

Table 8. Top 10 Participants in the Ad Hoc Track Relevant in Context Task

Participant	gP[5]	gP[10]	gP[25]	gP[50]	MAgP
udalian-16	0.2566	0.2318	0.1888	0.1511	0.1552
qutau-18	0.2618	0.2223	0.1802	0.1454	0.1489
rmit-1	0.2483	0.2335	0.1792	0.1379	0.1358
uamsterdam-4	0.2403	0.2121	0.1647	0.1275	0.1323
unigordon-7	0.2531	0.2205	0.1680	0.1283	0.1302
utwente-5	0.2067	0.1838	0.1512	0.1187	0.1233
inria-5	0.2483	0.2335	0.1861	0.1358	0.1147
justsystem-14	0.2072	0.1732	0.1342	0.1023	0.1107
mines-7	0.2120	0.1913	0.1527	0.1185	0.1081
maxplanck-8	0.2168	0.1879	0.1356	0.1050	0.1077

4.3 Relevant in Context Task

We now discuss the results of the Relevant in Context Task in which non-overlapping results (elements or passages) need to be returned grouped by the article they came from. The task was evaluated using generalized precision where the generalized score per article was based on the retrieved highlighted text. The official measure for the task was mean average generalized precision (MAgP).

Table 8 shows the top 10 participating groups (only the best run per group is shown) in the Relevant in Context Task. The first column lists the participant, see Table 6 for the full name of group, and see Appendix 6 for the precise run label. The second to fifth column list generalized precision at 5, 10, 25, 50 retrieved articles. The sixth column lists mean average generalized precision.

Here we briefly summarize the information available about the experiments conducted by the top five groups (based on MAgP).

Dalian University of Technology. Using the CO query. See the description for the Focused Task above. Although submitted as CO run, image finding topics received special handling promoting elements with paths containing image of figure to the top of the ranking. Cluster the returned elements per document, and remove overlap top-down.

Queensland University of Technology. Using the CO query: plural/singular expansion was used on the query, as well as removal of words preceded by a minus sign. GPX [4] was used to rank elements, based on a leaf-node index and $tf \cdot icf$ (term frequency times inverted collection frequency) weighting modified by i) the number of unique terms, ii) the proximity of query-term matches, and iii) boosting of query-term occurrences in the name field. All leaf-node-scores were normalized by their length, and the overall article's similarity score was added. The score of elements was calculated directly from the content of the nodes, obviating the need for score propagation with decaying factors.

RMIT University. Using the CO query. This is a baseline article run using Zettair [16] with the Okapi similarity measure with default settings. The title from each topic was automatically translated as an input query to Zettair. The similarity of an article to a query determines its final rank.

University of Amsterdam. Using the CO query. Having an index with only the "container" elements – elements that frequently contain an entire highlighted passage at INEX 2006 – basically corresponding to the main layout structure. A language model was used with a standard length prior and an incoming links prior, after list-based removal of overlapping elements the final results are clustered per article on a first-come, first-served basis.

Robert Gordon University. Using the CO query. An element's score was computed by a mixture language model combining estimates based on element full-text and a "summary" of it (i.e., extracted titles, section titles, and figure captions nested inside the element). A prior was used according to an element's location in the original text, and the length of its path. For the post-processing, they filter out redundant elements by selecting the highest scored element from each of the paths. Elements are reordered so that results from the same article are grouped together.

Based on the information from these and other participants:

- Solid article ranking seems a prerequisite for good overall performance, with third best run, *rmit-1*, and the seventh best run, *inria-5*, retrieving only full articles.
- The use of the structured query does not appear to promote overall performance: all five groups submitting a CAS query run had a superior CO query run.

4.4 Best in Context Task

We now discuss the results of the Best in Context Task in which documents were ranked on topical relevance and a single best entry point into the document was identified. The Best in Context Task was evaluated using generalized precision but here the generalized score per article was based on the distance to the assessor's best-entry point. The official measure for the task was mean average generalized precision (MAgP).

Table 9. Top 10 Participants in the Ad Hoc Track Best in Context Task

Participant	gP[5]	gP[10]	gP[25]	gP[50]	MAgP
rmit-2	0.3551	0.3280	0.2554	0.1931	0.1919
qutau-19	0.3256	0.2736	0.2138	0.1734	0.1831
uwaterloo-3	0.2600	0.2467	0.2181	0.1716	0.1817
udalian-7	0.2512	0.2416	0.2024	0.1601	0.1759
unigordon-2	0.3405	0.2906	0.2278	0.1761	0.1742
uamsterdam-16	0.3325	0.2917	0.2292	0.1788	0.1731
justsystem-7	0.2904	0.2714	0.2054	0.1611	0.1661
inria-8	0.3551	0.3280	0.2610	0.1952	0.1633
maxplanck-6	0.2005	0.2053	0.1735	0.1348	0.1350
utwente-2	0.2562	0.2246	0.1821	0.1430	0.1339

Table 9 shows the top 10 participating groups (only the best run per group is shown) in the Best in Context Task. The first column lists the participant, see Table 6 for the full name of group, and see Appendix 6 for the precise run label. The second to fifth column list generalized precision at 5, 10, 25, 50 retrieved articles. The sixth column lists mean average generalized precision.

Here we briefly summarize the information available about the experiments conducted by the top five groups (based on MAgP).

RMIT University. Using the CO query. This is the exact same run as the article run for the Relevant in Context Task. See the description for the Relevant in Context Task above.

Queensland University of Technology. Using the CO query. See the description for the Relevant in Context Task above. The best scoring element was selected.

University of Waterloo. Using the CO query. See the description for the Focused Task above. Based on the Focused run, duplicated articles were removed in a post-processing step.

Dalian University of Technology. Using the CO query. See the description for the Focused Task and Relevant in Context above. Return the element which has the largest score per document.

Robert Gordon University. Using the CO query. See the description for the Relevant in Context Task above. For the best-in-context task, the element with the highest score for each of the documents is chosen.

Based on the information from these and other participants:

- As for the Relevant in Context Task, we see again that solid article ranking is very important. In fact, the full article run *rmit-2* is the most effective system. Also the eighth best participant, *inria-8*, is retrieving only full articles.
- Using the start of the whole article as a best-entry-point, as done by the top scoring article run, appears to be a reasonable strategy.
- With the exception of *uamsterdam-16*, which used a filter based on all CAS target elements in the topic set, all best runs per group use the CO query.

4.5   Significance Tests

We tested whether higher ranked systems were significantly better than lower ranked system, using a t-test (one-tailed) at 95%. Table 10 shows, for each task, whether it is significantly better (indicated by "$\star$") than lower ranked runs. For example, For the Focused Task, we see that the early precision (at 1% recall) is a rather unstable measure and none of the runs are significantly different. Hence we should be careful when drawing conclusions based on the Focused Task results. For the Relevant in Context Task, we see that the top run is significantly better than ranks 3 through 10, the second best run better than ranks 4 through 10, the third ranked system better than ranks 6 through 10, and the fourth and

Table 10. Statistical significance (t-test, one-tailed, 95%)

(a) Focused Task

	1	2	3	4	5	6	7	8	9	10
udalian-15		-	-	-	-	-	-	-	-	-
mines-2			-	-	-	-	-	-	-	-
uwaterloo-0				-	-	-	-	-	-	-
cmu-0					-	-	-	-	-	-
maxplanck-3						-	-	-	-	-
utampere-5							-	-	-	-
udoshisha-0								-	-	-
qutau-8									-	-
inria-2										-
rmit-0										

(b) Relevant in Context Task

	1	2	3	4	5	6	7	8	9	10
udalian-16		-	★	★	★	★	★	★	★	★
qutau-18			-	★	★	★	★	★	★	★
rmit-1				-	-	★	★	★	★	★
uamsterdam-4					-	-	★	★	★	★
unigordon-7						-	★	★	★	★
utwente-5							-	-	-	-
inria-5								-	-	-
justsystem-14									-	-
mines-7										-
maxplanck-8										

(c) Best in Context Task

	1	2	3	4	5	6	7	8	9	10
rmit-2			-	-	-	★	★	★	★	★
qutau-19			-	-	-	-	-	-	★	★
uwaterloo-3				-	-	-	-	-	★	★
udalian-7					-	-	-	-	★	★
unigordon-2						-	-	-	★	★
uamsterdam-16							-	-	★	★
justsystem-7								-	★	★
inria-8									★	★
maxplanck-6										-
utwente-2										

fifth ranked systems better than ranks 7 through 10. For the Best in Context Task, we see that the top run is significantly better than ranks 5 through 10, the second to eighth ranked systems are significantly better than those at rank 9 and 10.

5   Analysis of Run and Topic Types

In this section, we will discuss relative effectiveness of element and passage retrieval approaches, and on the relative effectiveness of systems using the keyword and structured queries.

5.1   Elements Versus Passages

We received some, but few, submissions using passage results. We will look at the relative effectiveness of element and passage runs.

As we saw above, in Section 4, for all three tasks the best scoring runs used elements as the unit of retrieval. All nine official passage submissions were from the same participant. Table 11 shows their best passage runs for the three ad

Table 11. Ad Hoc Track: Passage runs

(a) Focused Task

Participant	iP[0.00]	iP[0.01]	iP[0.05]	iP[0.10]	MAiP
uotago-3	0.4850	0.4716	0.3423	0.2639	0.0902

(b) Relevant in Context Task

Participant	gP[5]	gP[10]	gP[25]	gP[50]	MAgP
uotago-1	0.1625	0.1529	0.1213	0.0955	0.1033

(c) Best in Context Task

Participant	gP[5]	gP[10]	gP[25]	gP[50]	MAgP
uotago-6	0.1377	0.1415	0.1194	0.0994	0.1064

Table 12. CAS query target elements over all 130 topics

Target Element	Frequency
*	51
article	29
section	28
figure	9
p	5
image	5
title	1
(section|p)	1
body	1

hoc tasks. As it turns out, the passage run *otago-3* would have been the 12th ranked participant (out of 26) for the Focused Task; *otago-1* would have been the 11th ranked group (out of 18) for the Relevant in Context Task; and *otago-6* would have been the 13th ranked group (out of 19) for the Best in Context Task.

This outcome is consistent with earlier results using passage-based element retrieval, where passage retrieval approaches showed comparable but not superior behavior to element retrieval approaches [5, 6].

It is hard to draw any conclusions for several reasons. First, the passage runs took no account of document structure with passages frequently starting and ending mid-sentence. Second, with only a single participant it is not clear whether the approach is comparable or the participant's runs are only comparable. Third, this is the first year passage retrieval has run at INEX and so the technology is less mature than element retrieval.

We hope and expect that the test collection and the passage runs will be used for further research into the relative effectiveness of element and passage retrieval approaches.

5.2   CO Versus CAS

We now zoom in on the relative effectiveness of the keyword (CO) and structured (CAS) queries. As we saw above, in Section 4, the best two runs for the Focused task used the CAS query, and one of the top 10 runs for the Best in Context Task used the CAS query.

All topics have a CAS query since artificial CAS queries of the form

```
//*[about(., keyword title)]
```

were added to topics without CAS title. Table 12 show the distribution of target elements. In total 111 topics had a CAS query formulated by the authors. Some authors already used the generic CAS query above. There are only 86 topics with a non-trivial CAS query.[3]

---

[3] Note that some of the wild-card topics (using the "*" target) in Table 12 had non-trivial about-predicates and hence have not been regarded as trivial CAS queries.

Table 13. Ad Hoc Track CAS Topics: CO runs (left-hand side) versus CAS runs (right-hand side)

(a) Focused Task

Participant	iP[0.00]	iP[0.01]	iP[0.05]	iP[0.10]	MAiP	Participant	iP[0.00]	iP[0.01]	iP[0.05]	iP[0.10]	MAiP
mines-2	0.6207	0.5426	0.3848	0.3016	0.1285	udalian-15	0.5503	0.5159	0.4481	0.4050	0.1795
udoshisha-0	0.5472	0.5190	0.3995	0.3454	0.1588	maxplanck-3	0.5780	0.4919	0.3834	0.3402	0.1397
cmu-0	0.6047	0.5184	0.4213	0.3679	0.1475	justsystem-3	0.5238	0.4798	0.3736	0.3087	0.1175
uwaterloo-0	0.5397	0.5140	0.4384	0.4079	0.1938	udoshisha-1	0.5337	0.4519	0.3466	0.2969	0.1319
qutau-8	0.5225	0.5124	0.4808	0.4594	0.2120	uamsterdam-10	0.4840	0.4413	0.3835	0.3443	0.1671
udalian-2	0.5343	0.5045	0.4429	0.4077	0.1903	unitoronto-0	0.4921	0.4079	0.3148	0.2680	0.1059
rmit-0	0.5115	0.5024	0.4734	0.4340	0.2351	qutau-9	0.4072	0.4033	0.3895	0.3590	0.1614
inria-2	0.5096	0.5007	0.4724	0.4315	0.2258	unigranada-0	0.3981	0.2644	0.1006	0.0637	0.0229
justsystem-0	0.5292	0.4998	0.4207	0.3599	0.1331						
unigordon-1	0.5189	0.4922	0.4297	0.3918	0.1977						

(b) Relevant in Context Task

Participant	gP[5]	gP[10]	gP[25]	gP[50]	MAgP	Participant	gP[5]	gP[10]	gP[25]	gP[50]	MAgP
qutau-18	0.2798	0.2286	0.1846	0.1482	0.1654	udalian-14	0.2525	0.2217	0.1800	0.1419	0.1578
udalian-4	0.2570	0.2345	0.1871	0.1442	0.1622	uamsterdam-13	0.2473	0.2180	0.1626	0.1237	0.1351
rmit-1	0.2505	0.2356	0.1719	0.1299	0.1455	qutau-10	0.2218	0.1892	0.1507	0.1178	0.1150
uamsterdam-4	0.2428	0.2137	0.1637	0.1242	0.1416	justsystem-15	0.2005	0.1687	0.1224	0.0952	0.1136
unigordon-7	0.2708	0.2296	0.1640	0.1214	0.1407	maxplanck-5	0.2293	0.1926	0.1448	0.1022	0.1048
utwente-5	0.2068	0.1821	0.1455	0.1094	0.1313						
inria-5	0.2505	0.2356	0.1810	0.1312	0.1259						
justsystem-14	0.2145	0.1765	0.1334	0.0981	0.1219						
maxplanck-8	0.2294	0.1921	0.1353	0.1042	0.1185						
mines-7	0.2119	0.1783	0.1181	0.0864	0.1137						

(c) Best in Context Task

Participant	gP[5]	gP[10]	gP[25]	gP[50]	MAgP	Participant	gP[5]	gP[10]	gP[25]	gP[50]	MAgP
rmit-2	0.3552	0.3262	0.2436	0.1842	0.2013	udalian-17	0.2523	0.2442	0.2095	0.1705	0.1800
uwaterloo-3	0.2869	0.2658	0.2259	0.1744	0.1986	uamsterdam-16	0.3233	0.2879	0.2233	0.1728	0.1768
qutau-19	0.3415	0.2777	0.2221	0.1818	0.1964	justsystem-9	0.3065	0.2712	0.2077	0.1710	0.1652
udalian-7	0.2609	0.2491	0.2105	0.1665	0.1899	qutau-3	0.2805	0.2366	0.1679	0.1299	0.1529
unigordon-2	0.3616	0.2950	0.2220	0.1683	0.1854	maxplanck-1	0.2726	0.2466	0.1965	0.1374	0.1281
justsystem-7	0.3109	0.2931	0.2183	0.1689	0.1792	unigranada-6	0.1930	0.1821	0.1548	0.1277	0.1139
uamsterdam-7	0.2706	0.2634	0.2123	0.1676	0.1760	irit-4	0.0337	0.0329	0.0316	0.0219	0.0170
inria-8	0.3552	0.3262	0.2521	0.1877	0.1735						
maxplanck-6	0.2088	0.2188	0.1790	0.1417	0.1451						
utwente-2	0.2532	0.2134	0.1592	0.1216	0.1366						

The CAS topics numbered 415, 416, 418-424, 426-432, 434-440, 442-448, 454, 459, 461, 463, 464, 466, 470, 472, 474, 476-491, 493-498, 500, 501, 507, 508, 511, 515, and 525-543. As it turned out, 77 of these CAS topics were assessed. The results presented here are restricted to only these 77 CAS topics.

Table 13 lists the top 10 participants measured using just the 77 CAS topics and for the Focused Task (a), the Relevant in Context Task (b), and the Best in Context Task (c). For the Focused Task the best two CAS runs outperform the CO runs, as they did over the full topic set. For the Relevant in Context Task, the best CAS run would have ranked fourth among CO runs. For the Best in Context Task, the best two CAS runs would rank sixth and seventh among the CO runs.

We look in detail at the Focused Task runs, where CAS submissions were competitive. Overall, the CAS submissions appear to perform similarly on the subset of 77 CAS topics to the whole set of topics. This was unexpected as these topics do contain real structural hints. The 77 CAS topics constitute three-quarters of the full topic set, making it reasonable to get such a result. However, there are some notable performance characteristics among the CO submissions:

- Some runs (like *maxplanck-3*) perform equally well as over all topics.
- Some runs (like *rmit-0* and *udoshisha-0*) perform much better than over all topics. A possible explanation is the larger number of article-targets among the CAS queries.
- Some runs (like *utampere-5*) perform less well than over all topics.

We should be careful to draw conclusions based on these observations, since the early precision differences between the runs tend not to be significant.

6 Discussion and Conclusions

In this paper we provided an overview of the INEX 2007 Ad Hoc Track that contained three tasks: For the *Focused Task* a ranked-list of non-overlapping results (elements or passages) was required. For the *Relevant in Context Task* non-overlapping results (elements or passages) grouped by the article that they belong to were required. For the *Best in Context Task* a single starting point (element's starting tag or passage offset) per article was required. We discussed the results for the three tasks, and analysed the relative effectiveness of element and passage runs, and of keyword (CO) queries and structured queries (CAS).

When examining the relative effectiveness of CO and CAS we found that the best Focused Task submissions use the CAS query, showing that structural hints can help promote initial precision. This provides further evidence that structured queries can be a useful early precision enhancing device [8]. Although, when restricting to non-trivial CAS queries, we see no real gain for the CAS submissions relative to the CO submissions.

An unexpected finding is that article retrieval is a reasonably effective at XML-IR: an article-only run scored the eighth best group for the Focused Task; the third best for the Relevant in Context Task; and the top ranking group for the Best in Context Task. This demonstrates the importance of the article ranking in the "in context" tasks. The chosen measures were also not unfavorable towards article-submissions:

- For the Relevant in Context Task, the F-score per document equally rewards precision and recall. Article runs have excellent recall, and in the case of Wikipedia, where articles tend to be focused on a single topic, acceptable precision.
- For the Best in Context Task, the window receiving scores was 1,000 characters which, although more strict than the measures at INEX 2006, remains too lenient.

Given the efforts put into the fair comparison of element and passage retrieval approaches, the number of passage submissions was disappointing. The passage runs that were submitted ignored document structure—perhaps the identification based on the XML structure turned out to be difficult, or perhaps the technology is just not yet mature. Although we received only passage results from a single participant, and should be careful to avoid hasty conclusions, we saw that the passage based approach was better than average, but not superior to element based approaches. This outcome is consistent with earlier results using passage-based element retrieval [5, 6]. The comparative analysis of element and passage retrieval approaches was the aim of the track, hoping to shed light on the value of the document structure as provided by the XML mark-up. Although few official submissions used passage retrieval approaches, we hope and expect that the resulting test collection will prove its value in future use. After all, the main aim of the INEX initiative is to create bench-mark test-collections for the evaluation of structured retrieval approaches.

Acknowledgments

Eternal thanks to Benjamin Piwowarski for completely updating the X-RAI tools to ensure that all passage offsets can be mapped exactly.

Jaap Kamps was supported by the Netherlands Organization for Scientific Research (NWO, grants # 612.066.513, 639.072.601, and 640.001.501), and by the E.U.'s 6th FP for RTD (project MultiMATCH contract IST-033104).

References

[1] Clarke, C.L.A.: Range results in XML retrieval. In: Proceedings of the INEX 2005 Workshop on Element Retrieval Methodology, pp. 4–5, Glasgow, UK (2005)

[2] Clarke, C.L.A., Kamps, J., Lalmas, M.: INEX 2007 retrieval task and result submission specification. In: Pre-Proceedings of INEX 2007, pp. 445–453 (2007)

[3] Denoyer, L., Gallinari, P.: The Wikipedia XML Corpus. SIGIR Forum 40, 64–69 (2006)

[4] Geva, S.: GPX – gardens point XML IR at INEX 2005. In: Fuhr, N., Lalmas, M., Malik, S., Kazai, G. (eds.) INEX 2005. LNCS, vol. 3977, pp. 204–253. Springer, Heidelberg (2006)

[5] Huang, W., Trotman, A., O'Keefe, R.A.: Element retrieval using a passage retrieval approach. In: Proceedings of the 11th Australasian Document Computing Symposium (ADCS 2006), pp. 80–83 (2006)

[6] Itakura, K.Y., Clarke, C.L.A.: From passages into elements in XML retrieval. In: Proceedings of the SIGIR 2007 Workshop on Focused Retrieval, University of Otago, Dunedin New Zealand, pp. 17–22 (2007)

[7] Kamps, J., Koolen, M.: On the relation between relevant passages and XML document structure. In: Proceedings of the SIGIR 2007 Workshop on Focused Retrieval, University of Otago, Dunedin New Zealand, pp. 28–32 (2007)

[8] Kamps, J., Marx, M., de Rijke, M., Sigurbjörnsson, B.: Articulating information needs in XML query languages. Transactions on Information Systems 24, 407–436 (2006)

[9] Kamps, J., Pehcevski, J., Kazai, G., Lalmas, M., Robertson, S.: INEX 2007 evaluation measures. In: Fuhr, N., Lalmas, M., Trotman, A. (eds.) INEX 2006. LNCS, vol. 4518. Springer, Heidelberg (2007)

[10] Kekäläinen, J., Järvelin, K.: Using graded relevance assessments in IR evaluation. Journal of the American Society for Information Science and Technology 53, 1120–1129 (2002)

[11] Lalmas, M., Piwowarski, B.: INEX 2007 relevance assessment guide. In: Pre-Proceedings of INEX 2007, pp. 454–463 (2007)

[12] Ogilvie, P., Callan, J.: Parameter estimation for a simple hierarchical generative model for xml retrieval. In: Fuhr, N., Lalmas, M., Malik, S., Kazai, G. (eds.) INEX 2005. LNCS, vol. 3977, pp. 211–224. Springer, Heidelberg (2006)

[13] Trotman, A., Geva, S.: Passage retrieval and other XML-retrieval tasks. In: Proceedings of the SIGIR 2006 Workshop on XML Element Retrieval Methodology, University of Otago, Dunedin New Zealand, pp. 43–50 (2006)

[14] Trotman, A., Larsen, B.: INEX 2007 guidelines for topic development. In: Pre-Proceedings of INEX 2007, pp. 436–444 (2007)

[15] Wumpus. The Wumpus search engine (2007), http://www.wumpus-search.org

[16] Zettair. The Zettair search engine (2007), http://www.seg.rmit.edu.au/zettair/

Appendix: Full Run Names

Run	Label
cmu-0	p40_nophrasebase
inria-2	p11_ent-ZM-Focused
inria-5	p11_ent-ZM-RiC
inria-8	p11_ent-ZM-BiC
irit-4	p49_xfirm.cos.01_BIC
justsystem-0	p41_VSM_CO_01
justsystem-14	p41_VSM_CO_09
justsystem-15	p41_VSM_CAS_10
justsystem-3	p41_VSM_CAS_04
justsystem-7	p41_VSM_CO_14
justsystem-9	p41_VSM_CAS_16
maxplanck-1	p25_TOPX-CAS-exp-BIC
maxplanck-3	p25_TOPX-CAS-Focused-all
maxplanck-5	p25_TOPX-CAS-RIC
maxplanck-6	p25_TOPX-CO-all-BIC
maxplanck-8	p25_TOPX-CO-all-exp-RIC
mines-2	p53_EMSE.boolean.Prox200NF.0012
mines-7	p53_EMSE.boolean.Prox200NRm.0010
qutau-10	p9_RIC_05
qutau-18	p9_RIC_07
qutau-19	p9_BIC_07
qutau-3	p9_BIC_04
qutau-8	p9_FOC_03
qutau-9	p9_FOC_04
rmit-0	p32_zet-okapi-Focused
rmit-1	p32_zet-okapi-RiC
rmit-2	p32_zet-okapi-BiC
uamsterdam-10	p36_inex07_contain_beta1_focused_clp_10000_cl_cas_pool_filter
uamsterdam-13	p36_inex07_contain_beta1_focused_clp_10000_cl_cas_pool_filter_ric_hse
uamsterdam-16	p36_inex07_contain_beta1_focused_clp_10000_cl_cas_pool_filter_bic_hse
uamsterdam-4	p36_inex07_contain_beta1_focused_clp_10000_cl_ric_hse
uamsterdam-7	p36_inex07_contain_beta1_focused_clp_10000_cl_bic_hse
udalian-14	p26_DUT_03_Relevant
udalian-15	p26_DUT_03_Focused
udalian-16	p26_DUT_01_Relevant
udalian-17	p26_DUT_03_Best
udalian-2	p26_DUT_01_Focused_3
udalian-4	p26_DUT_02_Relevant
udalian-7	p26_DUT_02_Best
udoshisha-0	p22_Kikori-CO-Focused
udoshisha-1	p22_Kikori-CAS-Focused
unigordon-1	p35_Focused-LM
unigordon-2	p35_BestInContext-LM
unigordon-7	p35_RelevantInContext-LM
unigranada-0	p4_CID_pesos_15_util_2
unigranada-6	p4_CID_pesos_15_bic
unitoronto-0	p60_4-sr
uotago-1	p10_DocsNostem-PassagesNoStem-StdDevNo
uotago-3	p10_DocsNostem-PassagesStem-StdDevNo-Focused
uotago-6	p10_DocsNostem-PassagesStem-StdDevNo-BEP
utampere-5	p55_Foc k=0.3, v=4.5, cont=2.3
utwente-2	p45_articleBic
utwente-5	p45_star_logLP_RinC
uwaterloo-0	p37_FOER
uwaterloo-3	p37_BICERGood

INEX 2007 Evaluation Measures

Jaap Kamps[1], Jovan Pehcevski[2], Gabriella Kazai[3], Mounia Lalmas[4],
and Stephen Robertson[3]

[1] University of Amsterdam, The Netherlands
kamps@science.uva.nl
[2] INRIA Rocquencourt, France
jovan.pehcevski@inria.fr
[3] Microsoft Research Cambridge, United Kingdom
{gabkaz,ser}@microsoft.com
[4] Queen Mary, University of London, United Kingdom
mounia@dcs.qmul.ac.uk

Abstract. This paper describes the official measures of retrieval effectiveness that are employed for the Ad Hoc Track at INEX 2007. Whereas in earlier years all, but only, XML elements could be retrieved, the result format has been liberalized to arbitrary passages. In response, the INEX 2007 measures are based on the amount of highlighted text retrieved, leading to natural extensions of the well-established measures of precision and recall. The following measures are defined: The Focused Task is evaluated by interpolated precision at 1% recall (iP[0.01]) in terms of the highlighted text retrieved. The Relevant in Context Task is evaluated by mean average generalized precision ($MAgP$) where the generalized score per article is based on the retrieved highlighted text. The Best in Context Task is also evaluated by mean average generalized precision ($MAgP$) but here the generalized score per article is based on the distance to the assessor's best-entry point.

1 Introduction

Focused retrieval investigates ways to provide users with direct access to relevant information in retrieved documents, and includes tasks like question answering, passage retrieval, and XML element retrieval [18]. Since its launch in 2002, INEX has studied different aspects of focused retrieval by mainly considering XML element retrieval techniques that can effectively retrieve information from structured document collections [7]. The main change in the Ad Hoc Track at INEX 2007 was to allow the retrieval of arbitrary document parts, which can represent XML elements or passages [3]. That is, a retrieval result can be either an XML element (a sequence of textual content contained within start/end tags), or an arbitrary passage (a sequence of textual content that can be either contained within an element, or can span across a range of elements). In this paper, we will use the term "document part" to refer to both XML elements and arbitrary passages. These changes address requests to liberalize the retrieval format to ranges of elements [2] and to arbitrary passages [16]. However, this

N. Fuhr et al. (Eds.): INEX 2007, LNCS 4862, pp. 24–33, 2008.

simple change had dear consequences for the measures as used up to now at INEX [6, 9, 10, 13, 14]. By allowing arbitrary passages, we loose the "natural" retrieval unit of elements that was the basis for earlier measures. At INEX 2007 we have adopted an evaluation framework that is based on the amount of highlighted text in relevant documents (similar to the HiXEval measures [15]). In this way we build directly on highlighting assessment procedure used at INEX, and define measures that are natural extensions of the well-established measures of precision and recall used in traditional information retrieval [1].

This paper is organized as follows. In Section 2, we briefly describe the ad hoc retrieval tasks at INEX 2007, and the resulting relevance assessments. Then in three separate sections, we discuss the evaluation measures used for each of the INEX 2007 tasks: the Focused Task (Section 3); the Relevant in Context Task (Section 4); and the Best in Context Task (Section 5). We finish with a some discussion and conclusions in Section 6.

2   Ad Hoc Retrieval Track

In this section, we briefly summarize the ad hoc retrieval tasks, and the resulting relevance judgments.

2.1   Ad Hoc Retrieval Tasks

The INEX 2007 Ad Hoc Track investigated the following three retrieval tasks as defined in [3]. First, there is the Focused Task.

Focused Task. This task asks systems to return a ranked list of non-overlapping, most focused document parts that represent the most appropriate units of retrieval. For example, in the case of returning XML elements, a paragraph and its container section should not both be returned. For this task, from all the estimated relevant (and possibly overlapping) document parts, systems are required to choose those non-overlapping document parts that represent the most appropriate units of retrieval.

The second task corresponds to an end-user task where focused retrieval answers are grouped per document, in their original document order, providing access through further navigational means. This assumes that users consider documents as the most natural units of retrieval, and prefer an overview of relevance in their original context.

Relevant in Context. This task asks systems to return non-overlapping relevant document parts clustered by the unit of the document that they are contained within. An alternative way to phrase the task is to return documents with the most focused, relevant parts highlighted within.

The third task is similar to Relevant in Context, but asks for only a single best point to start reading the relevant content in an article.

Best in Context. This task asks systems to return a single document part per
document. The start of the single document part corresponds to the best
entry point for starting to read the relevant text in the document.

Given that passages can be overlapping in sheer endless ways, there is no meaningful equivalent of the *Thorough Task* as defined in earlier years of INEX.

Note that there is no separate passage retrieval task, and for all the three
tasks arbitrary passages may be returned instead of elements. For all the three
tasks, systems could either use the title field of the topics (content-only topics)
or the cas-title field of the topics (content-and-structure topics). Trotman and
Larsen [17] provide a detailed description of the format used for the INEX 2007
topics.

2.2   Relevance Assessments

Since 2005, a highlighting assessment procedure is used at INEX to gather relevance assessments for the INEX retrieval topics [12]. In this procedure, assessors
from the participating groups are asked to highlight sentences representing the
relevant information in a pooled set of documents of the Wikipedia XML document collection [4]. After assessing an article with relevance, a separate best
entry point judgment is also collected from the assessor, marking the point in
the article that represents the best place to start reading.

The Focused and Relevant in Context Tasks will be evaluated against the text
highlighted by the assessors, whereas the Best in Context Task will be evaluated
against the best-entry-points.

3   Evaluation of the Focused Task

3.1   Assumptions

In the Focused Task, for each INEX 2007 topic, systems are asked to return a
ranked list of the top 1,500 non-overlapping most focused relevant document parts.
The retrieval systems are thus required not only to rank the document parts according to their estimated relevance, but to also decide which document parts are
the most focused non-overlapping units of retrieval.

We make the following evaluation assumption about the Focused Task: *The
amount of relevant information retrieved is measured in terms of the length of
relevant text retrieved.* That is, instead of counting the number of relevant documents retrieved, in this case we measure the amount of relevant (highlighted)
text retrieved.

3.2   Evaluation Measures

More formally, let p_r be the document part assigned to rank r in the ranked list
of document parts L_q returned by a retrieval system for a topic q (at INEX 2007,

$|L_q| = 1,500$ elements or passages). Let $rsize(p_r)$ be the length of highlighted (relevant) text contained by p_r in characters (if there is no highlighted text, $rsize(p_r) = 0$). Let $size(p_r)$ be the total number of characters contained by p_r, and let $Trel(q)$ be the total amount of (highlighted) relevant text for topic q. $Trel(q)$ is calculated as the total number of highlighted characters across all documents, i.e., the sum of the lengths of the (non-overlapping) highlighted passages from all relevant documents.

Measures at selected cutoffs. Precision at rank r is defined as the fraction of retrieved text that is relevant:

$$P[r] = \frac{\sum_{i=1}^{r} rsize(p_i)}{\sum_{i=1}^{r} size(p_i)} \tag{1}$$

To achieve a high precision score at rank r, the document parts retrieved up to and including that rank need to contain as little non-relevant text as possible.

Recall at rank r is defined as the fraction of relevant text that is retrieved:

$$R[r] = \frac{\sum_{i=1}^{r} rsize(p_i)}{Trel(q)} \tag{2}$$

To achieve a high recall score at rank r, the document parts retrieved up to and including that rank need to contain as much relevant text as possible.

An issue with the precision measure $P[r]$ given in Equation 1 is that it can be biased towards systems that return several shorter document parts rather than returning one longer part that contains them all (this issue has plagued earlier passage retrieval tasks at TREC [20]). Since the notion of ranks is relatively fluid for passages, we opt to look at precision at recall levels rather than at ranks. Specifically, we use an interpolated precision measure $iP[x]$, which calculates interpolated precision scores at selected recall levels:

$$iP[x] = \begin{cases} \max_{1 \le r \le |L_q|} (P[r] \wedge R[r] \ge x) & \text{if } x \le R[|L_q|] \\ \\ 0 & \text{if } x > R[|L_q|] \end{cases} \tag{3}$$

where $R[|L_q|]$ is the recall over all documents retrieved. For example, $iP[0.01]$ calculates interpolated precision at the 1% recall level for a given topic.

Over a set of topics, we can also calculate the interpolated precision measure, also denoted by $iP[x]$, by calculating the mean of the scores obtained by the measure for each individual topic.

Overall performance measure. In addition to using the interpolated precision measure at selected recall levels, we also calculate overall performance scores based on the measure of average interpolated precision AiP. For an INEX topic, we calculate AiP by averaging the interpolated precision scores calculated at 101 standard recall levels $(0.00, 0.01, \ldots, 1.00)$:

$$AiP = \frac{1}{101} \cdot \sum_{x=0.00,0.01,\ldots,1.00} iP[x] \tag{4}$$

Performance across a set of topics is measured by calculating the mean of the AiP values obtained by the measure for each individual topic, resulting in mean average interpolate precision ($MAiP$). Assuming there are n topics:

$$MAiP = \frac{1}{n} \cdot \sum_{t} AiP(t) \tag{5}$$

3.3   Results Reported at INEX 2007

For the Focused Task we report the following measures over all INEX 2007 topics:

- Mean interpolated precision at four selected recall levels: $iP[x]$, $x \in [0.00, 0.01, 0.05, 0.10]$; and
- Mean interpolated average precision over 101 recall levels ($MAiP$).

The official evaluation for the Focused Task is an early precision measure: interpolated precision at 1% recall ($iP[0.01]$).

4   Evaluation of the Relevant in Context Task

4.1   Assumptions

The Relevant in Context Task is a variation on document retrieval, in which systems are first required to rank documents in a decreasing order of relevance and then identify a set of non-overlapping, relevant document parts. We make the following evaluation assumption: *All documents that contain relevant text are regarded as (Boolean) relevant documents.* Hence, at the article level, we do not distinguish between relevant documents.

4.2   Evaluation Measures

The evaluation of the Relevant in Context Task is based on the measures of generalized precision and recall [11], where the per document score reflects how well the retrieved text matches the relevant text in the document. The resulting measure was introduced at INEX 2006 [8, 13].

Score per document. For a retrieved document, the text identified by the selected set of non-overlapping retrieved parts is compared to the text highlighted by the assessor. More formally, let d be a retrieved document, and let p be a document part in d. We denote the set of all retrieved parts of document d as $\mathcal{P}_d$. Let $Trel(d)$ be the total amount of highlighted relevant text in the document d. $Trel(d)$ is calculated as the total number of highlighted characters in a document, i.e., the sum of the lengths of the (non-overlapping) highlighted passages.

We calculate the following for a retrieved document d:

- Document precision, as the fraction of retrieved text (in characters) that is highlighted (relevant):

$$P(d) = \frac{\sum\limits_{p \in \mathcal{P}_d} rsize(p)}{\sum\limits_{p \in \mathcal{P}_d} size(p)} \qquad (6)$$

The $P(d)$ measure ensures that, to achieve a high precision value for the document d, the set of retrieved parts for that document needs to contain as little non-relevant text as possible.

- Document recall, as the fraction of highlighted text (in characters) that is retrieved:

$$R(d) = \frac{\sum\limits_{p \in \mathcal{P}_d} rsize(p)}{Trel(d)} \qquad (7)$$

The $R(d)$ measure ensures that, to achieve a high recall value for the document d, the set of retrieved parts for that document needs to contain as much relevant text as possible.

- Document F-Score, as the combination of the document precision and recall scores using their harmonic mean [19], resulting in a score in [0,1] per document:

$$F(d) = \frac{2 \cdot P(d) \cdot R(d)}{P(d) + R(d)} \qquad (8)$$

For retrieved non-relevant documents, both document precision and document recall evaluate to zero.

We may choose either precision, recall, the F-score, or even other aggregates as document score (S(d)). For the Relevant in Context Task, we use the F-score as the document score:

$$S(d) = F(d) \qquad (9)$$

The resulting $S(d)$ score varies between 0 (document without relevant text, or none of the relevant text is retrieved) and 1 (all relevant text is retrieved without retrieving any non-relevant text).

Scores for ranked list of documents. Given that the individual document scores $(S(d))$ for each document in a ranked list $\mathcal{L}$ can take any value in $[0, 1]$, we employ the evaluation measures of generalized precision and recall [11].

More formally, let us assume that for a given topic there are in total $Nrel$ relevant documents, and let $IsRel(d_r) = 1$ if document d at document-rank r contains highlighted relevant text, and $IsRel(d_r) = 0$ otherwise. Let $Nrel$ be the total number of document with relevance for a given topics.

Over the ranked list of documents, we calculate the following:

- generalized precision ($gP[r]$), as the sum of document scores up to (and including) document-rank r, divided by the rank r:

$$gP[r] = \frac{\sum_{i=1}^{r} S(d_i)}{r} \tag{10}$$

- generalized Recall ($gR[r]$), as the number of relevant documents retrieved up to (and including) document-rank r, divided by the total number of relevant documents:

$$gR[r] = \frac{\sum_{i=1}^{r} IsRel(d_i)}{Nrel} \tag{11}$$

Based on these, the average generalized precision AgP for a topic can be calculated by averaging the generalized precision scores obtained for each natural recall points, where generalized recall increases:

$$AgP = \frac{\sum_{r=1}^{|\mathcal{L}|} IsRel(d_r) \cdot gP[r]}{Nrel} \tag{12}$$

For non-retrieved relevant documents a generalized precision score of zero is assumed.

The mean average generalized precision ($MAgP$) is simply the mean of the average generalized precision scores over all topic.

4.3 Results Reported at INEX 2007

For the Relevant in Context Task we report the following measures over all topics:

- Non-interpolated mean generalized precision at four selected ranks: $gP[r]$, $r \in [5, 10, 25, 50]$; and
- Non-interpolated mean average generalized precision ($MAgP$).

The official evaluation for the Relevant in Context Task is the overall mean average generalized precision ($MAgP$) measure, where the generalized score per article is based on the retrieved highlighted text.

5 Evaluation of the Best in Context Task

5.1 Assumptions

The Best in Context Task is another variation on document retrieval where, for each document, a single best entry point needs to be identified. We again assume that all documents with relevance are equally desirable.

5.2   Evaluation Measures

The evaluation of the Best in Context Task is also based on the measures of generalized precision and recall [11], where the per document score reflects how well the retrieved entry point matches the best entry point in the document. Note that at INEX 2006 a different, and more liberal, distance measure was used [13].

Score per document. The document score $S(d)$ for this task is calculated with a distance similarity measure, $s(x, b)$, which measures how close the system-proposed entry point x is to the ground-truth best entry point b given by the assessor. Closeness is assumed to be an inverse function of distance between the two points. The maximum value of 1 is achieved when the two points match, and the minimum value is zero.

We use the following formula for calculating the distance similarity measure:

$$s(x, b) = \begin{cases} \frac{n - d(x,b)}{n} & \text{if } 0 \leq d(x, b) \leq n \\ \\ 0 & \text{if } d(x, b) > n \end{cases} \tag{13}$$

where the distance $d(x, b)$ is measured in characters, and n is the number of characters representing the visible part of the document that can fit on a screen (typically, $n = 1,000$ characters).

We use the $s(x, b)$ distance similarity score as the document score for the Best in Context Task:

$$S(d) = s(x, b) \tag{14}$$

The resulting $S(d)$ score varies between 0 (non-relevant document, or the distance between the system-proposed entry point and the ground-truth best entry point is more than n characters) and 1 (the system-proposed entry point is identical to the ground-truth best entry point).

Scores for ranked list of documents Completely analogous to the Relevant in Context Task, we use generalized precision and recall to determine the score for the ranked list of documents. For details, see the above discussion of the Relevant in Context Task in Section 4.

5.3   Results Reported at INEX 2007

For the Best in Context Task we report the following measures over all topics (using $n = 1,000$)

- Non-interpolated mean generalized precision at four selected ranks: $gP[r]$, $r \in [5, 10, 25, 50]$; and
- Non-interpolated mean average generalized precision ($MAgP$).

The official evaluation for the Best in Context Task is the overall mean average generalized precision ($MAgP$) measure with the generalized score per article is based on the distance to the best-entry point.

6   Discussion and Conclusions

This paper described the official measures of retrieval effectiveness that are employed for the Ad Hoc Track at INEX 2007. The main innovation at INEX 2007 was a liberalization of the allowed retrieval results. Whereas in earlier years all, but only, XML elements could be retrieved, the result format was extended to ranges of elements and arbitrary passages. In order to allow for a fair comparison of the effectiveness of both element-based and passage-based runs, all INEX 2007 measures were based on the amount of highlighted text retrieved, leading to natural extensions of the well-established measures of precision and recall.

The following three measures have been defined: The Focused Task is evaluated by interpolated precision at 1% recall (iP[0.01]) in terms of the highlighted text retrieved. The Relevant in Context Task is evaluated by mean average generalized precision ($MAgP$) where the generalized score per article is based on the retrieved highlighted text. The Best in Context Task is also evaluated by mean average generalized precision ($MAgP$) but here the generalized score per article is based on the distance to the assessor's best-entry point.

Given that the Focused Task measure is defined in terms of recall rather than ranks, it is less straightforward to relate the measure to user's reading effort. As it turned out, the precision at 1% recall was indeed measuring very early precision—usually obtained after one or a few results. That is, given the total length of highlighted or relevant text per topic, and the reasonable precision of the initial results of retrieval systems, the targeted recall was reached within the first few results. Further research is needed to establish whether the chosen recall level corresponds well enough to the intuitions underlying the Focused Task.

The Best in Context Task measure used a window of 1,000 characters around the assessor's best entry point to award a generalized precision score per document, which turned out to be quite lenient. That is, given the total length of Wikipedia articles, and the large fraction of best entry points that are placed relatively early in the article, the generalized precision score is reflecting to a large degree the "article retrieval" component also already awarded in the generalized recall scores. Further research is needed to establish whether the chosen window of characters corresponds well enough to the intuitions underlying the Best in Context Task.

The results of the INEX 2007 Ad Hoc track are detailed in the track overview paper [5].

Acknowledgements

We thank Benjamin Piwowarski and James A. Thom for their valuable comments on earlier drafts of this paper.

Jaap Kamps was supported by the Netherlands Organization for Scientific Research (NWO, grants # 612.066.513, 639.072.601, and 640.001.501), and by the E.U.'s 6th FP for RTD (project MultiMATCH contract IST-033104).

References

[1] Baeza-Yates, R., Ribeiro-Neto, B. (eds.): Modern Information Retrieval. ACM Press/Addison Wesley Longman, New York, Harlow (1999)

[2] Clarke, C.L.A.: Range results in XML retrieval. In: Proceedings of the INEX 2005 Workshop on Element Retrieval Methodology, pp. 4–5, Glasgow, UK (2005)

[3] Clarke, C.L.A., Kamps, J., Lalmas, M.: INEX 2007 retrieval task and result submission specification. In: Pre-Proceedings of INEX 2007, pp. 445–453 (2007)

[4] Denoyer, L., Gallinari, P.: The Wikipedia XML corpus. SIGIR Forum 40(1), 64–69 (2006)

[5] Fuhr, N., Kamps, J., Lalmas, M., Malik, S., Trotman, A.: Overview of the INEX 2007 ad hoc track. In: Fuhr, N., Lalmas, M., Trotman, A. (eds.) INEX 2006. LNCS, vol. 4518. Springer, Heidelberg (2007)

[6] Gövert, N., Kazai, G.: Overview of the INitiative for the Evaluation of XML retrieval (INEX) 2002. In: Proceedings of the First Workshop of the INitiative for the Evaluation of XML retrieval (INEX), pp. 1–17. ERCIM Publications (2003)

[7] INEX. INitiative for the Evaluation of XML Retrieval (2007), http://inex.is.informatik.uni-duisburg.de/

[8] Kamps, J., Lalmas, M., Pehcevski, J.: Evaluating Relevant in Context: Document retrieval with a twist. In: Proceedings of the 30th Annual International ACM SIGIR Conference on Research and Development in Information Retrieval, pp. 723–724. ACM Press, New York (2007)

[9] Kazai, G.: Report of the INEX 2003 metrics work group. In: INEX 2003 Workshop Proceedings, pp. 184–190 (2004)

[10] Kazai, G., Lalmas, M.: INEX 2005 evaluation measures. In: Fuhr, N., Lalmas, M., Malik, S., Kazai, G. (eds.) INEX 2005. LNCS, vol. 3977, pp. 16–29. Springer, Heidelberg (2006)

[11] Kekäläinen, J., Järvelin, K.: Using graded relevance assessments in IR evaluation. Journal of the American Society for Information Science and Technology 53(13), 1120–1129 (2002)

[12] Lalmas, M., Piwowarski, B.: INEX 2007 relevance assessment guide. In: Pre-Proceedings of INEX 2007, pp. 454–463 (2007)

[13] Lalmas, M., Kazai, G., Kamps, J., Pehcevski, J., Piwowarski, B., Robertson, S.: INEX 2006 evaluation measures. In: Fuhr, N., Lalmas, M., Trotman, A. (eds.) INEX 2006. LNCS, vol. 4518, pp. 20–34. Springer, Heidelberg (2007)

[14] Malik, S., Lalmas, M., Fuhr, N.: Overview of INEX 2004. In: Fuhr, N., Lalmas, M., Malik, S., Szlávik, Z. (eds.) INEX 2004. LNCS, vol. 3493, pp. 1–15. Springer, Heidelberg (2005)

[15] Pehcevski, J., Thom, J.A.: HiXEval: Highlighting XML retrieval evaluation. In: Fuhr, N., Lalmas, M., Malik, S., Kazai, G. (eds.) INEX 2005. LNCS, vol. 3977, pp. 43–57. Springer, Heidelberg (2006)

[16] Trotman, A., Geva, S.: Passage retrieval and other XML-retrieval tasks. In: Proceedings of the SIGIR 2006 Workshop on XML Element Retrieval Methodology, Seattle, USA, pp. 43–50 (2006)

[17] Trotman, A., Larsen, B.: INEX 2007 guidelines for topic development. In: Pre-Proceedings of INEX 2007, pp. 436–444 (2007)

[18] Trotman, A., Geva, S., Kamps, J. (eds.): Proceedings of the SIGIR 2007 Workshop on Focused Retrieval, University of Otago, Dunedin New Zealand (2007)

[19] van Rijsbergen, C.J.: Information Retrieval. Butterworths, London (1979)

[20] Wade, C., Allan, J.: Passage retrieval and evaluation. Technical report, CIIR, University of Massachusetts, Amherst (2005)

XML Retrieval by Improving
Structural Relevance Measures Obtained from
Summary Models

M.S. Ali, Mariano P. Consens, and Shahan Khatchadourian

University of Toronto
{sali,consens,shahan}@cs.toronto.edu

Abstract. In XML retrieval, there is often more than one element in
the same document that could represent the same focused result. So, a
key challenge for XML retrieval systems is to return the set of elements
that best satisfies the information need of the end-user in terms of both
content and structure. At INEX, there have been numerous proposals
for how to incorporate structural constraints and hints into ranking.
These proposals either boost the score of or filter out elements that have
desirable structural properties. An alternative approach that has not
been explored is to rank elements by improving their structural relevance.
Structural relevance is the expected relevance of a list of elements, based
on a graphical model of how users browse elements within documents. In
our approach, we use summary graphs to describe the process of a user
browsing from one part of a document to another.

In this paper, we develop an algorithm to structurally score retrieval
scenarios using structural relevance. The XML retrieval system identifies the candidate scenarios. We apply structural relevance with a given
summary model to identify the most structurally relevant scenario. This
results in improved system performance. Our approach provides a consistent way to apply different user models to ranking. We also explore
the use of score boosting using these models.

1 Introduction

INEX is a forum dedicated to research in information retrieval from collections
of XML documents. The INEX 2007 Ad-hoc Track highlights the comparison
of focused XML retrieval systems that return either elements or passages. Our
work here considers only elements. The main challenges in this INEX task are
to identify where relevant text appears in the collection; and to return the most
appropriate XML element(s) that contain the text [9]. Elements from the same
document may be returned in the same system output, but the elements must
not overlap.

For a given topic, the XML retrieval system finds candidate elements based
on which elements contain the topic keywords. For instance, consider returning
elements for the keywords "Herman Melville Moby-Dick" in the Wikipedia document shown in Figure 1. The keywords appear directly in the content of the

N. Fuhr et al. (Eds.): INEX 2007, LNCS 4862, pp. 34–48, 2008.

elements /article/name, /article/body/p, /article/body/section[1], and /article/body/section[2]. To return these specific results, we could return any of the the ancestor paths for each of the aforementioned elements. A basic challenge in structural retrieval (*i.e.,* using structural hints and constraints in XML retrieval) is to determine which subset of these elements would best satisfy the information needs of the end-user.

```
<article>
  <name id="xxx">Moby-Dick</name>
  <body>
    <figure>
      <image xlink:href="cover.png"/>
      <caption>Cover(1851): Moby-Dick: The Whale</caption>
    </figure>
    <p>Moby-Dick is a novel by Herman Melville.</p>
    <section><title>Historical background</title>
      Moby-Dick appeared in 1851, during an ...
    </section>
    <section><title>Major Themes</title>
      Moby-Dick is a highly symbolic work, but also includes chapters
      on natural history. Major themes include  ...
        <section><title>Symbolism</title>
          All of the members of the Pequod's crew have biblical-sounding,
          improbable, or descriptive names,...
        </section>
    </section>
  </body>
</article>
```

Fig. 1. Example of a Wikipedia document

Existing approaches to structural retrieval have relied on rote return structures and ad-hoc tuning parameters to score elements. A naive approach assumes that XML documents are structured as articles, and so only logical elements such as articles, sections and paragraphs are returned in the search results. Another approach is to allow users to specify structure, such as using NEXI which is a notation for expressing XML queries that includes structural constraints and hints [11]. NEXI can be used in conjunction with XPATH to retrieve strict XML structural paths according to what the user specifies in the query. Other approaches to structural retrieval, like XRANK [6] or Clarke's Re-ranking Algorithm [4] , use element weighting schemes to iteratively score and re-rank results to improve the final system output.

To reliably and consistently apply these approaches to new collections requires a significant effort and cost in user studies. These approaches rely on ad-hoc heuristics and their tuning parameters must be derived directly from empirical user studies. User studies are always ultimately required to understand how to best satisfy end-user information needs, but preliminary work at INEX has suggested that the best structural elements are a function of the document structure found in the collection documents and not what the user specifies as hints or constraints in a query [10]. In other words, end-users can recognize good structure when they see it, but they are not good at a priori expressing it. So, for existing approaches that rely on user studies, there is little hope of successfully applying the methods to other collections without the costly overhead of a user study.

This proposal is concerned mainly with the challenge of returning the most appropriate XML elements. In this paper, we introduce an alternative approach that relies instead on an analysis of the document structure. We use structural summaries [3,8] to simulate how a reviewer of a document may browse from element to element within the document based on different structural factors such as incoming/outgoing paths between elements, or the relative amount of content in elements. In [2], it is shown how given an entry point to an XML document and a summary model of the document, one can estimate which other elements the reviewer may see while browsing from the entry point. [2] shows how this can be used to evaluate a system output using the Structural Relevance (SR) measure. Here, we show how SR can be applied to structurally score XML retrieval results that do not have assessments. Thus, if a given system provides a set of candidate outputs, then SR can be used to find the best one to ultimately return to the end-user. Our proposal augments a keyword-based search system with structural scoring. The results show that this augmentation can significantly improve the effectiveness of the keyword-based system, and the approach suggests that it may provide a more reliable, efficient and consistent way to apply structural hints and constraints to new XML collections. The paper is structured as follows. First, we present the proposed approach. Next, we present the IR system and the results of our participation in the INEX ad-hoc track for different summary models. Finally, we conclude that improving structural relevance improves the effectiveness of search systems, but there remains further work in better understanding users' preferences in focused results and the relationship between summary models and formal user models.

2   Overall Approach

2.1   Structural Relevance

Structural relevance (SR) is the expected relevance of a system output based on the probability of the reviewer seeing specific elements from a document only once when browsing the document repeatedly from multiple entry points [2]. We call this probability the isolation of element e in the system output R. By conditioning the relevance value of e on the isolation of each element in the system output, the expected relevance value of the system output can be evaluated. This approach allows us to replace the number of relevant documents in traditional metrics like precision with a probabilistic equivalent.

The SR of the ranked list of the system output R is,

$$SR(R) = \sum_{i=1}^{k} rel(e_i) \cdot P(e_i; R[e_i]) \tag{1}$$

where the system output R is a top-k ranked list of elements $R = \{e_1, e_2, \ldots, e_k\}$, $R[e]$ is the ranked list up to the rank of element e, and $rel(e)$ is the assessed relevance value of element e. The isolation $P(e; R[e])$ is the probability that element e will be seen only once (and, if not, considered not relevant by the user) in the ranked list up to the rank of element e, $R[e]$. So, in calculating SR, the expected relevance value of each element is a Bernoulli trial of whether the element e will not be seen by the user by browsing the higher- or tie-ranked elements. Thus, $SR[R]$ is the sum of the expected relevance value of elements in R in k Bernoulli trials that are assumed to be independent trials.

In [2], the isolation of an element in the system output $P(e; R)$ is shown to be decomposable into isolation between elements $P(e; f)$, where the two elements e and f are (i) from the same document, and (ii) f is either higher-ranked or tie-ranked to e. $P(e; f)$ refers to the probability that element e will be seen given that a reviewer enters the document from element f.

So, to modify a traditional metric like precision, we define Structural Relevance in Precision (SRP) by substituting Equation 1 for the number of relevant elements, as

$$SRP(R) = \frac{\sum_{i=1}^{k} rel(e_i) \cdot P(e_i; R[e_i])}{k} \tag{2}$$

where k is the number of elements in the system output.

In summary, isolation is a Markovian model of a user browsing a document. In element retrieval, SR uses isolation between elements as a weak independence assumption to evaluate results. This allows SR to evaluate the relevance of elements retrieved from the same document based on the rank order of elements and the isolation between the elements by conditioning the relevance of each element in the results with the overall isolation of each element in the system output.

2.2 Structural Summaries

In the previous section, we presented a brief overview of SR and discussed how it could be calculated using element-to-element probabilities. In order to make SR a practical measure to calculate, [2] provides an approximation of the isolation between elements $P(e; f)$ using structural summaries; a graph-based partitioning of elements in the collection. In this section, we show how a structural summary is calculated. In the next section, we show how summaries are used to approximate isolation between elements, and define different summary models for modeling different user behaviours.

A structural summary is a graph of the structure of the documents in a collection [3]. Each node in a summary graph represents a structural constraint

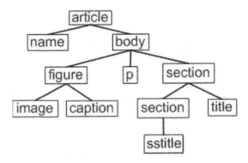

Fig. 2. Summary graph of Wikipedia document

Table 1. Weight of summary graph nodes by extent size

Path	Label	Weight (Extent Size)
/article	article	1
/article/name	name	1
/article/body	body	1
/article/body/figure	figure	1
/article/body/figure/image	img	1
/article/body/figure/caption	caption	1
/article/body/p	p	1
/article/body/section	sec	2
/article/body/section/title	title	2
/article/body/section/section	ss	1
/article/body/section/section/title	sstitle	1

on the collection that is a partition of the collection. So, in a summary, commonly occurring document structures in the collection define larger partitions than less common structures. Consider the document shown in Figure 1 and Figure 2 which shows the incoming path summary for the document. Each node in the summary graph represents a partition of the document based on the incoming label path defined from the document root article to the node. The extent of the partition is a conservative value that is calculated from the elements in the partition (such as the number of elements in the partition, or the number of characters contained in the elements in the partition). Table 1 shows the extents of the incoming summary of Figure 2 based on the number of elements in each partition.

2.3 Summary Models

In [2], the extent size of the structural summary are used to define the transition matrix for finding isolation. Table 2 shows a transition matrix based on the extents size as the weights for nodes (Table 1) of the incoming summary graph (Figure 2). For convenience, the incoming summary nodes are referred to by

Table 2. Node weights (by extent size) mapped into a transition matrix

Source	Destination											
	article	name	body	figure	img	caption	p	sec	title	ss	sstitle	π
article	0	1	1	0	0	0	0	0	0	0	0	0.091
name	1	0	0	0	0	0	0	0	0	0	0	0.045
body	1	0	0	1	0	0	1	2	0	0	0	0.227
figure	0	0	1	0	1	1	0	0	0	0	0	0.136
img	0	0	0	1	0	0	0	0	0	0	0	0.045
caption	0	0	0	1	0	0	0	0	0	0	0	0.045
p	0	0	1	0	0	0	0	0	0	0	0	0.045
sec	0	0	2	0	0	0	0	0	1	1	0	0.182
title	0	0	0	0	0	0	0	1	0	0	0	0.045
ss	0	0	0	0	0	0	0	1	0	0	1	0.091
sstitle	0	0	0	0	0	0	0	0	0	1	0	0.045

the labels in Table 1. We call these the summary node identifiers (SID). The transition weights between SID's is based on the extent of the graph nodes. Here, we assume that the transitions between summary nodes are equal and bidirectional. This allows us to consider the summary graph as a time-reversible Markov process. Let the SID's of two summary nodes be denoted as i and j, respectively. Let the transition weight between the nodes be denoted as w_{ij}, such that $w_{ij} = w_{ji}$. So, the stationary probabilities of the Markov model is computed as,

$$\pi_i = \frac{\sum_j w_{ij}}{\sum_i \sum_j w_{ij}} \tag{3}$$

Using the stationary probability, the isolation of summary node i is $P(i) = 1 - \pi_i$. The isolation of the node is the probability that, at any given time while browsing a document, the reviewer will *not* be in the partition defined by node i. The isolation between nodes in the summary can be used to approximate the isolation between elements in a document. Let e and f denote elements from the same document where f is the entry point and e is the element that represents the part of the document to not be seen from f. We denote the isolation between e and f using $P(e; f)$. The isolation of the elements is approximately the same as the inverse isolation for summary graph nodes if we assume that a user who is browsing will always enter the partition of element e from outside of the partition of the summary node of e. We denote the partition of the summary node of element e as (e). So, the isolation between elements e and f is approximately,

$$P(e; f) \approx \pi_{(e)} \tag{4}$$

For elements that are not from the same document, they are completely isolated so $P(e; f) = 1$. The isolation of an element from itself is zero, $P(e; e) = 0$. The important simplification that this approximation introduces is that the calculation of the isolation for an element is not dependent on any other elements in the system output. The final column in Table 2 shows the stationary probabilities for the summary graph of Table 1 based on the extents measured by incoming paths. From a user perspective, we interpret the probability π_i as the fraction of time that a user who uses a description of the document structure (*i.e.* the summary) to browse will spend π_i in partition i of the document.

The combination of a particular summary graph with node weights measured in a particular unit is called a *summary model*. By using different ways of weighting the nodes, we hypothesize that we can model different user behaviours via summary models. The weights can be measured using any nodal property that is conservative to an element, such as the extent size or number of bytes of content. For instance, instead of using the extent size (number of incoming paths) of each element, we could have used other measures such as the amount of content. The interpretation of any given summary model is beyond the scope of this paper. But, it is important to note that the summary acts as a single big parameter such that SR can be calculated efficiently for different summary models (*i.e.*, with different Markovian assumptions such as whether a reviewer will browse according to structural paths, node content or otherwise).

Finally, in this work, we define three summary models for the incoming summary; namely (i) extent size (*path*), (ii) amount of content in each element measured in characters (*content*), and (iii) the amount of content divided by the depth of the node in the summary graph (*depth*). In future work, we plan to investigate further how these models correspond to actual user behaviour.

2.4   Improving Structural Relevance of Results

Structural relevance (SR) is a measure of the isolation between elements. Roughly speaking, isolation is the degree to which the browsing of an element is independent of some other element. For instance, if a person browses into element A and will likely see B, but not vice versa, we say that element B is not isolated from A, but that A is isolated from B. In ranking, we use structural relevance in precision (SRP) as a structural score. By applying SRP, Equation 2, at every rank cut-off to a candidate system output, we can determine the effect of each element on the overall structural relevance of the system output.

To improve the structural relevance of results, let us assume that the IR system itself will provide a set of candidate system outputs Ω. For instance, in our system, Ω was composed of the set of different system outputs based on selecting the subsets of elements from the same documents that produce the highest structural scores. For any ranked list R composed of m distinct documents where R_i represents the elements in the ranked list R from document i such that $R = \bigcup_{i=1}^{m} R_i$, then the number of possible system outputs ℓ is,

$$\ell = |\Omega| = \prod_{i=1}^{m} |R_i|! \tag{5}$$

The algorithm shown in Figure 3 (below) evaluates each candidate system output in Ω to find the candidate with the highest structural score.

Algorithm. *FindMostIsolatedList*

Input: Summary of collection (π) and a weakly-ordered overlapped ranked list R.
Output: Ranked list with highest structural score R^*
1: let Ω be the set of possible outputs.
2: let ℓ be the number of outputs in Ω.
3: let R^* be the highest scoring output in Ω.
4: let $high$ be the highest SRP score found.
5: $high = 0$
6: **for** $j = 1$ to ℓ **do**
7:     let $R = \Omega^{(j)}$ be the j-th output in Ω
8:     let $score = SRP(R)$, /* see eq. 2 */
9:     **if** $score > high$ **then**
10:        $R^* = R$
11:    **end if**
12: **end for**

Fig. 3. Find the most structurally relevant output

In finding the most structurally relevant output, we assumed that all returned elements were relevant. We could have made different such as either using an expected search length (esl) [5] or randomly assigning elements as being relevant or not.

3   System Implementation

3.1   Lucene

A structural summary of all distinct paths in the Wikipedia XML collection was generated using code from DescribeX [3] which consisted of 55486 nodes (with aliasing on tags containing the substrings link, emph, template, list, item, or indentation). As the collection was summarized, modified Apache Lucene [1] code was used to index the tokens. The posting list also included character offsets. Tokens not excluded from the stop word filter were lower-cased and punctuation symbols were removed. Since the structural summary was generated at the same time as each document was summarized, payload information was associated with each token occurrence containing the summary partition in which the token appears.

To accommodate the structural hints in the INEX topics, separate indexes were built for each tag identified by the structural hint present within the set of INEX topics which included "article", "section", "p", "image", and "figure". For example, building an index for the "p" tag would index the first "p" element and its children, including nested "p" elements, until its respective closing tag. Thus, a file, with multiple non-overlapping indexed elements will create multiple documents within the index, is easily identified since the index stores the character offsets as previously mentioned. This results in having element-level documents that allows the calculation of idf scores for terms within elements. Table 3 shows the index sizes which includes the payload information.

Table 3. Index sizes using tag-level documents

Tag	Size
article	6.07GB
section	4.84GB
p	4.63GB
image	129MB
figure	192MB

Lucene's querying code was modified in order to accept INEX queries involving structural hints. The queries were encoded using boolean operators to represent token that were mandatory, optional, or to be excluded. Double quotes indicating adjacent tokens were removed since token positions were not indexed. Prior to running a query, the query was examined for any structural hints and the required indexes were searched as one using Lucene's regular application interface. If no structural hints were identified, the complete set of element indexes were used in the search.

3.2 Boosting

As previously mentioned, the collection was indexed at the element-level for "article", "section", "p", "image", and "figure" tags. In our experiments, we included runs with score boosting per term occurrence and using the average of the term scores as the overall document score. The boost used was the stationary probability (Equation 3) of the partition of the element in which the term occurs. The baseline payload score per occurrence was set to 1 and the boosted term score was the baseline plus the stationary probability. So, we implemented boosted runs for the three summary models considered in this paper; namely path, content and depth summary models.

3.3 Post-Processing

Lucene returned exhaustive results across indexes. The output of results from all indexes was combined into a a single weakly-ordered, overlapped ranked list. The different retrieval scenarios were then determined as described in Section 2.4. For overlapped elements, the first stage of post-processing would evaluate the effect of returning the different possible subsets of the overlapped elements (*i.e.,* finding elements that reduce SRP) with the caveats that (i) if two elements are originally ranked with one higher than the other, then the re-ranked result will maintain this relationship if both elements appear in the output, and (ii) for tied elements, any re-ordering of results is permissible. In our experiments, all elements in the results were assumed to be relevant. The most structurally relevant retrieval scenario was returned for the second stage of post-processing.

The second stage of post-processing was to remove overlap that would conflict with focused results. In focused results sibling elements are allowed in the results, but ancestor-descendant relationships between elements are not allowed. We implemented a simple rule that would choose the highest ranked ancestor-descendant element for final output, and removed all lower ranked elements from the final output.

4   Results

The graphs in Figure 4 show the overall results for the University of Toronto's submissions in the Focused Task for element retrieval in the INEX Ad-Hoc Track measured using HiXEval (left) and structural relevance in precision (right). HiX-Eval [7], the official INEX 2007 measure, was measured using average interpolated precision. Structural relevance in precision (SRP) was measured using a path summary model at rank cut-offs for $k = 100$ [2]. The two measures produce very similar system rankings (with an average rank position change of $+/-2.1$ positions out of 26 teams). Both measures suggest that there was a central "core" of systems that performed similarly (between 0.1 and 0.2 MAiP, and between 0.15 and 0.3 MASRP, respectively). In [2], it was shown that SRP using a path summary model agrees with XCG in ranking ad-hoc systems, and similarly HiX-Eval has been shown to also agree with XCG [7]. So, it is not surprising that these two measures agree with one and other in this respect. The results were produced using the algorithm presented in Section 2.4 using a path summary model of the Wikipedia collection.

We conducted further experiments to see whether we could improve our keyword-based XML retrieval system with structural scoring by boosting term occurrences in Lucene (see Section 3.2), and post-processing outputs from Lucene (see Section 2.4). We selected 10 test topics composed of 5 of our best performing topics (424,445,449,479, and 499) and 5 of our worst topics (414,503,511,519, and 542). All of the topics chosen had a non-zero precision $SRP > 0$. For these

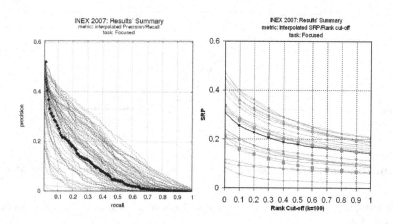

Fig. 4. Official INEX 2007 Results with HiXEval iP for focused task (left) and with SRP (right) for University of Toronto

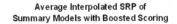

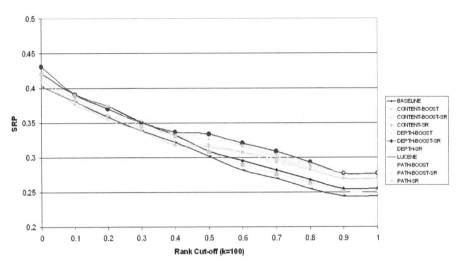

Fig. 5. Structural scoring using boosts and post-processing measured by MASRP

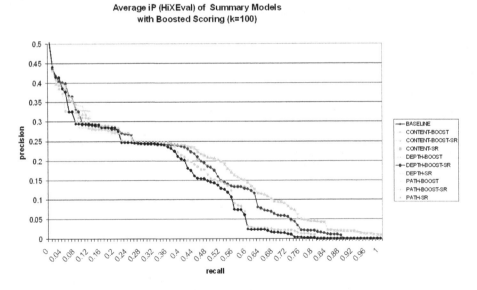

Fig. 6. Structural scoring using boosts and post-processing measured by MAiP

experiments, we used the path, depth and content summary models described in Section 2.3.

Altogether, we generated 11 new runs. The first run was the raw, overlapped results that were output directly from Lucene (*LUCENE*). Next, we resolved the overlap in the *LUCENE* run by returning only the highest ranked overlapped elements (*BASELINE*). Using the three different summary models (path, content and depth), we boosted the term occurrences in Lucene for each summary model as described in Section 3.2 and resolved the overlap in the Lucene output (*PATH-BOOST, CONTENT-BOOST*, and *DEPTH-BOOST*). Next, without boosting the Lucene output, we post-processed the Lucene output using each summary model as described in Section 2.4, and resolved the overlap from the post-processor output (*PATH-SR, CONTENT-SR*, and *DEPTH-SR*). Finally, we combined boosting, post-processing, and overlap resolution for our three final runs where the boost and post-processing both used the same summary model (*PATH-BOOST-SR, CONTENT-BOOST-SR*, and *DEPTH-BOOST-SR*). The results using HiXEval for 10 runs are shown in Figure 6. The *LUCENE* run was the output from Lucene with neither overlap resolution, boosting nor post-processing. The LUCENE run did not comply with the focused task because overlap was not removed from it (and, so, is not appropriate for measurement using the HiXEval measure for focused runs). The results using SRP for all 11 runs are shown in Figure 5. SRP makes no distinction between focused or unfocused constraints in the results, so the comparison of the focused to the unfocused run is valid.

In Table 4, we report the mean-average SRP and mean-average iP for the 11 experimental runs. According to both measures, our approach yielded improvements over *BASELINE*. Considering HiXEval, all runs outperformed the *BASELINE* run. The boosting of term occurrences was the most effective improvement. Boosting with the depth summary (*DEPTH-BOOST*) increased performance by 26% (from *BASELINE* $MAiP = 0.1442$ to *DEPTH-BOOST* $MAiP = 0.1817$). The post-processing of results using summary models showed only marginal improvements ($< 4\%$ improvements was seen for *PATH-SR, CONTENT-SR*, and *DEPTH-SR* runs). Moreover, the combination of both post-processing and boosting lessened the effect of the boosts alone (*e.g.*, there was a 15% improvement in the *DEPTH-BOOST-SR* run over the *BASELINE* run, as compared to a 26% improvement over *BASELINE* for the *DEPTH-BOOST* run).

The above analysis is not really fair because our approach to structural scoring used the output from Lucene (*LUCENE*) and not the baseline run (*BASELINE*). The Lucene run *LUCENE* cannot be accurately evaluated using the official INEX 2007 measure for the focused task because *LUCENE* is not a valid focused run. In saying that, intuitively from past experience at INEX, we know that it will have poor results because of the overlap, but, in improving systems, we content that it is critical that we can accurately evaluate invalid runs as well as valid runs.

Using SRP to evaluate performance, the post-processed-only runs did worse than *BASELINE* ($< 1\%$ lower performance for all summary models). But, all runs resulted in improvements over *LUCENE*. Boosting-only runs did not significantly improve results more than the *BASELINE* runs ($< 1\%$ for all summary models). But, the combination of boosts with post-processing, and in particular

Table 4. Overall performance for structural scoring using boosts and post-processing

RUN	MASRP	MAiP
BASELINE	0.3278	0.1442
CONTENT-BOOST	0.3271	0.1794
CONTENT-BOOST-SR	0.3374	0.1639
CONTENT-SR	0.3248	0.1499
DEPTH-BOOST	0.3298	**0.1817**
DEPTH-BOOST-SR	**0.3412**	0.1661
DEPTH-SR	0.3248	0.1499
LUCENE	0.3154	n/a
PATH-BOOST	0.3262	0.1804
PATH-BOOST-SR	0.3359	0.1653
PATH-SR	0.3248	0.1499

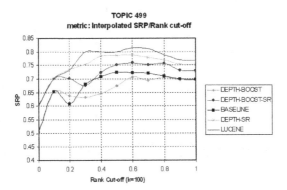

Fig. 7. Experimental runs for a best topic (Topic 499) using the depth summary model

the *DEPTH-BOOST-SR* run, showed a marked improvement in results (up to 8% improvement over *LUCENE* and 4% over *BASELINE*).

In Figures 7 and 8, we show the effects of the various stages of our approach using the depth summary model. Topic 499 (Figure 7) was one of our best submitted topics. Whereas, Topic 503 (Figure 8) was one of our worst submitted topics. Consider Topic 499, it was noted that *LUCENE* was the best run in SRP. Most of the elements that were returned for this run had been assessed as being relevant. So, using the heuristic of selecting only the highest ranking overlapped element for the *BASELINE* run, removed relevant elements from the output. It should be noted that, by boosting, there was a deterioration of results (both with and without post-processing). This result suggests that for highly relevant, but unfocused results, to find the best focused results using our approach would be to apply the post-processor and then remove the overlap. In Topic 499, the largest improvement from the *BASELINE* run was the *DEPTH-SR* run (an improvement of 10.7%).

In contrast, if we consider Topic 503 (Figure 8), then we see the exact opposite. The unfocused output of Lucene is unaffected by removing overlap and/or

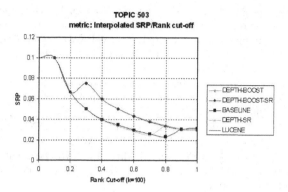

Fig. 8. Experimental runs for a worst topic (Topic 503) using the depth summary model

applying post-processing, but we note a marked improvement when we introduce boosting. The largest improvement between the *BASELINE* run was the *DEPTH-BOOST* run which was an improvement of 19.4%. Overall, this suggests that both the boost and post-processing are necessary in order to use our approach to improve unfocused keyword search systems such as Lucene for the Focused Task.

In general, we conclude that, in using our approach, the most effective strategy to improve keyword search for focused retrieval in XML is the combination of boosts and post-processing with the depth summary model. There remains an outstanding question as to how these results relate to evaluations based on empirical user models. But, these results strongly suggest that the use of SR for structural scoring is an effective way to improve keyword-based search systems.

5   Conclusion

We have presented a novel method for introducing structural scoring into element retrieval where the parameterization of our model allows for complex modeling of user behaviour based on summary models derived from the collection. Our approach does not make any assumptions about the collection, and can be easily applied to any XML collection. The experimental results suggest that our structural approach can improve results. Future research will involve comparing empirical user models to summary models, applying summary models other than incoming summaries, and using different summary models for personalizing search.

References

1. Apache Lucene Java (2008), http://lucene.apache.org
2. Ali, M., Consens, M.P., Lalmas, M.: Structural Relevance in XML Retrieval Evaluation. In: SIGIR 2007 Workshop on Focused Retrieval, Amsterdam, The Netherlands, 2007, July 27 (2007)

3. Ali, M.S., Consens, M., Khatchadourian, S., Rizzolo, F.: DescribeX: Interacting with AxPRE Summary Descriptions. In: ICDE 2008 (accepted, 2008)
4. Clarke, C.: Controlling overlap in content-oriented XML retrieval. In: SIGIR 2005: Proc. of the 28th Ann. Intl. ACM SIGIR Conf. on Res. and Dev. in IR, pp. 314–321. ACM Press, New York (2005)
5. Cooper, W.S.: Expected search length: A single measure of retrieval effectiveness based on weak ordering action of retrieval systems. J of the Amer. Soc. for Info. Scie. 19, 30–41 (1968)
6. Guo, L., Shao, F., Botev, C., Shanmugasundaram, J.: XRANK: Ranked keyword search over xml documents. In: ACM SIGMOD. ACM Press, New York (2003)
7. Pehcevski, J., Thom, J.A.: HiXEval: Highlighting XML retrieval evaluation. In: Fuhr, N., Lalmas, M., Malik, S., Kazai, G. (eds.) INEX 2005. LNCS, vol. 3977. Springer, Heidelberg (2006)
8. Polyzotis, N., Garofalakis, M.: Statistical synopses for graph-structured xml databases. In: SIGMOD 2002: Proceedings of the 2002 ACM SIGMOD international conference on Management of data, pp. 358–369. ACM Press, New York (2002)
9. Trotman, A., Geva, S.: Passage retrieval and other xml-retrieval tasks. In: Proc. SIGIR 2006 Workshop on XML Element Retrieval Methodology, pp. 43–50 (2006)
10. Trotman, A., Lalmas, M.: Why structural hints in queries do not help xml-retrieval. In: SIGIR 2006: Proceedings of the 29th annual international ACM SIGIR conference on Research and development in information retrieval, pp. 711–712. ACM Press, New York (2006)
11. Trotman, A., Sigurbjornsson, B.: Narrowed extended XPath I (NEXI). In: Proc. INEX Workshop, pp. 16–39 (2004)

TopX @ INEX 2007

Andreas Broschart[1], Ralf Schenkel[1], Martin Theobald[2], and Gerhard Weikum[1]

[1] Max-Planck-Institut für Informatik, Saarbrücken, Germany
{abrosch,schenkel,weikum}@mpi-inf.mpg.de
http://www.mpi-inf.mpg.de/departments/d5/
[2] Stanford University
theobald@stanford.edu
http://infolab.stanford.edu/

Abstract. This paper describes the setup and results of the Max-Planck-Institut für Informatik's contributions for the INEX 2007 Ad-Hoc Track task. The runs were produced with TopX, a search engine for ranked retrieval of XML data that supports a probabilistic scoring model for full-text content conditions and tag-term combinations, path conditions as exact or relaxable constraints, and ontology-based relaxation of terms and tag names.

1 System Overview

TopX [2,5] aims to bridge the fields of database systems (DB) and information retrieval (IR). From a DB viewpoint, it provides an efficient algorithmic basis for top-k query processing over multidimensional datasets, ranging from structured data such as product catalogs (e.g., bookstores, real estate, movies, etc.) to unstructured text documents (with keywords or stemmed terms defining the feature space) and semistructured XML data in between. From an IR viewpoint, TopX provides ranked retrieval based on a relevance scoring function, with support for flexible combinations of mandatory and optional conditions as well as text predicates such as phrases, negations, etc. TopX combines these two aspects into a unified framework and software system, with emphasis on XML ranked retrieval.

Figure 1 depicts the main components of the TopX system. The *Indexer* parses and analyzes the document collection and builds the index structures for efficient lookups of tags, content terms, phrases, structural patterns, etc. TopX currently uses Oracle10g as a storage system, but the JDBC interface would easily allow other relational backends, too. An *Ontology* component manages optional ontologies with various kinds of semantic relationships among concepts and statistical weighting of relationship strengths.

At query run-time, the *Core Query Processor* decomposes queries (which can be either NEXI or XPath Full-Text) and invokes the top-k algorithms. It maintains

N. Fuhr et al. (Eds.): INEX 2007, LNCS 4862, pp. 49–56, 2008.

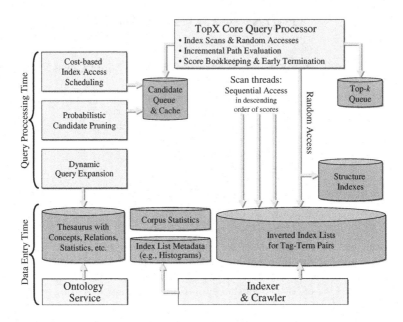

Fig. 1. TopX architecture

intermediate top-k results and candidate items in a priority queue, and it schedules accesses on the precomputed index lists in a multi-threaded architecture. Several advanced components provide means for run-time acceleration:

- The *Probabilistic Candidate Pruning* component [6] allows TopX to drop candidates that are unlikely to qualify for the top-k results at an early stage, with a controllable loss and probabilistic result guarantees.
- The *Index Access Scheduler* [1] provides a suite of scheduling strategies for sorted and random accesses to index entries.
- The *Incremental Path Evaluation* uses additional cost models to decide when to evaluate structural conditions like XML path conditions, based on specialized indexes for XML structure.
- The *Dynamic Query Expansion* component [4] maps the query keywords and/or tags to concepts in the available ontology and incrementally generates query expansion candidates.

As our INEX runs focused on result quality, not on efficiency, they were produced using only the Index Access Scheduler and Incremental Path Evaluation. TopX supports three different front-ends: a servlet with an HTML end-user interface (that was used for the topic development of INEX 2006 and 2007), a Web Service with a SOAP interface (that was used by the Interactive track), and a Java API (that was used to generate our runs).

2 Data Model and Scoring

We refer the reader to [2] for a thorough discussion of the scoring model. This section shortly reviews important concepts.

2.1 Data Model

We consider a simplified XML data model, where idref/XLink/XPointer links are disregarded. Thus every document forms a tree of nodes, each with a *tag* and a related *content*. We treat attributes nodes as children of the corresponding element node. The content of a node is either a text string or it is empty. With each node, we associate its *full-content* which is defined as the concatenation of the text contents of all the node's descendants in document order.

2.2 Content Scores

For content scores we make use of element-specific statistics that view the full-content of each element as a bag of words:

1) the *full-content term frequency*, $ftf(t, n)$, of term t in node n, which is the number of occurrences of t in the full-content of n;
2) the *tag frequency*, N_A, of tag A, which is the number of nodes with tag A in the entire corpus;
3) the *element frequency*, $ef_A(t)$, of term t with regard to tag A, which is the number of nodes with tag A that contain t in their full-contents in the entire corpus.

The score of an element e with tag A with respect to a content condition of the form T[about(., t)] (where T is either e's tag A or the tag wildcard operator *) is then computed by the following BM25-inspired formula:

$$score(e, \text{T[about(., t)]}) = \tag{1}$$
$$\frac{(k_1 + 1)\, ftf(t, e)}{K + ftf(t, n)} \cdot \log\left(\frac{N_A - ef_A(t) + 0.5}{ef_A(t) + 0.5}\right)$$

with $K =$
$$k_1\left((1 - b) + b\frac{\sum_{t'} ftf(t', e)}{avg\{\sum_{t'} ftf(t', e') \mid e' \text{ with tag } A\}}\right)$$

Note that this definition enforces a strict evaluation of the query condition, i.e., only elements whose tag matches the query tag get a non-zero score. For a query content condition with multiple terms, the score of an element satisfying the tag constraint is computed as the sum of the element's content scores for the corresponding content conditions, i.e.:

$$score(e, \text{T[about(., } t_1 \ldots t_m)\text{]}) = \sum_{i=1}^{m} score(e, \text{T[about(., } t_i)\text{]}) \tag{2}$$

Note that content-only (aka "CO") queries have not been in the focus when defining this scoring function, but keyword conditions as subconditions of structural queries. CO queries are therefore evaluated with the tag wildcard operator '*' which matches any tag, and hence the score for an element e (whose tag is A) with respect to a CO query //*[about(.,$t_1 \ldots t_m$)] is defined as

$$score(e, *[\text{about}(., \ t_1 \ldots t_m)]) = \sum_{i=1}^{m} score(e, A[\text{about}(., \ t_i)]) \quad (3)$$

Another option for scoring content-only queries would be to consider the whole pool of elements at once, removing any constraints on the tag of results in the scoring formula (note that this corresponds to a BM25 scoring function computed over all elements):

$$score'(e, //*[\text{about}(., \ \text{t})]) = \quad (4)$$
$$\frac{(k_1 + 1) \ ftf(t, e)}{K' + ftf(t, n)} \cdot \log \left(\frac{N - ef(t) + 0.5}{ef(t) + 0.5} \right)$$

$$\text{with } K' =$$
$$k_1 \left((1 - b) + b \frac{\sum_{t'} ftf(t', e)}{avg\{\sum_{t'} ftf(t', e')\}} \right)$$

Here, N is the number of elements in the collection, and $ef(t)$ is the number of elements that contain the term t. This variant of the scoring function is a straightforward application of text retrieval methods to XML, where each element is considered as independent 'document' for the scoring. We will report additional (non-official) experiments with this modified scoring function later.

TopX provides the option to evaluate queries either in conjunctive mode or in "andish" mode. In the first case, all terms (and, for content-and-structure queries, all structural conditions) must be met by a result candidate, but still different matches yield different scores. In the second case, a node is already considered a match if it satisfies at least one content condition in the target dimension specified in the NEXI/XPath query. Evaluating content-and-structure queries in this "andish" mode allows for results which do not strictly match all query conditions; however, even in this mode, tags specified in the query must be strictly matched.

Orthogonally to this, TopX can be configured to return two different granularities as results: in *document mode*, TopX returns the best documents for a query, whereas in *element mode*, the best target elements are returned, which may include several elements from the same document. For the INEX experiments in this year's AdHoc track, we used element mode with some additional post-processing for the Focused task, and document mode for the RelevantIn-Context and BestInContext tasks.

2.3   Structural Scores

Given a query with structural and content conditions, we transitively expand all structural query dependencies. For example, in the query //A//B//C[about(., t)] an element with tag C has to be a descendant of both A and B elements. Branching path expressions can be expressed analogously. This process yields a *directed acyclic graph* (DAG) with tag-term conditions as leaves, tag conditions as inner nodes, and all transitively expanded descendant relations as edges.

Our structural scoring model essentially counts the number of navigational (i.e., tag-only) conditions that are completely satisfied by a result candidate and assigns a small and constant score mass c for every such condition that is matched. This structural score mass is combined with the content scores. In our setup we have set $c = 1$, whereas content scores are normalized to $[0, 1]$, i.e., we emphasize the structural parts.

3   AdHoc Track Results

As the recent development of TopX has focused on efficiency issues, its scoring function used to rank results did not change from the experiments reported last year [3]. The discussion of the experimental results in this section therefore focuses on differences introduced by the new metrics used for INEX 2007.

For each subtask, we submitted the following four runs:

- CO-{subtask}-all: a CO run that considered the terms in the title of a topic without phrases and negations, allowing all tags for results.
- CO-{subtask}-ex-all: a CO run that considered terms as well as phrases and negations (so-called *expensive predicates*), again without limiting tags of results.
- CAS-{subtask}-all: a CAS run that considered the castitle of a topic if it was available, and the title otherwise. The target tag was evaluated strictly, whereas support conditions were optional; phrases and negations were ignored.
- CAS-{subtask}-ex-all: a CAS run that additionally considered phrases and negations.

All runs were created without stemming and without stopword removal in the collection (but removing stopwords from the queries).

3.1   Focused Task

Our runs for the focused task were produced by first producing a run with all results (corresponding to the *Thorough* task in previous years) and then post-processing the run to remove any overlap. For each such run, we kept an element e if there was no other element e' from the same document in the run that had a higher score than e and had a path that overlapped with e's path.

Table 1. Official results for the Focused Task: interpolated precision at different recall levels (ranks are in parentheses) and mean average interpolated precision

run	iP[0.00]	iP[0.01]	iP[0.05]	iP[0.10]	MAiP
TOPX-CAS-Focused-all	0.5780 (3)	0.5066 (10)	0.4006 (16)	0.3430 (26)	0.1307 (30)
TOPX-CAS-Focused-ex-all	0.5321 (21)	0.4740 (23)	0.3644 (40)	0.3067 (38)	0.1228 (35)
TOPX-CO-Focused-all	0.5300 (22)	0.4777 (21)	0.3879 (28)	0.3275 (31)	0.1227 (36)
TOPX-CO-Focused-ex-all	0.5434 (13)	0.4893 (16)	0.3999 (18)	0.3289 (30)	0.1231 (34)

Official Submissions. This simple, syntactic post-processing yielded good results for the CAS runs (shown in Table 1). Especially for the early recall levels, TopX performed well with peak rank 3 for iP[0.0] and rank 10 in the official result (iP[0.01]). Interestingly, the CAS run that considered phrases and negation did slightly worse than its counterpart without expensive predicates, whereas the CO run with phrases and negation did better than the plain CO run. Compared to 2006, the results are surprising as our CO runs were much better than our CAS runs then.

Additional Runs. In addition to our official runs, we evaluated (1) the modified scoring function *score'*, (2) the effect of stemming, (3) the effect of limiting results to *article* elements only, and (4) the effect of conjunctive query evaluation on the article level.

Table 2. Additional results for the Focused Task: interpolated precision at different recall levels and mean average interpolated precision (virtual ranks are in parentheses)

run	iP[0.00]	iP[0.01]	iP[0.05]	iP[0.10]	MAiP
score'	0.5593 (8)	0.5100 (9)	0.3880 (27)	0.3310 (31)	0.1343 (30)
score', article-only	0.4832 (47)	0.4723 (26)	0.4431 (8)	0.4120 (4)	0.2175 (2)
score', stems	0.5710 (4)	0.5296 (1)	0.4230 (13)	0.3751 (15)	0.1583 (22)
score', stems, article-only	0.4762 (47)	0.4671 (30)	0.4467 (8)	0.4174 (3)	0.2268 (1)
score', stems, article-only, conj	0.4400 (52)	0.4293 (47)	0.4022 (16)	0.3806 (13)	0.1737 (15)

Table 2 shows the results for these settings. Note that we limited the set of elements to those with tags article, body, section, p, normallist and item for efficiency reasons; results with the complete set of elements did not differ much. It is evident that the modified scoring is more effective for these queries than the original scoring used for the original runs. Additionally, stemming further improves results, with a peak (virtual) rank of 1 for iP[0.01]. Runs where results were limited to article elements only delivered best results for late recall points, and again stemming helps to further improve result quality. No element-based run of any participating group had a better MAiP value than our article-only run with stemming (only RMIT's article-only run achieved a comparable performance). Switching to conjunctive queries did not improve results.

Table 3. Results for the RelevantInContext Task: generalized precision/recall at different ranks and mean average generalized precision (ranks are in parentheses)

run	gP[5]	gP[10]	gP[25]	gP[50]	MAgP
TOPX-CO-all-RIC	0.2081 (30)	0.1963 (22)	0.1443 (28)	0.1106 (26)	0.1071 (29)
TOPX-CO-ex-all-RIC	0.2168 (27)	0.1879 (28)	0.1356 (30)	0.1050 (31)	0.1077 (28)
TOPX-CAS-RIC	0.2201 (26)	0.1941 (24)	0.1486 (27)	0.1086 (28)	0.1003 (35)
TOPX-CAS-ex-RIC	0.1612 (42)	0.1528 (42)	0.1205 (41)	0.0893 (41)	0.0842 (39)

Table 4. Results for the BestInContext Task: generalized precision/recall at different ranks and mean average generalized precision (ranks are in parentheses)

run	gP[5]	gP[10]	gP[25]	gP[50]	MAgP
TOPX-CO-all-BIC	0.2005 (44)	0.2053 (42)	0.1735 (38)	0.1320 (37)	0.1348 (36)
TOPX-CO-ex-all-BIC	0.2078 (43)	0.1940 (43)	0.1637 (42)	0.1646 (41)	0.1324 (33)
TOPX-CAS-BIC	0.2591 (25)	0.2294 (30)	0.1874 (29)	0.1330 (36)	0.1287 (38)
TOPX-CAS-ex-BIC	0.2338 (36)	0.2167 (37)	0.1767 (37)	0.1294 (38)	0.1280 (39)

3.2   RelevantInContext Task

To produce the runs for the RelevantInContext task, we ran TopX in document mode. This yielded a list of documents ordered by the highest score of any element within the document, together with a list of elements and their scores for each document.

The relative results (Table 3) are worse than 2006, with peak rank of 22 at 10 documents. Comparing the absolute values of the metrics for 2006 and 2007 shows that TopX even performed slightly better this year than last year (with a peak MAgP of 0.0906 in 2006 and 0.1077 in 2007, both measured with the 2007 metrics), but the other participants seemingly improved their systems a lot.

3.3   BestInContext Task

To compute the best entry point for a document, we post-processed the RelevantInContext runs by simply selecting the element with highest score from each document and ordered them by score. The results (Table 4) show that this did not work as well as 2006, with a peak rank of 25 this year (compared to a peak rank of 1 for 2006). Unlike the RelevantInContext task, the absolute performance values are also a lot lower (0.2096 in 2006 vs. 0.1348 in 2007). Especially CO runs performed much worse than expected in general, even though they performed better than our CAS runs or mean average generalized precision. We attribute this to the fact that we evaluated target tags strictly in CAS runs, so we limited our choice of best entry points to elements with these tags.

4   Conclusion

This paper presented the results of the runs produced for the INEX 2007 AdHoc Track with the TopX search engine. This year, runs using CAS topics performed better than runs with CO topics, and TopX performed especially well for the Focused task. Additional experiments showed that a modified scoring function and stemming can further improve results for CO queries in the Focused tasks, and that it is difficult to beat an article-only run.

References

1. Bast, H., Majumdar, D., Theobald, M., Schenkel, R., Weikum, G.: IO-Top-k: Index-optimized top-k query processing. In: Proceedings of the 32nd International Conference on Very Large Data Bases (VLDB 2006), pp. 475–486 (2006)
2. Theobald, M., Bast, H., Majumdar, D., Schenkel, R., Weikum, G.: Topx: efficient and versatile top-k query processing for semistructured data. VLDB J. (accepted for publication 2008)
3. Theobald, M., Broschart, A., Schenkel, R., Solomon, S., Weikum, G.: Topx – adhoc track and feedback task. In: Proceedings of the 5th International Workshop of the Initiative for the Evaluation of XML Retrieval (INEX 2006), pp. 233–242 (2006)
4. Theobald, M., Schenkel, R., Weikum, G.: Efficient and self-tuning incremental query expansion for top-k query processing. In: Proceedings of the 28th Annual International ACM SIGIR Conference on Research and Development in Information Retrieval, pp. 242–249 (2005)
5. Theobald, M., Schenkel, R., Weikum, G.: An efficient and versatile query engine for TopX search. In: Proceedings of the 31st International Conference on Very Large Data Bases (VLDB 2005), pp. 625–636 (2005)
6. Theobald, M., Weikum, G., Schenkel, R.: Top-k query evaluation with probabilistic guarantees. In: Proceedings of the 30th International Conference on Very Large Data Bases (VLDB 2004), pp. 648–659 (2004)

The Garnata Information Retrieval System at INEX'07

Luis M. de Campos, Juan M. Fernández-Luna, Juan F. Huete,
Carlos Martín-Dancausa, and Alfonso E. Romero

Departamento de Ciencias de la Computación e Inteligencia Artificial
E.T.S.I. Informática y de Telecomunicación, Universidad de Granada,
18071 – Granada, Spain
{lci,jmfluna,jhg,cmdanca,aeromero}@decsai.ugr.es

Abstract. This paper exposes the results of our participation at INEX'07 in the AdHoc track and the comparison of these results with respect to the ones obtained last year. Three runs were submitted to each of the Focused, Relevant In Context and Best In Context tasks, all of them obtained with Garnata, our Information Retrieval System for structured documents. As in the past year, we use a model based on Influence Diagrams, the CID model. The result of our participation has been better than the last year and we have reached an acceptable position in the ranking for the three tasks. In the paper we describe the model, the system and we show the differences between our systems at INEX'06 and INEX'07, which make possible to get a better performance.

1 Introduction

This is the second year that members of the research group "Uncertainty Treatment in Artificial Intelligence" at the University of Granada submit runs to the INEX official tasks, although before 2006 we also contributed to INEX with the design of topics and the assessment of relevance judgements. Like in the past year, we have participated in the Ad hoc Track with an experimental platform to perform structured retrieval using Probabilistic Graphical Models [5,8,10], called Garnata [4].

This year we have improved the version of Garnata that we used at INEX'06 in two ways, and we have also adapted it to cope with the three, non thorough tasks proposed this year, namely Focused, Relevant in Context and Best in Context. For each of these tasks, we have submitted three runs, all of them using Garnata with a different set of parameters. The results of this second participation are considerably better than those of the past year, where we were in the last positions of the ranking. Nevertheless, we are still quite far from the first positions, so there is still room for improvement, and more research and experimentation need to be carried out.

The paper is organised as follows: the next section describes the probabilistic graphical models underlying Garnata. Sections 3 and 4 give details about the new characteristics/improvements incorporated into the system and the adaptation

N. Fuhr et al. (Eds.): INEX 2007, LNCS 4862, pp. 57–69, 2008.

of Garnata to generate outputs valid for the three tasks, respectively. In Section 5 we discuss the experimental results. The paper ends with the conclusions and some proposals for future work with our system.

2 Probabilistic Graphical Models in the Garnata System

The Garnata IRS is based on probabilistic graphical models, more precisely an influence diagram and the corresponding underlying Bayesian network. In this section we shall describe these two models and how they are used to retrieve document components from a document collection through probabilistic inference (see [2,3] for more details). Alternative probabilistic graphical models for structured information retrieval can also be found in the literature [6,7,9]. We assume a basic knowledge about graphical models.

2.1 The Underlying Bayesian Network

We consider three different kinds of entities associated to a collection of structured documents, which are represented by the means of three different kinds of random variables: *index terms*, *basic structural units*, and *complex structural units*. These variables are in turn represented in the Bayesian network through the corresponding *nodes*. Term nodes form the set $\mathcal{T} = \{T_1, T_2, \ldots, T_l\}$; $\mathcal{U}_b = \{B_1, B_2, \ldots, B_m\}$ is the set of basic structural units, those document components which only contain terms, whereas $\mathcal{U}_c = \{S_1, S_2, \ldots, S_n\}$ is the set of complex structural units, that are composed of other basic or complex units. For those units containing both text and other units, we consider them as complex units, and the associated text is assigned to a new basic unit called *virtual unit*, see the example in Figure 1[1]. The set of all structural units is therefore $\mathcal{U} = \mathcal{U}_b \cup \mathcal{U}_c$.

The binary random variables associated with each node T, B or S take its values from the sets $\{t^-, t^+\}$, $\{b^-, b^+\}$ or $\{s^-, s^+\}$ (the term/unit is not relevant or is relevant), respectively. A unit is considered relevant for a given query if it satisfies the user's information need expressed by this query. A term is relevant in the sense that the user believes that it will appear in relevant units/documents.

Regarding the arcs of the model, there will be an arc from a given node (either term or structural unit) to the particular structural unit the node belongs to. The hierarchical structure of the model determines that each structural unit $U \in \mathcal{U}$ has *only one* structural unit as its child: the unique structural unit containing U (except for the leaf nodes, i.e. the complete documents, which have no child). We shall denote $U_{hi(U)}$ the single child node associated with node U (with $U_{hi(U)} = $ null if U is a leaf node).

To assess the numerical values for the required probabilities $p(t^+)$, $p(b^+ |pa(B))$ and $p(s^+|pa(S))$, for every node in $\mathcal{T}$, $\mathcal{U}_b$ and $\mathcal{U}_c$, respectively, and every

---

[1] Of course this type of unit is non-retrievable and it will not appear in the XPath route of its descendants, it is only a formalism that allows us to clearly distinguish between units containing only text and units containing only other units.

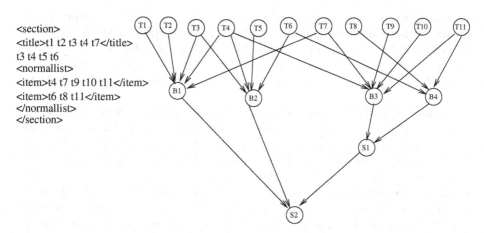

Fig. 1. Sample XML text and the corresponding Bayesian network. Ti represent index terms; the basic unit B1 corresponds with the tag `<title>`, and B3 and B4 with the tag `<item>`; the complex units S1 and S2 correspond with the tags `<normallist>` and `<section>` respectively; B2 is a virtual unit used to store the text within S2 which is not contained in any other unit inside it.

configuration $pa(X)$ of the corresponding parent sets $Pa(X)$, we use the canonical model proposed in [1], which supports a very efficient inference procedure. These probabilities are defined as follows:

$$\forall B \in \mathcal{U}_b, \quad p(b^+|pa(B)) = \sum_{T \in R(pa(B))} w(T, B), \tag{1}$$

$$\forall S \in \mathcal{U}_c, \quad p(s^+|pa(S)) = \sum_{U \in R(pa(S))} w(U, S), \tag{2}$$

where $w(T, B)$ is a weight associated to each term T belonging to the basic unit B and $w(U, S)$ is a weight measuring the importance of the unit U within S. In any case $R(pa(U))$ is the subset of parents of U (terms for B, and either basic or complex units for S) relevant in the configuration $pa(U)$, i.e., $R(pa(B)) = \{T \in Pa(B) \mid t^+ \in pa(B)\}$ and $R(pa(S)) = \{U \in Pa(S) \mid u^+ \in pa(S)\}$. These weights can be defined in any way with the only restrictions that

$$w(T, B) \geq 0, \quad w(U, S) \geq 0, \quad \sum_{T \in Pa(B)} w(T, B) \leq 1, \text{ and } \sum_{U \in Pa(S)} w(U, S) \leq 1.$$

2.2   The Influence Diagram Model

The Bayesian network is now enlarged by including decision nodes, representing the possible alternatives available to the decision maker, and utility nodes, thus transforming it into an influence diagram. For each structural unit $U_i \in \mathcal{U}$,

R_i represents the decision variable related to whether or not to return U_i to the user (with values r_i^+ and r_i^-, meaning 'retrieve U_i' and 'do not retrieve U_i', respectively), and the utility node V_i measures the value of utility for the corresponding decision. We shall also consider a *global utility node* Σ representing the joint utility of the whole model (we assume an additive behavior of the model).

In addition to the arcs between the nodes present in the Bayesian network, a set of arcs pointing to utility nodes are also included, employed to indicate which variables have a direct influence on the desirability of a given decision. In order to represent that the utility function of V_i obviously depends on the decision made and the relevance value of the structural unit considered, we use arcs from each structural unit node U_i and decision node R_i to the utility node V_i. Moreover, we include also arcs going from $U_{hi(U_i)}$ to V_i, which represent that the utility of the decision about retrieving the unit U_i also depends on the relevance of the unit which contains it (of course, for those units U where $U_{hi(U)} = $ null, this arc does not exist). The utility functions associated to each utility node V_i are therefore $v(r_i, u_i, u_{hi(U_i)})$, with $r_i \in \{r_i^-, r_i^+\}$, $u_i \in \{u_i^-, u_i^+\}$, and $u_{hi(U_i)} \in \{u_{hi(U_i)}^-, u_{hi(U_i)}^+\}$.

Finally, the utility node Σ has all the utility nodes V_i as its parents. These arcs represent the fact that the joint utility of the model will depend on the values of the individual utilities of each structural unit. Figure 2 displays the influence diagram corresponding to the previous example.

2.3   Inference and Decision Making

Our objective is, given a query, to compute the expected utility of retrieving each structural unit, and then to give a ranking of those units in decreasing order of expected utility (at this moment we assume a thorough task, i.e. structural units in the output may overlap. In Section 4 we shall see how overlapping may be removed). Let $\mathcal{Q} \subseteq \mathcal{T}$ be the set of terms used to express the query. Each term $T_i \in \mathcal{Q}$ will be instantiated to t_i^+; let q be the corresponding configuration of the variables in $\mathcal{Q}$. We wish to compute the expected utility of each decision given q. As we have assumed a global additive utility model, and the different decision variables R_i are not directly linked to each other, we can process each one independently. The expected utilities for retrieving each U_i can be computed by means of:

$$EU(r_i^+ \mid q) = \sum_{\substack{u_i \in \{u_i^-, u_i^+\} \\ u_{hi(U_i)} \in \left\{ u_{hi(U_i)}^-, u_{hi(U_i)}^+ \right\}}} v(r_i^+, u_i, u_{hi(U_i)}) \, p(u_i, u_{hi(U_i)} \mid q) \qquad (3)$$

Although the bidimensional posterior probabilities $p(u_i, u_{hi(U_i)} \mid q)$ in eq. (3) could be computed exactly, it is much harder to compute them that the unidimensional posterior probabilities $p(u_i \mid q)$, which can be calculated very efficiently due to the specific characteristics of the canonical model used to define the conditional probabilities and the network topology. So, we approximate the bidimensional

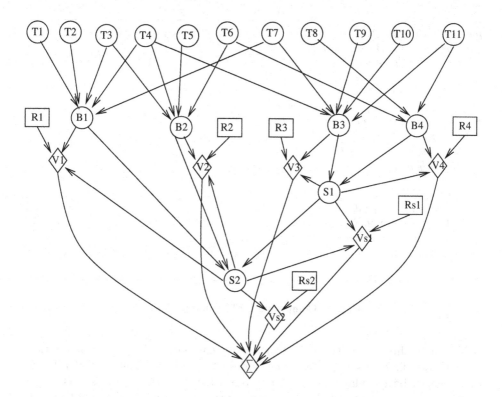

Fig. 2. Influence diagram for the example in Figure 1

probabilities as $p(u_i, u_{hi(U_i)}|q) = p(u_i|q) \times p(u_{hi(U_i)}|q)$. The computation of the unidimensional probabilities is based on the following formulas [2,3]:

$$\forall B \in \mathcal{U}_b, \quad p(b^+|q) = \sum_{T \in Pa(B) \setminus Q} w(T, B)\, p(t^+) + \sum_{T \in Pa(B) \cap R(q)} w(T, B), \quad (4)$$

$$\forall S \in \mathcal{U}_c, \quad p(s^+|q) = \sum_{U \in Pa(S)} w(U, S)\, p(u^+|q). \quad (5)$$

Figure 3 shows an algorithm that efficiently computes these probabilities, derived from eqs. (4) and (5), traversing only the nodes of the graph that require updating. It is assumed that the prior probabilities of all the nodes are stored in prior[X]; the algorithm uses variables prob[U] which, at the end of the process, will store the corresponding posterior probabilities. Essentially, the algorithm starts from the terms in Q and carries out a width graph traversal until it reaches the basic units that require updating, thus computing $p(b^+|q)$. Then, starting from these modified basic units, it carries out a depth graph traversal to compute $p(s^+|q)$, only for those complex units that require updating.

```
for each item T in Q
    for each unit B child of T
        if (prob[B] exists)
            prob[B] += w(T,B)*(1-prior[T]);
        else { create prob[B];
                prob[B] = prior[B]+w(T,B)*(1-prior[T]); }
for each basic unit B s.t. prob[B] exists {
    U = B; prod = prob[B]–prior[B];
    while (U_{hi(U)} is not NULL) {
        S = U_{hi(U)};
        prod *= w(U,S);
        if (prob[S] exists)
            prob[S] += prod;
        else { create prob[S];
                prob[S] = prior[S]+prod; }
        U = S; }
}
```

Fig. 3. Computing $p(b^+|q)$ and $p(s^+|q)$

The algorithm that initialises the process by computing the prior probabilities prior[U] (as the terms $T \in \mathcal{T}$ are root nodes, the prior probabilities prior[T] do not need to be calculated, they are stored directly in the structure) is quite similar to the previous one, but it needs to traverse the graph starting from all the terms in $\mathcal{T}$.

3   Changes from the Model Presented at INEX 2006

The two changes with respect to the model used at INEX'06 are related to the parametric part of the Garnata model. We explain first these parameters uses at INEX'06, before describing the changes made.

3.1   Parameters in Garnata

The parameters that need to be fixed in order to use Garnata are the prior probabilities of relevance of the terms, $p(t^+)$, the weights $w(T, B)$ and $w(U, S)$ used in eqs. (4) and (5), and the utilities $v(r_i^+, u_i, u_{hi(U_i)})$.

For the prior probabilities Garnata currently uses an identical probability for all the terms, $p(t^+) = p_0, \forall T \in \mathcal{T}$, with $p_0 = \frac{1}{|\mathcal{T}|}$.

The weights of the terms in the basic units, $w(T, B)$, follow a normalized tf-idf scheme:

$$w(T, B) = \frac{tf(T, B) \times idf(T)}{\sum_{T' \in Pa(B)} tf(T', B) \times idf(T')} \tag{6}$$

The weights of the units included in a complex unit, $w(U, S)$, measure, to a certain extent, the proportion of the content of the unit S which can be attributed to each one of its components:

$$w(U, S) = \frac{\sum_{T \in An(U)} tf(T, An(U)) \times idf(T)}{\sum_{T \in An(S)} tf(T, An(S)) \times idf(T)} \tag{7}$$

where $An(U) = \{T \in \mathcal{T} \mid T \text{ is an ancestor of } U\}$, i.e., $An(U)$ is the set of terms that are included in the structural unit U.

The utilities which are necessary to compute the expected utility of retrieving structural units, $EU(r_i^+ \mid q)$, namely $v(r_i^+, u_i, u_{hi(U_i)})$, are composed of a component which depends on the involved unit and another component independent on the specific unit and depending only on which one of the four configurations, $(u_i^-, u_{hi(U_i)}^-)$, $(u_i^-, u_{hi(U_i)}^+)$, $(u_i^+, u_{hi(U_i)}^-)$ or $(u_i^+, u_{hi(U_i)}^+)$, is being considered:

$$v(r_i^+, u_i, u_{hi(U_i)}) = nidf_Q(U_i) \times v(u_i, u_{hi(U_i)}) \tag{8}$$

with $v(u_i^-, u_{hi(U_i)}^-) = v^{--}$, $v(u_i^-, u_{hi(U_i)}^+) = v^{-+}$, $v(u_i^+, u_{hi(U_i)}^-) = v^{+-}$ and $v(u_i^+, u_{hi(U_i)}^+) = v^{++}$.

The part depending on the involved unit is defined as the sum of the inverted document frequencies of those terms contained in U_i that also belong to the query Q, normalized by the sum of the idfs of the terms contained in the query (a unit U_i will be more useful, with respect to a query Q, as more terms indexing U_i also belong to Q):

$$nidf_Q(U_i) = \frac{\sum_{T \in An(U_i) \cap Q} idf(T)}{\sum_{T \in Q} idf(T)} \tag{9}$$

Regarding the other component of the utility function independent on the involved unit, at INEX 2006 we used the following values

$$v^{--} = v^{-+} = v^{++} = 0, \; v^{+-} = 1$$

3.2 Changing Weights

We have modified the weights of the units included in a complex unit, $w(U, S)$, in order to also take into account, not only the proportion of the content of S which is due to U, but also some measure of the importance of the type (tag) of unit U within S. For example, the terms contained in a `collectionlink` (generally proper nouns and relevant concepts) or `emph2` should be cuantified higher than terms outside those units. Units labeled with `title` are also very informative, but units with `template` are not.

So, we call $I(U)$ the *importance of the unit* U, which depends of the type of tag associated to U. These values constitute a global set of free parameters, specified at indexing time. The new weights $nw(U, S)$, are then computed from the old ones in the following way:

$$nw(U, S) = \frac{I(U) \times w(U, S)}{\sum_{U' \in Pa(S)} I(U') \times w(U', S)} \tag{10}$$

Table 1. Importance of the different types of units used in the official runs

Tag	Weight file 8	Weight file 11	Weight file 15
name	20	100	200
title	20	50	50
caption	10	10	30
collectionlink	10	10	30
emph2	10	30	30
emph3	10	30	30
conversionwarning	0	0	0
languagelink	0	0	0
template	0	0	0

Table 2. Relative utility values of the different types of units used in the official runs

Tag	Utility file 1	Utility file 2	Utility file 3
conversionwarning	0	0	0
name	0.75	0.75	0.85
title	0.75	0.75	0.85
collectionlink	0.75	1.5	0.75
languagelink	0	0	0
article	2	2.5	2.5
section	1.5	1	1.25
p	1.5	1	1.5
body	1.5	1	2
emph2	1	1.5	1
emph3	1	1.5	1

We show in Table 1 the three different importance schemes used in the official runs. Unspecified importance values are set to 1 (notice that by setting $I(U) = 1, \forall U \in \mathcal{U}$, we get the old weights).

3.3   Changing Utilities

This year the formula of the utility values for a unit U is computed by considering another factor called *relative utility value*, $RU(U)$, which depends only on the kind of tag associated to that unit, so that:

$$v(r_i^+, u_i, u_{hi(U_i)}) = nidf_Q(U_i) \times v(u_i, u_{hi(U_i)}) \times RU(U_i) \tag{11}$$

It should be noticed that this value $RU(U)$ is different from the importance $I(U)$: a type of unit may be considered very important to contribute to the relevance degree of the unit containing it and, at the same time, is considered not very useful to retrieve this type of unit itself. For example, this may be the case of units having the tag `<title>`: in general a title alone may be not very useful for a user as the answer to a query, probably the user would prefer to get the content of the structural unit having this title; however, terms in a title tends to be highly representative of the content of a document part, so that the importance of the title should be greater than the importance derived simply of the proportion of text that the title contains (which will be quite low). The sets of utility values used in the official runs are displayed in Table 2.

In all the cases, the default value for the non-listed units is 1.0. We have also considered the case where all the relative utility values are set to 1.0 (which is equivalent to not to use relative utilities at all).

4 Adapting Garnata to the INEX 2007 Ad Hoc Retrieval Tasks

For each query, Garnata generates a list of document parts or structural units, ordered by relevance value (expected utility), as the output. So, this output is compatible with the thorough task used in previous editions but not with the three adhoc tasks for INEX 2007, *Focused, Relevant in Context* and *Best in Context*. To cope with these tasks, we still use Garnata but after we filter its output in a way which depends on the kind of task:

Focused task: The output must be an ordered list of structural units where overlapping has been eliminated. So, we must supply some criterion to decide, when we find two overlapping units in the output generated by Garnata, which one to preserve in the final output. The criterion we have used is to keep the unit having the greatest relevance value and, in case of tie, we keep the more general unit (the one containing a larger amount of text).

Relevant in Context task: In this case the output must be an ordered list of documents and, for each document, a set of non-overlapping structural units, representing the relevant text within the document (i.e., a list of non-overlapping units clustered by document). Therefore, we have to filter the output of Garnata using two criteria: how to select the non-overlapping units for each document, and how to rank the documents. To manage overlapping units we use the same criterion considered for the focused task. To rank the documents, we have considered three criteria to assign a relevance value to the entire document: the relevance value of a document is equal to: (1) the maximum relevance value of its units; (2) the relevance value of the "/article[1]" unit; (3) the sum of the relevance values of all its units. Some preliminary experimentation pointed out that the maximum criterion performed better, so we have used it in the official runs.

Best in Context task: The output must be an ordered list composed of a single unit per document. This single document part should correspond to the best entry point for starting to read the relevant text in the document. Therefore, we have to provide a criterion to select one structural unit for each document and another to rank the documents/selected units. This last criterion is the same considered in the relevant in context task (the maximum relevance value of its units). Regarding the way of selecting one unit per document, the idea is to choose some kind of *centroid* structural unit: for each unit U_i we compute the sum of the distances from U_i to each of the other units U_j in the document, the distance between U_i and U_j being measured as the number of links in the path between units U_i and U_j in the XML tree times the relevance value of unit U_j; then we select the unit having minimum sum of distances. In this way we try to select a unit which is nearest to the units having high relevance values.

5   Results of Our Model at INEX 2007

We have obtained the following results in the three tasks, using the combinations of weight and utility configurations displayed in Tables 3, 4 and 5.

As we can see in these results, the configuration of utilities with the value 3 is the most appropriate to get the best results in the different tasks, although we can not fix a specific configuration of weights that obtain the same results.

Finally, we show the graphics of the different tasks, where we can see the comparison of our results (red lines) with the results of the other organizations.

We have come to the conclusion that our system gets better results than the year before, so we have reached a middle position in the ranking (except for the focused task, where the results are worse) as we can see in the graphics and in the tables.

Table 3. Results for the Focused task

Weight file	Utility file	Ranking
8	3	62/79
15	2	70/79
15	none	71/79

Table 4. Results for the Relevant in Context task

Weight file	Utility file	Ranking
15	3	44/66
8	3	45/66
11	1	47/66

Table 5. Results for the Best in Context task

Weight file	Utility file	Ranking
8	3	40/71
15	None	45/71
15	2	47/71

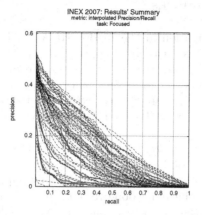

Fig. 4. Results for the Focused task

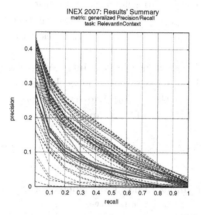

Fig. 5. Results for the Relevant in Context task

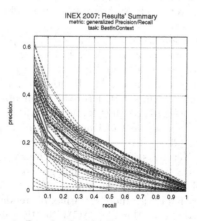

Fig. 6. Results for the Best in Context task

6   Concluding Remarks

In this year, our participation in the AdHoc track has been more productive than the one presented last year. In 2006, we only applied for one of the four AdHoc tasks (Thorough), and in 2007 we have sent results for all the tasks of the track. Besides, on 2006 we got a very bad ranking (lying on the percentile 91). The best runs of this year are clearly better than the one obtained last year (corresponding to percentiles 78 [Focused], 66 [Relevant in Context] and 56 [Best in Context]).

Results in the Relevant in Context and Best in Context tasks are at the end of the second-third of the ranking, but in Focused they are in a lower position. So, the filter used for Focused should be considerably improved.

On the other hand, we have not done yet a deep experimentation of different configurations for both the importance and the utility values. The parameters values used during INEX'07 were randomly selected configurations that obtained good results with the queries and relevance assessments of INEX'06. We think that the behaviour of our model could be clearly improved with a more systematic experimentation finding an optimal configuration of the parameters.

Acknowledgments. This work has been jointly supported by the Spanish Consejería de Innovación, Ciencia y Empresa de la Junta de Andalucía, Ministerio de Educación y Ciencia and the research programme Consolider Ingenio 2010, under projects TIC-276, TIN2005-02516 and CSD2007-00018, respectively.

References

1. de Campos, L.M., Fernández-Luna, J.M., Huete, J.F.: The BNR model: foundations and performance of a Bayesian network-based retrieval model. Int. J. Appr. Reason. 34, 265–285 (2003)
2. de Campos, L.M., Fernández-Luna, J.M., Huete, J.F.: Using context information in structured document retrieval: An approach using Influence diagrams. Inform. Process. Manag. 40(5), 829–847 (2004)
3. de Campos, L.M., Fernández-Luna, J.M., Huete, J.F.: Improving the context-based influence diagram for structured retrieval. In: Losada, D.E., Fernández-Luna, J.M. (eds.) ECIR 2005. LNCS, vol. 3408, pp. 215–229. Springer, Heidelberg (2005)
4. de Campos, L.M., Fernández-Luna, J.M., Huete, J.F., Romero, A.E.: Garnata: An information retrieval system for structured documents based on probabilistic graphical models. In: Proceedings of the Eleventh International Conference of Information Processing and Management of Uncertainty in Knowledge-Based Systems (IPMU), pp. 1024–1031 (2006)
5. Jensen, F.V.: Bayesian Networks and Decision Graphs. Springer, Heidelberg (2001)
6. Metzler, D., Croft, W.B.: Combining the language model and inference networks approaches to retrieval. Inform. Process. Manag. 40(5), 735–750 (2004)
7. Myaeng, S.H., Jang, D., Kim, M., Zhoo, Z.: A flexible model for retrieval of SGML documents. In: Proceedings of the 21st ACM–SIGIR Conference, pp. 138–145. ACM Press, New York (1998)

8. Pearl, J.: Probabilistic Reasoning in Intelligent Systems: Networks of Plausible Inference. Morgan and Kaufmann, San Mateo (1988)
9. Piwowarski, B., Gallinari, P.: A Bayesian network for XML information retrieval: searching and learning with the INEX collection. Inform. Retrieval 8(4), 655–681 (2005)
10. Shachter, R.: Probabilistic inference and influence diagrams. Oper. Res. 36(5), 527–550 (1988)

Dynamic Element Retrieval in the Wikipedia Collection

Carolyn J. Crouch, Donald B. Crouch,
Nachiket Kamat, Vikram Malik, and Aditya Mone

Department of Computer Science
University of Minnesota Duluth
Duluth, MN 55812, (218) 726-7607
ccrouch@d.umn.edu

Abstract. This paper describes the successful adaptation of our methodology for the dynamic retrieval of XML elements to a semi-structured environment. Working with text that contains both tagged and untagged elements presents particular challenges in this context. Our system is based on the Vector Space Model; basic functions are performed using the Smart experimental retrieval system. Dynamic element retrieval requires only a single indexing of the document collection at the level of the basic indexing node (i.e., the paragraph). It returns a rank-ordered list of elements identical to that produced by the same query against an all-element index of the collection. Experimental results are reported for both the 2006 and 2007 Ad-hoc tasks.

1 Introduction

Our work for INEX 2007 focuses on solving some of the interesting problems which arose for dynamic element retrieval when the experimental collection changed from IEEE to Wikipedia. Dynamic element retrieval—i.e., the dynamic retrieval of elements at the desired degree of granularity—has been the focus of our investigations at INEX for some time [3, 7]. We have demonstrated that our method works well for structured text (e.g., the INEX IEEE collection) and that it in fact produces a result virtually identical to that produced by the search of the same query against the corresponding all-element index [1]. The challenge in 2007 is to adapt our methods to the particular issues presented by a semi-structured collection such as Wikipedia.

The well structured IEEE collection lends itself quite naturally to representation by Fox's Extended Vector Space Model [5], which allows for the incorporation of objective identifiers (such as date of publication) along with the normal content identifiers associated with the document. Wikipedia documents, on the other hand, are semi-structured. They contain untagged text which is distributed throughout the documents. These documents can be nicely represented within the Vector Space Model [11]; retrieval then takes place using Smart [10] against an all-element index composed of articles, sections, and paragraphs (or terminal nodes). But semi-structured text poses particular problems for dynamic element retrieval, which requires that all the terminal nodes of a document be identifiable. Since the process requires the execution-time building of document trees of interest to the query, all of the terminal nodes or text-bearing elements of the tree must be present in order for their parent elements to be generated properly.

Thus the impact of untagged text on dynamic element retrieval is twofold. During parsing, it must be identified, so that it may subsequently be used in generating the

N. Fuhr et al. (Eds.): INEX 2007, LNCS 4862, pp. 70–79, 2008.

document schemas utilized by dynamic element retrieval as it builds the document trees. How do we manage untagged text while building the document tree, so as to guarantee that each parent element in the hierarchy (and thus the document) is properly constructed? Untagged text is not identifiable or retrievable in itself; it exists only as a component of its parent element. Thus a means of incorporating untagged text in the document schemas (so that it can later play its role in the generation of the document trees) must be devised.

The second impact of untagged text on dynamic element retrieval relates to its inclusion in the set of terminal nodes used to seed the process. Since the method requires an initial retrieval against the terminal node index to identify the documents of interest to the query (i.e., those whose trees will be built), we must determine the value of untagged text in this context. In other words, is the untagged text distributed throughout a document (or interspersed between tagged elements) important from the retrieval viewpoint?

Experiments to answer this and other, related questions were performed during the past year. In [2], we reported our successful method for dynamic query weighting, conceived by Ganapathibhotla and further described in [4] and [1]. But because new collections require tuning to determine appropriate parameter values for term weighting and because our hardware had difficulties handling the significant increase in size from IEEE to Wiki, we were unable to report specific results at that time. This year, using first the 2006 INEX test collection and evaluation metrics, we establish that dynamic element retrieval can be effectively applied to semi-structured collections, producing a result identical to that produced by the equivalent all-element retrieval. Our results are competitive with respect the Ad-hoc Thorough and Focused Overlap Off subtasks. We then apply our methods to the 2007 Ad-hoc subtasks, using the 2007 data set and evaluation metrics, and examine both all-element or baseline retrieval and dynamic element retrieval in this context.

2   Dynamic Element Retrieval and the Wikipedia Collection

In this section, we give a brief overview of our system, including dynamic element retrieval as implemented in a structured environment. We discuss the particular problems that arise with semi-structured text and report the adaptations required to accommodate these structural changes.

2.1   Dynamic Element Retrieval

Salton's Vector Space Model, upon which our system is based, is a foundational model in information retrieval, its methods instantly recognizable. The importance of a term within a vector is indicated by its term weight; many methods are available, but we use Singhal's *Lnu-ltu* term weighting in our system. Details of this weighting scheme may be found in [12, 13]. It is of particular interest in element retrieval where the elements vary considerably in length, depending on type (e.g., paragraph versus section and body). *Lnu-ltu* weighting attempts to deal with the ranking issues resulting from disparity in vector length. (See [1] for a more detailed discussion of this issue.) A brief overview of the system follows.

We selected the *paragraph*—in our view, the smallest meaningful unit of text—as our basic indexing unit in the earliest stages of our investigations. The term *paragraph* in this context means all the leaf nodes of a document tree. Thus it is used to refer to figure captions, lists, section titles, tables, abstracts—all the content-bearing terms that partition the document into mutually exclusive parts. (Although some of these elements may not be leaf nodes according to their DTDs, they are treated as leaf nodes in this context because their child nodes are too small to meaningful units in themselves.)

We first produce a paragraph (leaf node) parse of the documents. Paragraphs and queries are indexed; retrieval then takes place by running the *ltu*-weighted query against the *Lnu*-weighted paragraph vectors. The result is a list of *elements* in rank order. Every element in this list having a non-zero correlation with the query represents a terminal node in the body of a document with some relationship to the query.

Consider now the *n* top-ranked elements in this list. Our method of dynamic element retrieval builds a tree representation for each document having an element in this list. Each tree is built based on a schema of the document (produced as a by-product of parsing). Given its set of terminal nodes in the form of term-frequency vectors, a document tree is built, bottom-up, according to its schema. The content of each internal node is based solely on the content of its children. As each element vector is produced, it is *Lnu*-weighted and correlated with the query, which is itself *ltu*-weighted. After all element vectors, including the body element, have been generated, weighted and correlated with the query, the process continues with the next document. The resulting set of element vectors (i.e., all the elements from each document with a terminal node in the set of *n* top-ranked elements retrieved by the query) are then sorted and the top-ranked elements are reported.

2.2 Tagged vs. Untagged Text

The method described in above works beautifully for structured text, wherein all elements are tagged and thus uniquely identifiable. Dynamic element retrieval depends on having all the terminal nodes of a document tree present in the paragraph index. The initial paragraph retrieval then gives us a good indication of documents of interest to the query because all paragraphs that correlate highly with it are identified. (These paragraphs identify their parent documents, which may contain other elements of interest.) And the documents themselves are properly constructed because each terminal node in the document tree is present when the tree is constructed as per its schema.

The Wikipedia collection, on the other hand, contains untagged text which is distributed throughout its documents at the body and section levels. This untagged text cannot be retrieved except as a component of its parent element. Yet it must be recognizable during parsing and represented in the document schema so that the document tree, when generated, can be generated properly with untagged text included at the parent level. And if it is important for retrieval purposes, it must be available during the initial retrieval against the paragraph index, when documents of interest to the query are identified. Two questions now arise. First, how do we manage untagged text within the process of dynamic element retrieval? Second, is the presence of untagged text important from the retrieval viewpoint? That is, if such text is omitted from the terminal node index and thus not involved in identifying documents of potential interest, does retrieval suffer as a result?

With reference to the first question, above: We handle untagged text by consolidating it at the parent level under its own tag (*<mt>*). A body or section may have one or multiple pieces of untagged text within it; all such text is merged and becomes part of the *<mt>* element associated with the parent. We now have a structured document; *<mt>* elements are included in the document schemas in the same fashion as paragraph elements. Dynamic element retrieval now proceeds as indicated in Section 2.1 with one exception--all elements carrying the *<mt>* tag are discarded before the final set of elements associated with a document is returned.

Consider now the second question: Is untagged text important for retrieval purposes? Experiments designed to investigate this issue are reported in [6]. Results clearly show that untagged text is important in this context. That is, the initial leaf node parse (described in Section 2.1) must include *<mt>* elements as well as paragraphs to produce an improved result. This combined set of terminal nodes is then indexed, retrieval takes place, and the set of n top-ranked elements used to seed dynamic element retrieval now contains both paragraph and *<mt>* elements. If we fail to include untagged text at this stage, considerable text is lost for the purpose of identifying documents of potential relevance to the query. Dynamic element retrieval uses an initial retrieval against a set of terminal nodes in order to identify interesting documents. A set of terminal nodes which includes *<mts>* may well identify a different set of documents than that identified by a similar set in which the *<mts>* are not present. (See [6] for details.)

2.3 Terminal Node Expansion

One factor of interest here is what we refer to as *terminal node expansion.* Dynamic element retrieval returns a terminal node (e.g., paragraph) based on its correlation with the query. A terminal node may (and frequently does) contain children which are present not by virtue of their contribution to meaning but rather for formatting purposes (e.g., tags such as italics or bold). We process terminal nodes by retaining the text enclosed within such tags but removing the tags themselves. Thus any such tags embedded within the text of a terminal node are not present for evaluation purposes. That is, their xpaths are not reported. This significantly impacts results for the Thorough and Focused Tasks. The omission of these xpaths in fact has a considerable impact on evaluation scores as [8] shows. In the tables below, we report the best results obtained for each experiment. If those results were achieved by performing terminal node expansion after element retrieval, the table so indicates.

2.4 What About n?

The parameter n is used in dynamic element retrieval to seed the tree generation algorithm. It represents the number of elements (or paragraphs) input to the algorithm, which constructs a document tree for each paragraph in the input set. So for a set of n paragraphs, at most n trees will be built. There are many interesting aspects of this process. What is perhaps of most interest with respect to results presented here is that dynamic element retrieval is able, over a range of values for n, to produce results which are better than those produced by all-element retrieval. This holds true for both the INEX IEEE and Wikipedia collections. (For details and rationale, see [9].)

3 Experiments with the INEX 2006 Collection

The results of our experiments in dynamic element retrieval for the INEX 2006 Ad-hoc tasks (i.e., Thorough, Focused Overlap On, Focused Overlap Off, Best-in-Context, and All-in-Context) are reported below. In each case, slope and pivot values for the *Lnu-ltu* weighting scheme are .12 and 38, respectively. Relevance assessment is v.5. For each value of n (where n = 5, 10, 25, 50, 100, 250, 500, and 1000), dynamic element retrieval returns a ranked list of elements which is evaluated using the specified metric and compared against the baseline (all-element) retrieval using the same value of slope and pivot. The topic or query set used in all experiments reported here is CO. Terminal node expansion, if utilized, is specified.

3.1 Thorough Task with Terminal Node Expansion

Consider now Table 1, which reports results of the base case (all-element retrieval) vs. dynamic element retrieval as n ranges from 5 to 1000. All-element retrieval produces a MAep value of 0.0414 at 1500. Dynamic element retrieval, however, produces its best results at n = 50, where MAep at 1500 is 0.0474 and the intermediate MAep values [almost all] equal or exceed the baseline. We evaluated results for dynamic element retrieval for two cases, which are identical other than one case includes terminal node expansion and the other does not. For the Thorough Task, terminal node expansion produces in every instance a substantially improved result. The value of MAep @1500 for the base case (at 0.0414) would put these results at rank 1 when compared with the official results for this task; dynamic element retrieval at n =50 improves substantially on the base case.

Table 1. INEX 2006 Thorough Results, Dynamic vs. All-Element Retrieval

	MAep					
n	@10	@20	@50	@100	@500	@1500
all-el	0.0054	0.0088	0.0148	0.0202	0.0351	0.0414
1000	0.0055	0.0088	0.0148	0.0202	0.0351	0.0414
500	0.0055	0.0088	0.0148	0.0203	0.0351	0.0415
250	0.0055	0.0088	0.0148	0.0203	0.0351	0.0429
100	0.0055	0.0088	0.0148	0.0202	0.036	0.0466
50	0.0055	0.0088	0.0147	0.0202	0.0398	0.0474
25	0.0054	0.0088	0.0148	0.0209	0.0399	0.0439
10	0.0054	0.0088	0.0162	0.0238	0.0364	0.0372
5	0.0054	0.0093	0.0167	0.0236	0.0312	0.0312

3.2 Focused Task

Consider first the Focused Overlap Off Task. For this task, terminal node expansion is also important, with results much improved over the corresponding case without expansion. Table 2 shows that even the base case exceeds the best official values reported for nxCG@25 and nxCG@50. Dynamic element retrieval at n = 5 produces substantial improvement over the base case and exceeds the best case official results for all values of nxCG except nxCG@5, where our result ranks second.

Results drop dramatically when we look at the Focused Task with Overlap On. Terminal node expansion (as might be expected) is actually detrimental to this task. Hence Table 3 shows the results of base case retrieval vs. dynamic element retrieval. Base case retrieval produces the best overall results for this task which rank poorly when compared to the official results.

Table 2. INEX 2006 Focused Overlap Off, Dynamic vs. All-Element Retrieval

	nxCG			
n	@5	@10	@25	@50
all-el	0.4358	0.4086	0.3428	0.3030
1000	0.4341	0.4094	0.3428	0.3030
500	0.4296	0.4090	0.3425	0.3024
250	0.4241	0.4040	0.3388	0.3001
100	0.4190	0.4023	0.3366	0.2981
50	0.4190	0.4023	0.3372	0.2991
25	0.4190	0.4041	0.3410	0.3112
10	0.4391	0.4103	0.3801	0.3333
5	0.4361	0.4254	0.4014	0.3488

Table 3. INEX 2006 Focused Overlap On, Dynamic vs. All-Element Retrieval

	nxCG			
n	@5	@10	@25	@50
all-el	0.3241	0.2787	0.2220	0.1815
1000	0.3185	0.2720	0.2147	0.1733
500	0.3185	0.2720	0.2145	0.1734
250	0.3192	0.2720	0.2149	0.1733
100	0.3192	0.2720	0.2150	0.1738
50	0.3209	0.2720	0.2177	0.1765
25	0.3227	0.2757	0.2202	0.1782
10	0.3252	0.2751	0.1992	0.1450
5	0.3217	0.2370	0.1641	0.1033

3.3 Best-in-Context Task

For this task, we first use dynamic element retrieval to return the set of elements associated with a query. Dynamic element retrieval returns, for each document of interest to the query, all elements having a positive correlation with it. Given the set of all such elements associated with the document, we select a single element (in this case, the most highly correlating element) to return.

Table 4. INEX 2006 Best-in-Context (Dynamic Element Retrieval)

	BEPD				
n	@0.01	@0.1	@1.0	@10.0	@100.0
1500	0.1443	0.2246	0.3512	0.5414	0.7357
500	0.1254	0.1985	0.3131	0.4865	0.6658
250	0.1072	0.1701	0.2690	0.4199	0.5792
100	0.0746	0.1233	0.1979	0.3113	0.4358
50	0.0552	0.0936	0.1519	0.2395	0.3370
25	0.0351	0.0620	0.1025	0.1657	0.2381
10	0.0215	0.0376	0.0604	0.0971	0.1442
5	0.0135	0.0250	0.0388	0.0605	0.0891

Table 4 shows the results of this process. The best results were returned at $n = 1500$; these results would rank 18 (out of 77) when compared to the official results. It is clear that increasing the value of n here has the potential to produce an improved result. (For further insight into how these results may be improved, see Section 4.2.)

3.4 All-in-Context Task

We perform this task by first retrieving against a document index (i.e., an index of Wiki documents) and then using dynamic element retrieval to produce and rank the elements within each document for return to the user. The results of this task are reported in Table 5. Our results, compared to the top submissions on the INEX website, rank 25 out of 56.

Table 5. INEX 2006 All-in-Context (Dynamic Element Retrieval)

n	gp[5]	gp[10]	[gp25]	gp[50]	MAgP
1500	0.3036	0.2453	0.1723	0.1260	0.1254
500	0.3036	0.2453	0.1723	0.1260	0.1254
250	0.3036	0.2453	0.1723	0.1260	0.1246
100	0.3036	0.2453	0.1723	0.1260	0.1167
50	0.3036	0.2453	0.1723	0.1260	0.1037
25	0.3036	0.2453	0.1723	0.0861	0.0853
10	0.3036	0.2453	0.0981	0.0490	0.0615
5	0.3036	0.1518	0.0607	0.0303	0.0430

3.5 Synopsis: 2006 Results

We conclude from our experiments with the 2006 Ad-hoc tasks that our Thorough and Focused Overlap Off results, with terminal node expansion in each case, are excellent. Focused Overlap On results, on the other hand, fall dramatically. The Best-in-Context method performs relatively well, but All-in-Context results are mediocre at best. More analysis is needed to determine the rationale for lack of performance in the Focused Overlap On and the All-in-Context tasks.

4   Experiments with the INEX 2007 Collection

In this section, we include the results produced by our methods for the three INEX 2007 Ad-hoc tasks, namely, Focused, Best-in-Context, and Relevant-in-Context. To produce its best results, our system needs tuning to establish appropriate values for term weighting with respect to a metric. The results reported here were obtained using parameters tuned to the 2006 metric. We expect some improvement when 2007 tuning is complete.

4.1   Focused Task

The INEX 2007 Focused Task is identical to the 2006 Focused Overlap On task. The result of all-element vs. dynamic element retrieval is shown in Table 6. Based on iP[0.01], base case results rank at 37 (out of 79 official entries); best results for dynamic element retrieval at $n = 25$ are slightly better.

Table 6. INEX 2007 Focused, Dynamic vs. All-element Retrieval

n	iP[0.00]	iP[0.01]	iP[0.05]	iP[0.10]	MAiP
all-el	0.5075	0.4554	0.3390	0.2655	0.0883
1000	0.5259	0.4590	0.3182	0.2449	0.0769
500	0.5249	0.4583	0.3197	0.2482	0.0808
250	0.5251	0.4585	0.3228	0.2532	0.0837
100	0.5247	0.4586	0.3294	0.2638	0.0849
50	0.5222	0.4582	0.3390	0.2618	0.0846
25	0.5233	0.4642	0.3340	0.2530	0.0810
10	0.5160	0.4613	0.3182	0.2489	0.0694
5	0.5086	0.4336	0.2868	0.2395	0.0631
1	0.4008	0.3121	0.2178	0.1565	0.0430

4.2   Best-in-Context Task

Table 7 shows results obtained for the 2007 Best-in-Context Task. For this task (as indicated by analysis of the corresponding 2006 data), we increased n above the value normally utilized (to 2000, in this case). The resultant value of MAgP produces a rank of 35 (out of 71) when compared to the official results.

There are two obvious reasons for failing to produce improved results. The task involves producing the best entry point (BEP) for each document judged relevant to the query. One potential problem area lies in ensuring that each such document lies in the set produced by dynamic element retrieval. The second and more difficult problem is ensuring that the xpath returned as BEP in each case is in fact the BEP identified by the assessor.

Table 7. INEX 2007 Best-in-Context (Dynamic Element Retrieval)

n	MAgP
2000	0.1231
1000	0.1178
500	0.1086
250	0.0976
200	0.0946
150	0.0899
100	0.0820
50	0.0648
25	0.0523
10	0.0303
5	0.0209

The first problem is relatively simple to handle. We are still analyzing the data for this task, but early indications are that the element we identify as BEP (i.e., the most highly correlating element) may in fact be not the BEP identified by the assessor but rather a parent of that element. Further analysis should provide insight into solving this problem.

4.3 Relevant-in-Context Task

Table 8 displays the results achieved by dynamic element retrieval with respect to the Relevant-in-Context task. The best MAgP value is produced by n at 25; it ranks 37 out of 66 with respect to the official runs. Further analysis of these results is ongoing.

Table 8. INEX 2007 Relevant-in-Context (Dynamic Element Retrieval)

n	MAgP
1000	0.0471
500	0.0577
250	0.0818
200	0.0830
100	0.0885
50	0.0897
25	0.0935
10	0.0587
5	0.0448

5   Conclusions

In 2007, our system achieved its most recent goal. Working in the semi-structured environment of the INEX Wikipedia collection, it retrieves elements dynamically and returns a rank-ordered list of elements equivalent to that retrieved by a search of the corresponding all-element index. We examined the impact of changing from structured to semi-structured text in this environment and adapted our methods accordingly. We have shown that untagged text is important both with respect to content and its impact on retrieval. Results reported herein with respect to both the 2006 and 2007 Ad-hoc tasks show clearly that dynamic element retrieval is able to identify what we might call *elements of interest*, i.e., elements of potential interest with respect to the query. Our efforts in the coming year will focus on (1) refining strategies for identifying or selecting a particular element from this set as required by the task, and (2) applying our current methodology to passage retrieval.

References

[1] Crouch, C.: Dynamic element retrieval in a structured environment. ACM Transactions on Information Systems 24(4), 437–454 (2006)

[2] Crouch, C., Crouch, D., Ganapathibhotla, M., Bakshi, V.: Dynamic element retrieval in a semi-structured collection. In: Fuhr, N., Lalmas, M., Trotman, A. (eds.) INEX 2006. LNCS, vol. 4518, pp. 82–88. Springer, Heidelberg (2007)

[3] Crouch, C., Khanna, S., Potnis, P., Daddapaneni, N.: The dynamic retrieval of XML elements. In: Fuhr, N., Lalmas, M., Malik, S., Kazai, G. (eds.) INEX 2005. LNCS, vol. 3977, pp. 268–281. Springer, Heidelberg (2006)

[4] Ganapathibhotla, M.: Query processing in a flexible retrieval environment. M.S. Thesis, Department of Computer Science, University of Minnesota Duluth, Duluth, MN (2006), http://www.d.umn.edu/cs/thesis/Ganapathibhotla.pdf

[5] Fox, E.A.: Extending the Boolean and vector space models of information retrieval with p-norm queries and multiple concept types. Ph.D. Dissertation, Department of Computer Science, Cornell University (1983)

[6] Kamat, N.: Impact of untagged text in dynamic element retrieval. M.S. Thesis, Department of Computer Science, University of Minnesota Duluth, Duluth, MN (2007), http://www.d.umn.edu/cs/thesis/kamat.pdf

[7] Khanna, S.: Design and implementation of a flexible retrieval system. M. S. Thesis, Department of Computer Science, University of Minnesota Duluth, Duluth, MN (2005), http://www.d.umn.edu/cs/thesis/khanna.pdf

[8] Malik, V.: Impact of terminal node processing on element retrieval. M.S. Thesis, Department of Computer Science, University of Minnesota Duluth, Duluth, MN (2007), http://www.d.umn.edu/cs/thesis/malik.pdf

[9] Mone, A.: Dynamic element retrieval for semi-structured documents. M.S. Thesis, Department of Computer Science, University of Minnesota Duluth, Duluth, MN (2007), http://www.d.umn.edu/cs/thesis/mone.pdf

[10] Salton, G. (ed.): The Smart Retrieval System—Experiments in Automatic Document Processing. Prentice-Hall, Englewood Cliffs (1971)

[11] Salton, G., Wong, A., Yang, C.S.: A vector space model for automatic indexing. Comm. ACM 18(11), 613–620 (1975)

[12] Singhal, A.: AT&T at TREC-6. In: The Sixth Text REtrieval Conf (TREC-6), pp. 215–225 (1998)

[13] Singhal, A., Buckley, C., Mitra, M.: Pivoted document length normalization. In: Proc. of the 19th Annual International ACM SIGIR Conference, pp. 21–29 (1996)

The Simplest XML Retrieval Baseline That Could Possibly Work*

Philipp Dopichaj

University of Kaiserslautern
Gottlieb-Daimler-Str.
67663 Kaiserslautern
Germany
dopichaj@informatik.uni-kl.de

Abstract. Five years of INEX have produced many competing XML element retrieval methods that make use of the document structure. So far, no clearly best method has been identified, and there is even no clear evidence what parts of the document structure can be used to improve retrieval quality. Little research has been done on simply using standard information retrieval techniques for XML retrieval. This paper aims at addressing this; it contains a detailed analysis of the BM25 similarity measure in this context, revealing that this can form a viable baseline method.

1 Introduction

In the five years since the inception of INEX, much research on XML element retrieval methods has been done by the participants. Through the use of the INEX test collections, it was possible to determine the retrieval quality of the competing retrieval engines. One thing all retrieval engines participating in INEX have in common is that they make use of the XML document structure in some way, based on the reasonable assumption that retrieval engines that use more of the information that is available can yield better results.

To our knowledge, this assumption has never been tested in detail. To close this gap, we provide a detailed analysis of the retrieval quality that can be achieved by simply using the standard BM25 similarity measure with minimum adaptations to XML retrieval.

1.1 Evaluation Metrics

Over the years, the evaluation metrics and retrieval tasks used for INEX have changed considerably. In this paper, we will only evaluate the *thorough* retrieval task; this task is the simplest of all INEX tasks, and the results for the other tasks are typically created by applying a postprocessing step to the *thorough* results.

---

* ... and it does!

N. Fuhr et al. (Eds.): INEX 2007, LNCS 4862, pp. 80–93, 2008.

We use the standard nxCG measure as used for INEX 2005 and 2006 [], and the official assessments from the corresponding workshop web sites[1].

We do not use the official evaluation software EvalJ[2], but our own reimplementation of the official measures; this was necessary because the overhead of calling an external process would have been too high. We made sure that our version of the evaluation gives the same results (although at a slightly higher numerical accuracy).

1.2 Test Collections

The INEX workshops used a collection of IEEE computer society[3] journal and transactions articles through 2005, where later versions of the collection are supersets of earlier versions (new volumes were added). From 2006 on, a conversion of the English version of Wikipedia was used [2]. The evaluations in this thesis will be based on the collections from 2004, 2005, and 2006. Figure 1 gives an overview of various characteristics of the document collections.

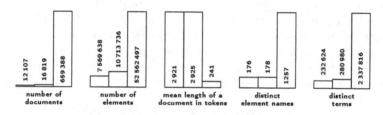

Fig. 1. Test collections statistics. The bars in each group are, from left to right, the IEEE 1.4 collection (2004), the IEEE 1.9 collection (2005), and the Wikipedia collection (2006). The token count excludes stop words.

For each year of the workshop, a new set of topics was created by the participants, consisting of a longer description of the information need and a query in NEXI format. The number of topics varied: in 2004, there were 40 CO topics (34 have been assessed), in 2005, there were 40 topics (29 assessed), and in 2006, there were 130 topics (114 assessed). For our evaluations, we will only use content-only topics.

The assessment procedure has changed against the years: In 2004, the assessors had to manually select both specificity and exhaustiveness on a scale from 0 to 2 for each element in the recall base. In 2005, a highlighting approach was introduced; the assessor used a virtual highlighter to mark relevant passages in the documents to denote specificity. In the next step, the exhaustiveness had to be set for each element as in 2004. From 2006 on, exhaustiveness was dropped from the assessments, only the highlighting approach to selectivity was retained.

---

[1] See http://inex.is.informatik.uni-duisburg.de/

[2] See http://evalj.sourceforge.net

[3] See http://www.computer.org

Note that we use nxCG for the evaluations on the INEX 2004 test collection, even though nxCG was not the official evaluation measure at the time. This is possible because the data that was collected for the assessments is compatible, and it makes the results presented in this paper more consistent and comparable. The results may not be as meaningful as the results for the other collections, but it is still interesting to see differences of behavior compared to the 2005 results, which are based on almost the same document collection.

1.3 Standard Similarity Measures

As mentioned before, we use the BM25 similarity measure as introduced by the Okapi project, as described by Robertson and Walker [10]. The core idea is the notion of *eliteness*, which denotes to what degree a document d is "elite" for term t. As with most information retrieval measures, eliteness is derived from the term frequency $\mathrm{tf}(t,d)$, and each term has a *global weight* w_i, which is derived from the term's document frequency $\mathrm{df}(t)$ and the total number of documents.

The conversion from the plain term frequency to the term eliteness probability can be adapted with the global parameter k_1; the formula ensures that the term eliteness is 0 if the term frequency is 0, and it asymptotically approaches 1 as the term frequency increases. This implies that the first few occurrences of a term make the greatest contribution to term eliteness – the function is steep close to 0. The eliteness of term t for document d, using a document-length normalization constant K (see below) is defined as:

$$\mathrm{eliteness}(t,d) = \frac{(k_1 + 1)\,\mathrm{tf}(t_i, d)}{K + \mathrm{tf}(t_i, d)} \cdot \underbrace{\log \frac{N - \mathrm{df}(t_i) + 0.5}{\mathrm{df}(t_i) + 0.5}}_{w_i} \tag{1}$$

An important feature of BM25 is *document-length normalization*. Based on the assumption that document length is caused either by needless verbosity – this implies normalization – or a more thorough treatment of the subject – this implies no normalization –, BM25 uses partial length normalization. The degree of normalization is controlled by a global parameter b.

$$K = k_1 \left((1 - b) + b \cdot \frac{\mathrm{len}(d)}{\mathrm{avg}(\mathrm{len}(d))} \right) \tag{2}$$

The final similarity of document d to the query q consisting of terms $t_1 \dots t_m$ is then accumulated as follows (we assume that there are no weights attached to query terms):

$$\mathrm{sim}(q,d) = \sum_{i=1}^{m} \mathrm{eliteness}(t_i, d) \tag{3}$$

For completeness, we will also examine the similarity measure used by the Apache Lucene project[4]. This similarity measure proved to be effective for our INEX 2005 submissions, with minor adaptations [3].

$$\mathrm{sim}(q,d) = \mathrm{coord}(q,d) \sum_{t \in q} \left[\sqrt{\mathrm{tf}(d,t)} \left(1 + \log\left(\frac{N}{\mathrm{df}(t)+1} \right) \right) \mathrm{lnorm}(d) \right] \quad (4)$$

$$\mathrm{lnorm}(d) = \frac{1}{\sqrt{\mathrm{len}(d)}} \quad (5)$$

$$\mathrm{coord}(q,d) = |\{t \in q : \mathrm{tf}(d,t) > 0\}| \quad (6)$$

The coordination factor $\mathrm{coord}(q,d)$ is the number of query terms in q that also occur in d. The intention is to reward documents that contain more of the query terms. The result is that documents that contain all the query terms will usually end up in the first ranks in the result list, which is usually the right thing to do.

1.4   Adaptation for XML Retrieval

The standard information retrieval similarity measures are based on the assumption that a document is atomic, that is, documents cannot be decomposed into sub-documents. This assumption is not valid for element retrieval, so minor adaptations have to be performed.

In particular, each document is split into its elements, and *every* element is stored in the index. The cost for indexing *all* elements may appear to be prohibitive, but with appropriate index structures, the overhead can be kept at an acceptable rate [5].

One change that this entails is the choice of the global frequency (in the original formulas, document frequency). Of course, it is still possible to use document frequency in element retrieval, but this is not the only option. In fact, if every element is indexed as if it were a document, the new concept of *element frequency* might well be a more logical choice.

There are other options [12, 8], but they require larger changes to the standard information retrieval techniques and index structures, so we will not consider them here.

2   Parameter Tuning for the Baseline Retrieval Engine

For both similarity measures, BM25 and Lucene, we will tune the parameters to suit XML retrieval; the default parameters are good for standard information retrieval, but will probably have to be adapted for this new scenario. The results for these similarity measures will then be compared to the best submitted results of the corresponding INEX workshop to put things in context.

---

[4] See http://lucene.apache.org

```
<section><title>Example document</title>
<p>A paragraph.</p>
<p>A paragraph with <it>inline</it> markup.</p>
</section>
```

(a) Input XML document.

XPath	Indexed contents
/section	Example document A paragraph. A paragraph with inline markup.
/section/title	Example document
/section/p[1]	A paragraph.
/section/p[2]	A paragraph with inline markup.
/section/p[2]/it	inline

(b) Indexed "documents".

Fig. 2. Example of XML document indexing

2.1   Lucene Similarity Measure

The Lucene similarity measure gave good results at least in 2005. In this section, we will evaluate two global weighting methods – element and document frequency – and a parameterizable version of Lucene's length normalization function:

– Standard length normalization:

$$\mathrm{lnorm_{luc}}(d) = \frac{1}{\sqrt{\mathrm{len}(d)}} \qquad (7)$$

– Standard length normalization with a constant value up to length l:

$$\mathrm{lnorm_{const}}(d) = \frac{1}{\sqrt{\max(\mathrm{len}(d), l)}} \qquad (8)$$

The following parameter combinations have to be tested, using $\mathrm{lnorm_{const}}$ (for 0, $\mathrm{lnorm_{const}}$ is effectively $\mathrm{lnorm_{luc}}$):

$$\underbrace{\{\mathrm{df}, \mathrm{ef}\}}_{\mathrm{gf}} \times \underbrace{\{0, 5, 10, \ldots, 195, 200\}}_{\mathrm{lnorm}}$$

In our INEX submissions, we used a non-linear adaptation of Lucene's function [3] – elements shorter than about 50 tokens basically get an RSV of 0. This length normalization function leads to inferior results in all experiments (in particular at higher ranks), so it is not included in the evaluation.

Tuning the length normalization is crucial to good performance, and what version is the best depends on the document collection. As figure 3 shows, for the IEEE collection, a soft threshold of 65 tokens yields the best results, whereas

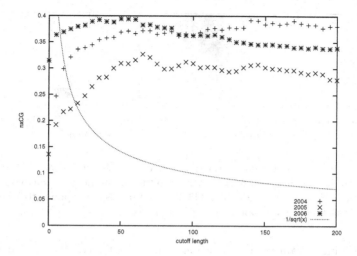

Fig. 3. Lucene retrieval quality (nxCG@10), using document frequency. For reference, a plot of the Lucene length normalization function is included in the plot.

for the Wikipedia collection, a lower value of about 50 is better. This can be explained by the different typical lengths of the documents in the collections: IEEE articles are much longer than Wikipedia articles, so the relevant parts are also longer (but this might also be a side effect of the assessment procedure).

For INEX 2004, the results are significantly worse than the best official results; it is unclear what the reason is. For INEX 2005, the Lucene similarity measure can exceed the best official submission at rank 10 (our own submission also using Lucene with a different length normalization function). For INEX 2006, the best Lucene results are about 10 percent worse than the best submitted results.

The results for the different global weighting functions are close to one another. This indicates that it does not matter whether document or element frequency is used with the Lucene similarity measure.

2.2 BM25 Similarity Measure

For BM25, length normalization is controlled by the parameters b and k_1. Permissible values for b are in the range $0 \ldots 1$, where 0 means "no length normalization" and 1 means "maximum influence of length normalization". The larger k_1 gets, the closer the local term weight gets to the raw term frequency.

According to Spärck Jones et al. [11], $b = 0.75$ and k between 1.2 and 2 work well on the TREC data, but it is unlikely that these parameter combinations can be transferred unchanged to XML retrieval. Theobald [12] uses $k_1 = 10.5$ and $b = 0.75$, but the TopX approach is sufficiently different from mine to warrant further exploration.

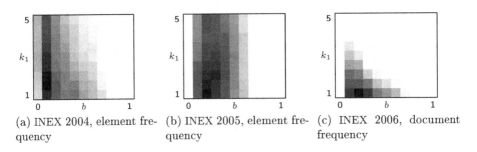

(a) INEX 2004, element frequency
(b) INEX 2005, element frequency
(c) INEX 2006, document frequency

Fig. 4. Parameter tuning for BM25; the darkness of each field corresponds to nxCG at cutoff rank 10. In each map, black corresponds to the maximum and white corresponds to 10 percent more than the minimum. The horizontal axis corresponds to b, from 0 to 1, and the vertical axis corresponds to k_1, from 1 to 5.

The following parameter combinations should be tested (the full range for b and a reasonable range for k_1):

$$\underbrace{\{0.0, 0.1, \ldots, 1.0\}}_{b} \times \underbrace{\{1, 1.5, 2, \ldots, 4.5, 5\}}_{k_1}$$

Figure 4 shows the results for the three test collections. It is obvious that a good choice of parameter b is much more critical than a good choice of k_1. In general, lower values of b work better than higher values, with the exception of $b = 0$ (that is, no length normalization). Compared to the best parameter values for traditional information retrieval ($b = 0.75$ and $k_1 = 1.2$), the best value of b for element retrieval is much lower (somewhere between 0.1 and 0.2), so the influence of length normalization is reduced.

Each parameter space has a global maximum; the parameters for this maximum are close for the different test collections, but not identical. In particular, it is surprising to see that the best parameters for 2004 and 2005 differ noticeably.

The reason is that in our usage scenario, length normalization also fulfills the purpose of selecting the right result granularity (should a chapter or a paragraph be ranked higher?). What happens is that for maximum length normalization ($b = 1$), very short elements are pushed to the front of the result lists, typically leading to a list of section titles or titles of cited works. This is obviously a bad result. With length normalization completely disabled ($b = 0$), there is a strong bias towards the longest elements, that is, complete articles or their bodies. For values of b between the extremes, the results are much more balanced; they are a mixture of sections, complete articles, and other elements. Although an occasional title does occur in the top ranks, this is the exception rather than the rule and does not do much harm. In fact, if all elements of fewer than ten terms are removed from the results, retrieval quality drops dramatically.

The best choice for the global frequency function depends on the document collection: Element frequency is best for the IEEE collection, whereas document frequency is better for the Wikipedia collection.

Using element frequency as the global frequency consistently leads to better results than using document frequency for the IEEE collection (2004 and 2005). Although this is consistent with the original formula, this result is somewhat surprising: Element frequency is not simple to interpret – terms that occur in deeply nested elements have a higher element frequency than terms that do not.

The explanation lies in a peculiarity of the BM25 formula: For terms that occur in more than half of all documents, the term weight w_i is negative so that the presence of these terms actually decreases the RSV:

$$w_i = \log \frac{N - \mathrm{df}(t_i) + 0.5}{\mathrm{df}(t_i) + 0.5} \tag{9}$$

To circumvent this problem, the term weight is generally set to 0 if it is negative, which means that these terms are treated as stop words.

In the IEEE collection, there are many terms that occur in more than half of the documents, so they cannot contribute to the RSV. There are, however, no terms for which the element frequency is high enough to obtain a negative weight, so this particular problem does not occur.

One might argue that terms that occur so frequently are useless for retrieval, but this is not necessarily the case for element retrieval: The terms "IEEE", "volume", and "computer" basically occur in all documents, so they have no discriminatory power at the document level. On the other hand, they may well be useful for element retrieval. For example, if a user searches for "IEEE conferences", elements that mention both terms are likely to be relevant, but elements that only mention "conferences" will have a high rate of false positives.

For the 2006 data, the behavior of element and document frequency is roughly identical, with document frequency being slightly better. This discrepancy is somewhat puzzling: what characteristic affects this? In the Wikipedia collection, the topics of the documents are more diverse, so there are no terms (apart from stop words) that occur in more than half of the documents, so the problem of negative term weights does not occur. The only outlier in this respect is the term "0", which occurs in almost all documents' header.

Figure 5 illustrates the effect of the global frequency for all tested combinations of b and k_1.

2.3 Comparison with the Official Submissions

So far, we have obtained the best BM25 parameter combinations for the various test collections, but it is still unclear how the results compare to the results of XML retrieval systems. It is hard to determine a single best official run, so we will compare the quality of the base retrieval engine with the maximum of all official submissions to that year's workshop. That is, for each rank, the nxCG value averaged over all topics for each submission is calculated, and we use the maximum as the comparison run; the resulting curve does not correspond to a real run, but it gives us an indication of where the baseline stands with respect to the others. Lucene results are excluded because they are exceeded in all cases by BM25 results.

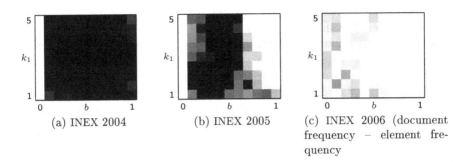

(a) INEX 2004   (b) INEX 2005   (c) INEX 2006 (document frequency – element frequency)

Fig. 5. Choice of global frequency for BM25. The heat maps show the difference between the results for element frequency and the results for document frequency; each square corresponds to one combination of b and k_1. White squares denote no change or better results for document frequency, all other shades of gray denote the degree of improvement when using element frequency.

From the INEX 2005 results, one can see that unmodified BM25 already yields high-quality results, even compared to the official submissions. This is somewhat alarming, as it shows that the methods tailored to XML retrieval fail to better the general-purpose algorithms.

Further tuning resulted in the values presented in table 1. For INEX 2005, there is a noticeable increase in retrieval quality, whereas for INEX 2006, the increase is less pronounced. For INEX 2004, the optimum result of the base retrieval engine is significantly worse than the best submitted run. This is surprising, considering that the 2004 and 2005 collections basically use the same document collection. It should be noted, however, that the assessment procedure has changed between these rounds of INEX. Figure 6 shows the results for the 2005 and 2006 collections compared to the maximum of the submissions for all ranks and shows that the good quality at rank 10 is not completely isolated.

In a real-world scenario, there are usually no relevance assessments available, so it is impossible to find the optimal parameter values. However, the values for the 2005 and 2006 test collections are close in magnitude although the collections

Table 1. Best parameters and evaluation results for the different test collections. In all cases, the Lucene similarity measure yielded worse results. The "base" column displays the value for the base engine, the "max" column displays the maximum of all official submissions in that year. The maximum from 2005 is our own submission.

	Parameters			nxCG@10	
Test collection	b	k_1	gf	base	max
INEX 2004	0.08	1.5	ef	0.4669	0.5099
INEX 2005	0.20	1.0	ef	0.3368	0.3037
INEX 2006	0.18	0.8	df	0.4332	0.4294

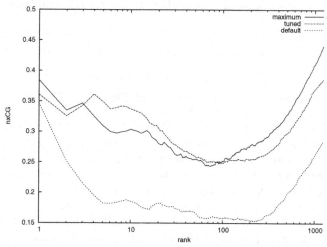

(a) INEX 2005, tuned is $b = 0.2, k_1 = 1$. The base retrieval engine is better than the best submissions up to about rank 100 (with the exception of the top ranks). Below that rank, performance gets significantly worse, possibly due to the pooling problems.

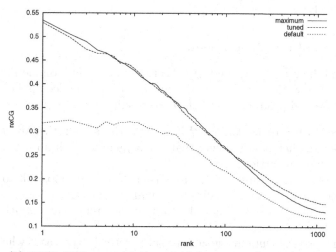

(b) INEX 2006, tuned is $b = 0.18, k_1 = 0.8$. Although the baseline does not quite reach the top status, it is close.

Fig. 6. Two of the base BM25 runs compared with the maximum run ("maximum"). The BM25 run with $b = 0.75$ and $k1 = 1.2$ ("default") shows what can be achieved without parameter tuning and the "tuned" BM25 run shows the best parameter combination for the test collection.

are very different; thus, one can assume that these values are good starting points for other collections.

3 Discussion

It is surprising to see how well a simple adaptation of standard information retrieval techniques can work for XML retrieval. Simply indexing all elements as if they were documents and applying BM25 with the right parameters can lead to better results than the best official submissions. One should keep in mind that the optimal parameters were determined after the fact by evaluating a large range of combinations on the assessed test data; the real submissions do not have the advantage of this fine-tuning.

On the other hand, the best parameters are very similar for the INEX 2005 IEEE collection and the Wikipedia collection, and minor deviations from the optimal results do not decrease retrieval quality much. Considering that these collections are very different from one another, it seems plausible to assume that using $b = 0.2$ and $k_1 = 1$ will work reasonably well in other situations. It is surprising that the best parameters are different for the INEX 2004 collection, which is almost identical to the 2005 collection. It is not clear what the reason is, but it should be kept in mind that we used an evaluation measure that was not official back then.

3.1 Realism of the Experiments

Keep in mind, however, that the test collections and evaluation metrics that are used at the INEX workshops do not entirely reflect the intended application area, and other potential problems may affect the results:

- Both the IEEE articles and the Wikipedia articles are rather short and self-contained so that it is unlikely that a fragment of such an article is more relevant than the article itself.
- The two collections differ in so many aspects that it is impossible to attribute the difference in retrieval quality to a single difference.
- The assessment process is not the same in different years, which makes it hard to do a comparison.
- Relevance assessments are generally subjective; in the cases where several people assessed the same topic, the assessments were quite different [13, 9].
- Runs that are evaluated, but were not included in the pooling process may suffer if they retrieve elements that are not in the pool. Whereas this effect has been shown to be minimal in the context of TREC [15], no study has been made in the context of INEX, but problems have been reported [14].
- The assessment interface differs from what a user of the retrieval system would see; it does not use ranking and is document-based, so the relation to real-world scenarios is unclear.

The last point needs further explanation: The unranked presentation of the results is inherent to the pooling approach that has successfully been used for traditional information retrieval evaluation for years. In the context of element retrieval, however, there is the problem that the pool does not reflect the retrieval results. Even if the pooled results only contain a single paragraph from a document, the assessor must assess the complete document. This in itself is a minor technical problem, but it seems likely that the assessment can be different from the assessment that would be obtained if the isolated paragraph were presented; if the paragraph is shown in the context of the document, the assessor may – consciously or not – use this context to rate the element's relevance.

3.2 Evaluation Metrics

It is clear that even the INEX organizers and participants have not yet reached consensus on how to evaluate the effectiveness of XML retrieval systems: Through the years, various metrics were adopted and abandoned, and even the basic retrieval tasks for the ad-hoc track are far from being fixed (INEX 2007 dropped the *thorough* task, which previously was the only task that had been done in every year). This is not avoidable, considering that XML retrieval is still a relatively young research area, but the lack of clear definitions makes it hard to do meaningful comparisons between systems.

In general, it is questionable whether the results from batch evaluations – as done in the INEX ad-hoc track – contribute to user satisfaction. Hersh et al. [7] compare several systems' performance on TREC data in batch and interactive experiments and come to the conclusion that there are significant differences in the results. In XML retrieval, the differences are likely to be even more pronounced, because the assessment user interface displays the results in a different fashion than an XML retrieval system would – the element results are shown in the context of the complete document. This is likely to affect the assessment: the users can take the surrounding material into account when judging the relevance of an element.

Buckley and Voorhees [1] discuss what it takes to draw conclusions with a sufficiently low error rate. The retrieval scenarios in this thesis are closest to their notion of web retrieval – it is very difficult to know how many relevant documents exist in total, so precision at a cutoff level of 10 to 20 should be used. In this scenario, precision is replaced by nxCG, but the reasoning is the same. To achieve a reasonable error rate, they suggest using 100 queries, which implies that only INEX 2006 data can be used to obtain reasonable conclusions (2004 and 2005 together have only 63 queries); unfortunately, the IEEE collection more closely matches the assumptions made in this thesis.

Overall, even document-based retrieval evaluation has problems, despite having a rather long tradition. For INEX, the problems are amplified by a number of new problems, partly specific to XML, partly due to the resources being much more limited than for TREC. Evaluations in INEX data are certainly far from worthless, but they should be interpreted with care.

4   Conclusions

We have shown that standard information retrieval techniques can yield surprisingly good results for XML element retrieval, even compared to techniques specifically designed for XML retrieval. This does not imply that the existing XML retrieval methods are inferior; this paper only examined retrieval quality as determined by the standard measures, storage size and speed have not been addressed. It is conceivable that other methods yield comparable retrieval quality with less overhead, or are less sensitive to parameter changes; this should definitely be examined in future research. It is hard to say what exactly the reasons are, but we hope that future research will reveal techniques for exploiting the document structure to achieve greater retrieval quality.

We propose that BM25 with suitable parameters should be used as a baseline to compare XML retrieval systems against. This may lead to painful conclusions at first – for example, we found that our work on structural patterns [4] does not work as well as previously though [6] –, but in the long run, we believe that it will lead to a higher acceptance of XML retrieval in the standard information retrieval community.

Note that the results reported in this paper only pertain to content-only retrieval and the *thorough* retrieval task. It is obviously impossible to directly use standard techniques for content-and-structure retrieval, because the standard methods do not support structural queries. For the other content-only tasks, like *focused* and *in context*, postprocessing steps on the baseline results can be used; in fact, most INEX participants already derive the results for the advanced tasks from the *thorough* results. Thus, the next logical step for further research is to combine existing approaches for the advanced tasks with the baseline retrieval methods presented here and examine what the results are.

References

[1] Buckley, C., Voorhees, E.M.: Evaluating evaluation measure stability. In: SIGIR 2000 proceedings, pp. 33–40. ACM Press, New York (2000)

[2] Denoyer, L., Gallinari, P.: The Wikipedia XML corpus. SIGIR Forum 40(1), 64–69 (2006)

[3] Dopichaj, P.: The University of Kaiserslautern at INEX 2005. In: INEX 2005 proceedings, pp. 196–210. Springer, Heidelberg (2006)

[4] Dopichaj, P.: Improving content-oriented XML retrieval by applying structural patterns. In: ICEIS 2007 proceedings. INSTICC, pp. 5–13 (2007a)

[5] Dopichaj, P.: Space-efficient indexing of XML documents for content-only retrieval. Datenbank-Spektrum 7(23) (November 2007b)

[6] Dopichaj, P.: Content-oriented retrieval on document-centric XML. PhD thesis, University of Kaiserslautern (2008), http://thesis.dopichaj.de

[7] Hersh, W., Turpin, A., Price, S., Chan, B., Kramer, D., Sacherek, L., Olson, D.: Do batch and user evaluations give the same results. In: SIGIR 2000 proceedings, pp. 17–24. ACM Press, New York (2000)

[8] Mass, Y., Mandelbrod, M., Amitay, E., Maarek, Y., Soffer, A.: JuruXML – an XML retrieval system at INEX 2002. In: INEX 2002 proceedings, pp. 73–80 (2002)

[9] Pehcevski, J., Thom, J.A.: HiXEval: Highlighting XML retrieval evaluation. In: Fuhr, N., Lalmas, M., Malik, S., Kazai, G. (eds.) INEX 2005. LNCS, vol. 3977, pp. 43–57. Springer, Heidelberg (2006)

[10] Robertson, S.E., Walker, S.: Some simple effective approximations to the 2-poisson model for probabilistic weighted retrieval. In: SIGIR 1994 proceedings, pp. 232–241. ACM Press, New York (1994),
http://portal.acm.org/citation.cfm?id=188490.188561

[11] Jones, K.S., Walker, S., Robertson, S.E.: A probabilistic model of information and retrieval:development and status. Technical report, Computer Laboratory, University of Cambridge (1998),
http://www.cl.cam.ac.uk/techreports/UCAM-CL-TR-446.html

[12] Theobald, M.: TopX – Efficient and Versatile Top-k Query Processing for Text, Structured, and Semistructured Data. PhD thesis, Universität des Saarlandes (2006)

[13] Trotman, A.: Wanted: Element retrieval users. In: Trotman, A., Lalmas, M., Fuhr, N. (eds.) Proceedings of the INEX 2005 Workshop on Element Retrieval Methodology, pp. 63–69 (2005), http://www.cs.otago.ac.nz/inexmw/

[14] Trotman, A., Pharo, N., Jenkinson, D.: Can we at least agree on something? In: Trotman, A., Geva, S., Kamps, J. (eds.) Proceedings of the SIGIR 2007 Workshop on Focused Retrieval, pp. 49–56 (2007),
http://www.cs.otago.ac.nz/sigirfocus/papers.html

[15] Zobel, J.: How reliable are the results of large-scale information retrieval experiments. In: SIGIR 1998 proceedings, pp. 307–314. ACM Press, New York (1998),
http://doi.acm.org/10.1145/290941.291014

Using Language Models and Topic Models for XML Retrieval

Fang Huang

School of Computing, The Robert Gordon University, Scotland
f.huang@rgu.ac.uk

Abstract. This paper exposes the results of our participation in the INEX 2007 ad hoc track. We implemented two different models: a mixture language model and a topic model. For the language model, we focused on the question of how shallow features of text display information in an XML document can be used to enhance retrieval effectiveness. Our language model combined estimates based on element full-text and the compact representation of the element. We also used non-content priors, including the location the element appears in the original document, and the length of the element path, to boost retrieval effectiveness. For the topic model, we looked at a recent statistical model called Latent Dirichlet Allocation[1], and explored how it could be applied to XML retrieval.

1 Introduction

In this paper, we describe our experiments in the INEX 2007 ad hoc track. With the rapidly widespread use of the eXtensible Markup Language (XML) on the internet, XML information retrieval (XML-IR) has been receiving growing research interest. A variety of approaches have been exploited to score XML elements' relevance to a user's query. Geva [3] described an approach based on the construction of a collection sub-tree that consists of all elements containing one or more of the query terms. Leaf nodes are assigned a score using a *tf.idf* variant, and scores are propagated upwards in the document XML tree, so that all ancestor elements are ranked. Ogilvie and Callan [7] proposed using hierarchical language models for ranking XML elements. An element's relevance is determined by weighted combining of several language models estimated, respectively, from the text of the element, its parent, its children, and the document. In our participation of INEX 2006, we[4] investigated which parts of a document or an XML element are more likely to attract a reader's attention, and proposed using these "attractive" parts to build a compact form of a document (or an XML element). We then used a mixture language model combining estimates based on element full-text, the compact form of it, as well as a range of non-content priors. The mixture language model presented in this paper is mainly based on our previous approach[4], but we made a few modifications to improve retrieval effectiveness.

N. Fuhr et al. (Eds.): INEX 2007, LNCS 4862, pp. 94–102, 2008.

We also experimented on how topic model, a recent unsupervised learning technique, can be use in XML retrieval. The specific model at the heart of this study is the Latent Dirichlet Allocation (LDA) model[1], a hierarchical Bayesian model employed previously to analyze text corpora and to annotate images[2]. The basic idea of a topic model is that documents are mixtures of topics, where a topic is a probability distribution over words. We used LDA to discover topics in the Wikipedia collection. Documents, XML elements, user queries and words were all represented as mixtures of probabilistic topics, and were compared to each other to calculate their relevance.

The remainder of this paper is organized as follows: Section 2 describes the mixture language model we used. Section 3 briefly introduces the LDA model and explains how LDA is used to model the relationships of documents in the Wikipedia collection. Our INEX experiments and submitted runs are presented in section 4. Section 5 discusses our results in the INEX 2007 official evaluation. The final part, section 6, concludes with a discussion and possible directions for future work.

2   The Retrieval Model

While current work in XML information retrieval focuses on exploiting the hierarchical structure of XML elements to implement more focused retrieval strategies, we believe that text display information together with some shallow features (e.g., an XML element's location in the original document) could be used to enhance retrieval effectiveness. This is based on the fact that when a human assessor reads an article, he (or she) usually can judge its relevance by skimming over certain parts of the documents. Intuitively, the titles, section titles, figures, tables, words underlined, and words emphasized in bold, italics or larger fonts are likely to be the most representative parts. In [4], we proposed to extract and put together all those most representative words to build a compact form of a document (or an XML element), and employed retrieval models that emphasized the importance of the compact form in identifying the relevance of an XML element. However, our results in the INEX 2006 evaluation showed that it did not perform as well as we expected. One reason might be that a compact form built like that contained some noise, as in the large, heterogeneous collection we used, not all the features we used are related to texts' importances. Based on this consideration, in this work, the compact form was generated by words only from titles, section titles, and figure captions. For the remainder of the paper, when we refer to the compact form of an XML element, we mean a collection of words extracted from the titles, section titles, and figure captions nested within that element.

The retrieval model we used is based on the language model, i.e., an element's relevance to a query is estimated by

$$P(e|q) \propto P(e) \cdot P(q|e) \tag{1}$$

where e is an XML element; q is a query consisting of the terms $t_1,...,t_k$; the prior, $P(e)$, defines the probability of element e being relevant in absence of a query; $P(q|e)$ is the probability of the query q, given element e.

2.1 Element Priors

The prior $P(e)$ defines the probability that the user selects an element e without a query. Elements are not equally important even though their contents are ignored. Several previous studies[5,9] reported that a successful element retrieval approach should be biased towards retrieving large elements. In INEX 2006, we conducted a preliminary experiment to investigate potential non-content features that might be used to boost retrieval effectiveness, and concluded that relevant elements tend to appear in the beginning parts of the text, and they are not likely to be nested in depth[4].

Based on these considerations, we calculate the prior of an element according to its location in the original document, and the length of its path.

$$P(e) = \frac{1}{5 + |e_{location}|} \cdot \frac{1}{3 + |e_{path}|} \tag{2}$$

where $e_{location}$ is the location value of element e; and e_{path} is the path length of e. Location was defined as the local order of an element ignoring its path. The path length of an element e equals to the number of elements in the path including e itself and those elements nesting e. For example, for an element /article[1]/body[1]/p[1] (the first paragraph in the document), the location value is 1 (the first paragraph), and the path length is 3.

2.2 Probability of the Query

Assuming query terms to be independent, $P(q|e)$ can be calculated according to a mixture language model:

$$P(q|e) = \prod_{i=1}^{k} (\lambda \cdot P(t_i|C) + (1 - \lambda) \cdot P(t_i|e)) \tag{3}$$

where λ is the so-called smoothing parameter; C represents the whole collection. $P(t_i|C)$ is the estimate based on the collection used to avoid sparse data problem.

$$P(t_i|C) = \frac{doc\_freq(t_i, e)}{\sum_{t' \in C} doc\_freq(t', C)} \tag{4}$$

The element language model, $P(t_i|e)$, defines where our method differs from other language models. In our language model, $P(t_i|e)$ is estimated by a linear combination of two parts:

$$P(t_i|e) = \lambda_1 \cdot P(t_i|e_{full}) + (1 - \lambda - \lambda_1) \cdot P(t_i|e_{compact}) \tag{5}$$

where λ_1 is a mixture parameter; $P(t_i|e_{full})$ is a language model for the full-text of element e; $P(t_i|e_{compact})$ is the estimate based on the compact representation of element e. Parameter λ and λ_1 play important roles in our model. Previous experiments[5,10] suggested that there was a correlation between the value of the smoothing parameter and the size of the retrieved elements. Smaller average sizes of retrieved elements require more smoothing than larger ones. In our experiments, the retrieval units, which are XML elements, are relatively small. We set the smoothing parameter $\lambda = 0.6$. And λ_1 was set to 0.3. In summary, the probability of a query is calculated by

$$P(q|e) = \prod_{i=1}^{k} (0.6(t_i|C) + 0.3(t_i|e_{full}) + 0.1(t_i|e_{compact})). \tag{6}$$

3 Using the Latent Dirichlet Allocation Model on Wikipedia Collection

Latent dirichlet allocation[1] is a generative probabilistic model for collections of discrete data such as text corpora. It assumes that each word of each document is generated by one of several "topics"; each topic is associated with a different conditional distribution over a fixed vocabulary. The same set of topics is used to generate the entire set of documents in a collection but each document reflects these topics with different relative proportions. Specifically, for a collection consists of words $w = w_1, w_2, ..., w_n$, where $w_i(1 \leq i \leq n)$ belongs to some documents, as in a word-document co-occurrence matrix. For each document d_i, we have a multinomial distribution over k topics, with parameters $\theta^{(d_i)}$, so for a word in document d_i, $P(z_i = j) = \theta_j^{(d_i)}$. The $j^{th}(1 \leq j \leq n)$ topic is represented by a multinomial distribution over the n words in the vocabulary, with parameters $\alpha^{(j)}$, so $P(w_i|z_i = j) = \alpha_{w_i}^{(j)}$. A Dirichlet prior is introduced for the topic distribution with parameters $\alpha_i(1 \leq i \leq k)$:

$$p(\theta|\alpha) = \frac{\Gamma(\sum_{i=1}^{k}\alpha_i)}{\prod_{i=1}^{k}\Gamma(\alpha_i)}\theta_1^{\alpha_1 - 1}...\theta_k^{\alpha_k - 1} \tag{7}$$

where the parameter α is a k-vector with components $\alpha_i > 0$, and $\Gamma(x)$ is the Gamma function. Thus, the probability of observing a document d_i is:

$$p(d_i|\alpha, \beta) = \int p(\theta|\alpha)(\prod_{n=1}^{N}\sum_{z_n} p(z_n|\theta)p(w_n|z_n, \beta))d\theta \tag{8}$$

where document d_i contains N words $w_n(1 \leq n \leq N)$. The number of parameters to estimate in this model is k parameters for the Dirichlet distribution and $n-1$ parameters for each of the k topic models. The estimation of parameters is done by variational inference algorithms.

We applied the LDA on the Wikipedia collection. All texts in the collection were lower-cased, stop-words removed using a stop-word list. After the preprocessing, each document was represented in a form of a word frequency vector.

A Gibbs sampling algorithm was then used to estimate parameters of LDA in our implementation. As the LDA model assumes that the dimensionality of the Dirichlet distribution (and thus the dimensionality of the topic variable z) is known and fixed, two topic models were learned in our experiments. The dimensionalities of them were 200 and 50, respectively. The content of words, documents, any XML elements, and user queries were then represented as vectors of topic probabilities. The similarity of a user query and an XML element were determined by cosine similarity between the two corresponding vectors.

4   INEX Experiments

In this section, we present our experiments in participating for the INEX 2007 ad hoc track.

4.1   Index

We created inverted indexes of the collection using Lucene[6]. Indexes were word-based. All texts were lower-cased, stop-words removed using a stop-word list, but no stemming. We considered paragraph elements to be the lowest possible level of granularity of a retrieval unit, and indexed text segments consisting at least one paragraph as a descendant element. For the remainder of the paper, when we refer to the XML elements considered in our investigation, we mean the segments that correspond to paragraph elements and to their ancestors. For each XML element, all text nested inside it was indexed. In addition to this, we added an extra field which corresponded to the compact representation of the element. As some studies[5,9] have already concluded that a successful element retrieval approach should be biased toward retrieving large elements, in the experiments, we indexed only those elements that consist of more than 200 characters (excluding stop words). The decision to measure in characters instead of words was based on the consideration that smaller segments such as "I like it." contains little information, while a sentence with three longer words tends to be more informative.

4.2   Query Processing

Our queries were created using terms only in the <title> parts of topics. Like the index, queries were word-based. The text was lower-cased and stop-words were removed, but no stemming was applied. '+', '-' and quotes in queries were simply removed. The modifiers "and" and "or" are ignored.

4.3   Submissions

We totally submitted 9 runs for the ad hoc track, three for each of the 3 tasks (Focused, Relevant-in-Context, and Best-in-Context). Table 1 lists a brief description of the runs.

Table 1. Ad-hoc runs submitted to INEX'07

RunID	Approach	INEX task
Focused-LM	mixture language model	Focused
Focused-TM-1	topic model with 200 topics	Focused
Focused-LDA	topic model with 50 topics	Focused
RelevantInContent-LM	mixture language model	Relevant-in-Context
RelevantInContent-TM-1	topic model with 200 topics	Relevant-in-Context
RelevantInContent-LDA	topic model with 50 topics	Relevant-in-Context
BestInContext-LM	mixture language model	Best-in-Context
BestInContext-TM-1	topic model with 200 topics	Best-in-Context
BestInContext-LDA	topic model with 50 topics	Best-in-Context

In our experiments, the top ranked elements were returned for further processing. For the Focused task, overlaps were removed by applying a post-filtering on the retrieved ranked list by selecting the highest scored element from each of the paths. In case of two overlapping elements with the same relevance score, the child element was selected. For the Relevant-in-Context task, we simply took the results for the Focused task, reordered the elements in the list such that results from the same article were grouped together in the same order they appeared in the original article. In the Best-in-Context task, the element with the highest score was chosen for each document. If there were two or more elements with the same highest score, the one that appears first in the original document was selected. For each of the runs, the top 1,500 ranked elements were returned as answers.

5 Evaluation and Results

The system's performance was evaluated against the INEX human relevance assessments. Details of the evaluation metrics can be found in [8]. Table 2 lists the result of our Focused runs, where $iP@j$, $j \in [0.00, 0.01, 0.05, 0.10]$, is the interpolated precision at j recall level cutoffs, and MAip is the mean average interpolated precision. Evaluation results of Relevant-in-Context runs and Best-in-Context runs are listed in Table 3 and Table 4, respectively. Here, $g[r]$, $r \in [5, 10, 25, 50]$, is non-interpolated generalized precision at r ranks; and MAgP is non-interpolated mean average generalized precision.

In general, our method based on mixture language model performed well compared to other submissions. Due to the pressure of time, we did not submit baseline runs for retrieval models based on full-text solely or without priors for comparison. Performances of Focused-LDA, RelevantInContent-LDA, and BestInContext-LDA are very poor. This is what we expected, as we used only 50 topics to model the collection in this group of runs. The results prompt us that 50 topics are not enough to describe the whole collection. This is reasonable, as the Wikipedia collection we used is a large heterogeneous corpus containing 659,388 documents with a large number of various topics. Furthermore, when we increased the number of topics,

Table 2. Results of Focused runs (totally 79 submissions)

RunID	iP@0.00		iP@0.01		iP@0.05		iP@0.10		MAiP	
	score	rank	score	rank	score	rank	score	rank	score	rank
Focused-LM	0.5120	33	0.4758	22	0.4118	13	0.3803	13	0.1894	8
Focused-TM-1	0.5346	18	0.4711	27	0.3788	34	0.3157	37	0.1301	31
Focused-LDA	0.0564	79	0.0277	79	0.0216	78	0.0188	78	0.0066	78

Table 3. Results of Relevant-in-Context runs (totally 66 submissions)

RunID	gP[5]		gP[10]		gp[25]		gp[50]		MAgP	
	score	rank	score	rank	score	rank	score	rank	score	rank
RelevantInContext-LM	0.2531	8	0.2205	12	0.1680	13	0.1283	14	0.1302	12
RelevantInContext-TM-1	0.2299	21	0.2064	18	0.1598	18	0.1270	17	0.1189	19
RelevantInContext-LDA	0.0100	66	0.0074	66	0.0122	65	0.0102	65	0.0081	63

Table 4. Results of Best-in-Context runs (totally 71 submissions)

RunID	gP[5]		gP[10]		gp[25]		gp[50]		MAgP	
	score	rank	score	rank	score	rank	score	rank	score	rank
BestInContext-LM	0.3405	5	0.2906	4	0.2278	4	0.1761	5	0.1742	8
BestInContext-TM-1	0.2273	39	0.2129	41	0.1775	35	0.1402	35	0.1308	35
BestInContext-LDA	0.0126	69	0.0091	69	0.0114	69	0.0099	69	0.0093	69

performances of Focused-TM-1, RelevantInContext-TM-1, and BestInContext-TM-1 (runs based on a topic model with 200 topics) are significantly improved. As the topic dimensionalities were randomly set as 50 and 200 in our experiments, we expect that retrieval results will be significantly improved given that we know the actually number of topic underlying the collection.

6   Conclusions and Future Work

We have presented, in this paper, our experiments of using language models and topic models for the INEX 2007 evaluation campaign. In our language model, we assumed important words could be identified according to the ways they were displayed in the text. We proposed to generate a compact representation of an XML element by extracting words appearing in titles, section titles, and figure captions the element nesting. Our retrieval methods emphasized the importance of these words in identifying relevance. We also integrated non-content priors that emphasized elements appeared in the beginning part of the original text, and elements that are not nested deeply. We used a mixture language model combining estimates based on element full-text, the compact form of it, as well as the non-content priors. In general, our system performed well compared to other submissions. However, due to the pressure of time, we could not submit baseline runs for comparisons of exactly how these priors and compact forms improve performances.

Our future work will focus on refining the retrieval models. Currently, the compact representation of an element is generated by words from certain parts of the text. However, the effectiveness of this method depends on the type of the documents. For example, in scientific articles, section titles (such as introduction, conclusion, etc) are not very useful for relevance judgment, whereas section titles in news reports are very informative. In the future, we will explore different patterns for generating compact representations depending on types of texts. This might involve genre identification techniques. We will investigate different priors' effectiveness and how different types of evidence can be combined to boost retrieval effectiveness.

We also explored how topic models can be used in XML retrieval. The LDA model was used to detect topics underlying the collection. We learned two topic models with topic numbers of 50 and 200, respectively. The evaluation results showed that runs based on the topic model with 200 topics achieved significantly better performances than runs based on a lower-dimensional topic space (50 topics). One assumption of the LDA model is that the dimensionality of the topic is known and fixed. In our experiments, dimensionalities were randomly set as 50 and 200. We expect the results will be better if we learn the number of topics underlying the collection. Our future work will focus on integrating text mining techniques to learn the number of topics before applying LDA model.

Acknowledgments

The Lucene-based indexer used this year was partly based on the indexing code developed for RGU INEX'06 by Stuart Watt and Malcolm Clark.

References

1. Blei, D., Ng, A., Jordan, M.: Latent Dirichlet Allocation. Journal of Machine Learning Research 3, 993–1022 (2003)
2. Blei, D., Jordan, M.: Modeling annotated data. In: Proceedings of the 26th Annual International ACM SIGIR Conference on Research and Development in Information Retrieval, pp. 127–134. ACM Press, New York (2003)
3. Geva, G.: Gardens point XML IR at INEX 2005. In: Proceedings of Fourth Workshop of the INitiative for the Evaluation of XML Retrieval (INEX 2005) (2006)
4. Huang, F., Watt, S., Harper, D., Clark, M.: Compact representations in XML retrieval. In: Fuhr, N., Lalmas, M., Trotman, A. (eds.) INEX 2006. LNCS, vol. 4518. Springer, Heidelberg (2007)
5. Kamps, J., Marx, M., de Rijke, M., Sigurbjornsson, B.X.: Retrieval: What to retrieve?. In: Proceedings of the 26th Annual International ACM SIGIR Conference on Research and Development in Information Retrieval (2003)
6. Lucene. The Lucene search engine (2005), http://jakarta.apache.org/lucene
7. Ogilvie, P., Callan, J.: Parameter estimation for a simple hierarchical generative model for XML retrieval. In: Proceedings of Fourth Workshop of the INitiative for the Evaluation of XML Retrieval (INEX 2005) (2006)

8. Pehcevski, J., Kamps, J., Kazai, G., Lalmas, M., Ogilvie, P., Piwowarski, B., Robertson, S.: INEX 2007 Evaluation Measures. In: INEX 2007 (2007)
9. Sigurbjornsson, B., Kamps, J., de Rijke, M.: An element-based approach to XML retrieval. In: INEX 2003 Workshop Proceedings (2004)
10. Zhai, C., Lafferty, J.: A study of smoothing methods for language models applied to ad hoc information retrieval. In: Proceedings of the 24th Annual International ACM SIGIR Conference on Research and Development in Information Retrieval (2001)

UJM at INEX 2007:
Document Model Integrating XML Tags

Mathias Géry, Christine Largeron, and Franck Thollard

Jean Monnet University, Hubert Curien Lab, Saint-Étienne, France
{Mathias.Gery,Christine.Largeron,
Franck.Thollard}@univ-st-etienne.fr

Abstract. Different approaches have been used to represent textual documents, based on boolean model, vector space model or probabilistic models. In text mining as in information retrieval (IR), these models have shown good results about textual documents modeling. They nevertheless do not take into account documents structure. In many applications however, documents are inherently structured (e.g. XML documents).

In this article[1], we propose an extended probabilistic representation of documents in order to take into account a certain kind of structural information: logical tags that represent the different parts of the document and formatting tags used to emphasized text. Our approach includes a learning step that estimates the weight of each tag. This weight is related to the probability for a given tag to distinguish the relevant terms.

1 Introduction

In Information Retrieval as in text mining many approaches are used to model documents. As stated in [1], these approaches can be organized in three families: models based on boolean model (for example fuzzy or extended boolean model); models based on vector space model; probabilistic models. The latter holds Bayesian networks, inference networks or belief networks. All these models appear to be appropriate to represent textual documents. They were successfully applied in categorization task or in information retrieval task.

However they all present the drawback of not taking into account the structure of the documents. It appears nevertheless that most of the available information either on the Internet or in textual databases are strongly structured. This is for example the case for scientific articles in which a title, an abstract, keywords, introduction, conclusion and other sections do not have the same importance. This is also true for the documents available on the Internet as they are written in languages (e.g. HTML or XML) that explicitly describe either the logical structure of the document (section, paragraph,...) and the formatting structure (*e.g.* font, size, color, ...).

For all these documents, the information provided by structure can be useful to emphasize some part of the textual documents. Consequently a given word does not have the same importance depending on its position in the article (*e.g.* in the title or in the

---

[1] This work has been partly funded by the Web Intelligence project (région Rhône-Alpes).

N. Fuhr et al. (Eds.): INEX 2007, LNCS 4862, pp. 103–114, 2008.

body) or if it is emphasized (bold font, etc.). Indeed, if the author of a web page deliberately writes a given word in a particular font, it could be thought that a particular information can be associated with the term and therefore that the term should be considered differently.

For all these reasons, recent works in information retrieval as in text mining, takes into account the structure of documents. This leads, in particular, to content oriented XML information retrieval (IR) that aims at taking advantage of the structure provided by the XML tree. Taking into account the structure can be done either at the indexing step or at the querying one. In the former [2,9,7], a structured document is indexed using a tree of logical textual elements. The terms weight in a given element is propagated through the structural relation, *i.e.* from leafs to the root or from root to leafs. In the latter [5], SQL query language has been adapted to the structured context in order to allow queries like "I look for a paragraph dealing with running, included in an article that deals with the New-York marathon and in which a photo of a marathon-man is present". The INEX competition (INitiative for Evaluation of XML Retrieval[2]) provides, since 2002, large collections of structured documents. Systems are evaluated through their ability to find relevant part of documents associated with XML element rather than the whole documents.

In this article, we propose to extend the probabilistic model in order to take into account the document structure (the logical structure and the formatting structure). Our approach is made up of two steps: the first one is a learning step, in which a weight is computed for each tag. This weight is estimated, based on the probability that a given tag distinguishes relevant terms. In the second step, the above weights are used to better estimate the probability for a document to be relevant for a given query.

An overview of our model is presented in the next section. A more formal one follows in section 3. The results obtained on the INEX 2006 & 2007 collections are then presented in section 4.

2   Integrating Tags into Document Modeling

In Information Retrieval, the probabilistic model [6] aims at estimating the relevance of a document for a given query through two probabilities: the probability of finding a relevant information and the probability of finding a non relevant information.

These estimates are based on the probability for a given term in the document to appear in relevant (or in non relevant) documents. Given a training collection in which the documents relevance according to some query is available, one can estimate the probability for a given term to belong to a relevant document (respectively non relevant document), given its distribution in relevant documents (respectively non relevant documents).

This probabilistic model leads to good results in textual information retrieval. Our goal here is to extend this model by taking into account the documents structure. Different kinds of "structure" can be considered. As an example, Fourel defined physical structure, layout structure, linguistic structure, discursive structure and logical structure

---

[2] See http://inex.is.informatik.uni-duisburg.de/2007/ for more details on the INEX competition.

[3]. In our model, we consider the structure defined through XML tags: logical structure (title, section, paragraph, etc.) and formatting structure (bold font, centered text, etc.).

Then, the structure is integrated in the probabilistic model at two levels:

1. The logical structure is used in order to select the XML elements (section, paragraph, table, etc.) that are considered at the indexing step. Given a query, these indexed elements are the only ones that can be ranked and returned to the user.
2. The formatting structure is then integrated into the probabilistic model, in order to improve terms weighting.

Integrating formatting tags needs a learning step in which a weight for each tag is computed. This weight is based on the probability, for a given tag, to distinguish relevant terms from non relevant ones. This is closely related to the classic probabilistic model, in which a weight for each term is estimated, based on the probability for the term to appear in relevant documents or in non relevant documents. But in our approach, tags are considered instead of terms and terms instead of documents. Thus the relevance is evaluated on documents parts (term by term) instead of whole documents, and the probability for a tag to distinguish relevant terms from non relevant ones is estimated. Accordingly, in the INEX collections, the relevance is defined on structural elements, i.e. XML elements and parts of them (i.e. sentences[3]).

During querying step, the probability for an element to be relevant is estimated based not only on the weights of the terms it contains, but also on the weights of the tags that labeled these terms.

A more formal presentation of our model is given in the next section.

3   A Probabilistic Model for the Representation of Structured Documents

3.1   Notations and Examples

Let $\mathcal{D}$ be a set of structured documents. We will consider here XML documents. Each logical element (article, section, paragraph, table, etc.) e_j of the XML tree will therefore be represented by a set of terms. We now present a running example in which three documents D_0, D_1 and D_2 are present:

D_0	D_1	D_2
\<article\>	\<article\>	\<article\>
\<p\> $t_1 t_2 t_3$ \</p\>	\<section\>	\<section\>
\<section\>	\<p\> $t_2 t_4$ \</p\>	\<p\>\<b\> t_5 \</b\>\</p\>
\<p\> $t_1 t_4$ \</p\>	\<p\> $t_2 t_5$ \</p\>	\<p\> $t_3 t_4$ \</p\>
\<p\> $t_2 t_5$ \</p\>	\</section\>	\<p\> $t_3 t_5$ \</p\>
\</section\>	\<p\> $t_2 t_1$ \</p\>	\</section\>
\</article\>	\</article\>	\</article\>

---

[3] In our model, we do not consider the relevance of sentences, but only the relevance of XML elements.

Each tag describing logical structure (*article*, *section*, *p*, etc.) defines elements that corresponds to a part of a document. Each element will be indexed. In the example, document D_2 is indexed by five elements: an *article* (tag <article>), a *section* (tag <section>) and three *paragraphs* (tag <p>).

We note:

- $E = \{e_j, j = 1, ..., l\}$, the set of the logical elements available in the collection (*article*, *section*, *p*, etc.);
- $T = (t_1, ..., t_i, ...t_n)$, a term index built from E;
- $B = \{b_1, ..., b_k, ..., b_m\}$, the set of tags.

Let E_j be a vector of random variables T_{ij} in $\{0, 1\}$:

$$E_j = (T_{10}, ..., T_{1k}, ..., T_{1m}, ..., T_{i0}, ..., T_{ik}, .., T_{im}, ..., T_{n0}, ..., T_{nk}, .., T_{nm})$$

with $\begin{cases} T_{ik} = 1 \text{ if the term } t_i \text{ appears in this element labeled by } b_k \\ T_{ik} = 0 \text{ if the term } t_i \text{ does not appear labeled by } b_k \\ T_{i0} = 1 \text{ if the term } t_i \text{ appears without being labeled by a tag in } B \\ T_{i0} = 0 \text{ if the term } t_i \text{ does not appear without being labeled} \end{cases}$

We note $e_j = (t_{10}, ..., t_{1k}, ..., t_{1m}, t_{i0}, ..., t_{ik}, .., t_{im}, t_{n0}, ..., t_{nk}, .., t_{nm})$ a realization of the random variable E_j.

In the previous example with three documents, we have $b_1 = article$, $b_2 = section$, $b_3 = p$, $b_4 = b$ and $T = \{t_1, ..., t_5\}$.

The element e_1: <p> t_1 t_2 t_3 </p> of D_0 can be represented by the vector:

$$\{t_{10}, t_{11}, t_{12}, t_{13}, t_{14}, t_{20}, t_{21}, ..., t_{53}, t_{54}\} = \{0, 1, 0, 1, 0, 0, 1, ..., 0, 0\}$$

since the term t_1 is labeled by *article* ($t_{11} = 1$), and p ($t_{13} = 1$) but neither by *section* ($t_{12} = 0$) nor by b ($t_{14} = 0$). We have $t_{10} = 0$ since the term does not appear without tag.

3.2   Term Based Score for an XML Element to Be Relevant

In the classic probabilistic model, the relevance of an element for a given query is function of the weights of the matching terms (*i.e.* terms of the query contained in the element). The weighting function BM25 [6], is broadly used to evaluate this weight, noted w_{ij}, of a term t_i in an element e_j. The term based relevance f_{term} of e_j is given by:

$$f_{term}(e_j) = \sum_{t_{ik} \in e_j} t_{ik} * w_{ij} \tag{1}$$

Given this classical model, the goal is now to propose an extension that will take into account the documents structure.

3.3   Tag Based Score for an XML Element to Be Relevant

In this section, we adapt the model introduced in [6] in order to take into account the documents structure described previously (cf. section 3.1). To do so, we not only consider term weights w_{ij}, but also tag weights.

In an information retrieval context, we want to estimate the relevance of an XML element e_j given a query. We thus want to estimate:

$P(R|e_j)$:   the probability of finding a relevant information (R) given an element e_j and a query.

$P(NR|e_j)$: the probability of finding a non relevant information (NR) given an element e_j and a query.

Let $f_1(e_j)$ be a document ranking function:

$$f_1(e_j) = \frac{P(R|e_j)}{P(NR|e_j)}$$

The higher $f_1(e_j)$, the more relevant the information presented in e_j. Using Bayes formula, we get:

$$f_1(e_j) = \frac{P(e_j|R) \times P(R)}{P(e_j|NR) \times P(NR)}$$

The term $\frac{P(R)}{P(NR)}$ being constant over the collection for a given query, it will not change the ranking of the documents. We therefore define f_2 – which is proportional to f_1 – as:

$$f_2(e_j) = \frac{P(e_j|R)}{P(e_j|NR)}$$

Using the Binary Independence Model assumption, we have:

$$P(E_j = e_j|R) = \prod_{t_{ik} \in e_j} P(T_{ik} = t_{ik}|R) \tag{2}$$

$$= \prod_{t_{ik} \in e_j} P(T_{ik} = 1|R)^{t_{ik}} \times P(T_{ik} = 0|R)^{1-t_{ik}} \tag{3}$$

In the same way, we get :

$$P(E_j = e_j|NR) = \prod_{t_{ik} \in e_j} (P(T_{ik} = 1|NR))^{t_{ik}} \times (P(T_{ik} = 0|NR))^{1-t_{ik}} \tag{4}$$

For sake of notation simplification, we note, for a given XML element:

$p_0 = P(T_{i0} = 0|R)$: the probability that t_i does not appear without being labeled, given a relevant element.

$p_{ik} = P(T_{ik} = 1|R)$: the probability that t_i appears labeled by b_k, given a relevant element.

$q_0 = P(T_{i0} = 0|NR)$: the probability that t_i does not appear without being labeled, given a non relevant element.

$q_{ik} = P(T_{ik} = 1|NR)$: probability that t_i appears labeled by b_k, given a non relevant element.

Using these notations in equations 3 and 4, we get:

$$P(e_j|R) = \prod_{t_{ik} \in e_j} (p_{ik})^{t_{ik}} \times (1 - p_{ik})^{1-t_{ik}},$$

$$P(e_j|NR) = \prod_{t_{ik} \in e_j} (q_{ik})^{t_{ik}} \times (1 - q_{ik})^{1-t_{ik}}.$$

The ranking function $f_2(e_j)$ can then be re-written:

$$f_2(e_j) = \frac{\prod_{t_{ik} \in e_j} (p_{ik})^{t_{ik}} \times (1 - p_{ik})^{1-t_{ik}}}{\prod_{t_{ik} \in e_j} (q_{ik})^{t_{ik}} \times (1 - q_{ik})^{1-t_{ik}}}$$

The *log* function being monotone increasing, taking the logarithm of the ranking function will not change the ranking. We can then define f_3 as:

$$
\begin{aligned}
f_3(e_j) &= \log(f_2(e_j)) \\
&= \sum_{t_{ik} \in e_j} (t_{ik} \log(p_{ik}) + (1 - t_{ik}) \log(1 - p_{ik}) - t_{ik} \log(q_{ik}) - (1 - t_{ik}) \log(1 - q_{ik}) \\
&= \sum_{t_{ik} \in e_j} t_{ik} \times \left(\log\left(\frac{p_{ik}}{1 - p_{ik}}\right) - \log(\frac{q_{ik}}{1 - q_{ik}}) \right) + \sum_{t_{ik} \in e_j} \log(\frac{1 - p_{ik}}{1 - q_{ik}})
\end{aligned}
$$

As before, the term $\sum_{t_{ik} \in e_j} \log(\frac{1-p_{ik}}{1-q_{ik}})$ is constant with respect to the collection (independent of t_{ik}). Not considering it will not change the ranking provided by $f_3(e_j)$:

$$f_{tag}(e_j) = \sum_{t_{ik} \in e_j} t_{ik} \log\left(\frac{p_{ik}(1 - q_{ik})}{q_{ik}(1 - p_{ik})}\right) \tag{5}$$

Thus, we obtain in this ranking function, a weight for each term t_i and each tag b_k. The weight of a term t_i labeled by b_k will be written w'_{ik}:

$$w'_{ik} = \log(\frac{p_{ik}(1-q_{ik})}{q_{ik}(1-p_{ik})})$$

Finally, in our probabilistic model that takes into account the document structure, the relevance of an XML element e_j, relatively to the tags, is defined through $f_{tag}(e_j)$:

$$f_{tag}(e_j) = \sum_{t_{ik} \in e_j} t_{ik} \times w'_{ik}$$

In practice, we have to estimate the probabilities p_{ik} and q_{ik}, $i \in \{1, .., n\}$, $k \in \{0, .., m\}$ in order to evaluate the element relevance. For that purpose, we used a learning set LS in which elements relevance for a given query is known. Given the set R (respectively NR) that contains the relevant elements (respectively non relevant ones) a contingency table can be built for each term t_i labeled by b_k:

	R	NR	$LS = R \cup NR$
$t_{ik} \in e_j$	r_{ik}	$n_{ik} - r_{ik}$	n_{ik}
$t_{ik} \notin e_j$	$R - r_{ik}$	$N - n_{ik} - R + r_{ik}$	$N - n_{ik}$
Total	R	$N - R$	N

with:

- r_{ik}: the number of times term t_i labeled by b_k is relevant in LS;
- $\sum_i r_{ik}$: the number of relevant terms labeled by b_k in LS.
- n_{ik}: the number of times term t_i is labeled by b_k in LS;
- $r'_{ik} = n_{ik} - r_{ik}$: the number of times term t_i labeled by b_k is not relevant in LS;
- R = $\sum_{ik} r_{ik}$: the number of relevant terms in LS;
- N-R = $\sum_{ik} r'_{ik}$: the number of non relevant terms in LS.

We can now estimate $\begin{cases} p_{ik} = P(t_{ik} = 1|R) &= \frac{r_{ik}}{R} \\ q_{ik} = P(t_{ik} = 1|NR) = \frac{n_{ik} - r_{ik}}{N - R} \end{cases}$

And w'_{ik} follows:

$$w'_{ik} = \log\left(\frac{\frac{r_{ik}}{R}\left(1 - \frac{n_{ik} - r_{ik}}{N - R}\right)}{\frac{n_{ik} - r_{ik}}{N - R}\left(1 - \frac{r_{ik}}{R}\right)}\right) = \log\left(\frac{r_{ik} * (N - n_{ik} - R + r_{ik})}{(n_{ik} - r_{ik}) * (R - r_{ik})}\right). \qquad (6)$$

3.4   Combining Term Based and Tag Based Scores

In order to estimate the relevance of an element e_j given a query, a global ranking function $fc(e_j)$ combining terms weights used in $f_{term}(e_j)$ and tags weights used in $f_{tag}(e_j)$, is introduced:

$$fc(e_j) = \sum_{t_{ik} \in e_j} w_{ij} \times C_k(w'_{ik})$$

where C is the function used to combine terms weights and tags weights.

We experiment different ways of combining terms weights and tags weights, in other words several functions C.

In the first one, called PSPM, the weight w_{ij} of each term t_i in e_j is multiplied with the weights w'_{ik} of the tags that label this term. More formally:

$$f_{PSPM}(e_j) = \sum_{t_{ik} \in e_j} w_{ij} \times \prod_{k/t_{ik}=1} w'_{ik}$$

We can note that some tags will reinforce the weight of the term ($w'_{ik} > 1$) while other will weaken it ($w'_{ik} \leq 1$).

The second model, called CSPM (for Closest Structured Probabilistic Model), only considers the weight w'_{ic} of the tag b_c that tags the term t_i and that is the closest to t_i.

$$f_{CSPM}(e_j) = \sum_{t_{ik} \in e_j} w_{ij} \times w'_{ic}$$

In the third model, called ASPM (for Average Structured Probabilistic Model) the weight w_{ij} of each term t_i in e_j is multiplied with the average of the weights w'_{ik} of the tags that label this term.

$$f_{ASPM}(e_j) = \sum_{t_{ik} \in e_j} w_{ij} \times \frac{\sum_{k/t_{ik}=1} w'_{ik}}{|\{k/t_{ik}=1\}|}$$

These strategies have been evaluated on the INEX 2006 & 2007 collections.

4   Experiments on INEX 2006 and 2007 Collection

4.1   INEX Collection

We used for our experimentations the INEX (Initiative for Evaluation of XML Retrieval) collection as it contains a significant amount of data together with the availability of relevant assessments.

The corpus contains 659,388 articles in English, from the free Wikipedia encyclopaedia. The documents are strongly structured as they are composed of 52 millions XML elements. Each XML article view as a tree contains, on average, 79 elements for an average depth of 6.72. Moreover, whole articles (textual content + XML structure) represent 4.5 Gb while the textual content weights only 1.6 Gb. The structural information thus represents more than twice the size of the textual one.

A set of queries is submitted by the participants during INEX 2006 competition (125 queries) and 2007 competition (130 queries). In order to evaluate information retrieval systems, the INEX campaign made available the relevance assessments corresponding to the 114 queries in 2006, and to the 107 queries in 2007.

4.2   Experimental Protocol

The corpus enriched by the INEX 2006 assessments is used as the LS training set in order to estimate the tags weights w'_{ik}.

The queries of INEX 2007 are then processed. The vector space model using BM25 weighting function is used as the baseline, without stemming nor stoplist. In order to understand the pro and cons of our structured document model, BM25 is also used as the term weighting function before integrating the tags weights.

We have evaluated our approach using the 107 assessed queries of INEX 2007. The evaluation measures used are the *precision* and *recall* measures as defined by [8].

The *interpolated average precision* (AiP), introduced by INEX, combines *precision* and *recall*, and provides an evaluation of the system results for each query. By averaging the AiP values on the set of queries, an overall measure of performance is defined [4]. This average is called *interpolated mean average precision* (MAiP).

4.3   Results

We have manually selected 14 tags in order to define the XML elements to consider. These logical structure tags will be the retrieval units, i.e. the tags considered during

Table 1. Tags frequencies (top 20)

collectionlink	16645121	normallist	1087545
item	5490943	row	954609
unknownlink	3847064	outsidelink	841443
cell	3814626	languagelink	739391
p	2689838	name	659405
emph2	2573195	body	659396
template	2396318	article	659389
section	1575519	conversionwarning	659388
title	1558235	br	378990
emph3	1484568	td	359908

the indexing step and therefore the tags the system will be able to return. These tags are *article, body, p, section, table, normallist, numberlist, title, row, td, tr, caption, definitionitem, th.*

Regarding the other tags (namely the formatting tags), we first selected the 61 tags that appear more than 300 times in the 659,388 documents (cf. table 1) and then manually removed the 6 we considered not relevant (*e.g.* br, hr, value, ...).

The weights of the 55 remaining tags were computed according to equation 6.

Table 2 presents the top 6 tags and their weights, together with the weakest 6 ones and their weights.

Table 2. Weight of the 6 strongest and 6 weakest tags

Top strongest weights		Top weakest weights	
h4	11,52	emph4	0,06
ul	2,92	tt	0,07
sub	2,34	font	0,08
small	2,21	big	0,08
strong	2,16	em	0,11
section	2,03	languagelink	0,12

We now compare the results obtained on the 107 queries of the INEX 2007 collection using our baseline and the three variants of our structured probabilistic model. Only BM25 and PSPM were submitted as official runs to INEX 2007, but all the results were computed using INEX evaluation programs (version 2, february 2008).

The results are synthesized either in table 3 or figure 1.

As can be seen, the BM25 baseline obtains a 5.32% MAiP, while PSPM obtains a 2.63% MAiP and ASPM 5.77% MAiP. The baseline is outperformed by our model ASPM, but produces better results than PSPM. Our interpretation of the latter is that multiplication impacts too strongly on small weights: two or three tags having small weights are enough to delete a term from the corpus, decreasing its weight to zero.

ASPM, that takes into account all tags by averaging their weights, performs slightly better than BM25 baseline. We can also notice that ASPM also performs better than

Table 3. MAiP of the three models evaluated on the 2007 collection

Model	@0.00	@0.01	@0.05	@0.10	@0.90	@1.00	MAiP	Rank
BM25 (baseline)	**0.4195**	**0.3221**	**0.2142**	**0.1530**	0.0004	0.0000	0.0532	63th
PSPM: all tags (weights product)	0.2266	0.1813	0.1100	0.0729	0.0000	0.0000	0.0263	72th
CSPM: closest tags only	0.1426	0.1426	0.1405	0.1271	0.0027	0.0000	0.0529	
ASPM: all tags (average weights)	0.1611	0.1611	0.1584	0.1455	**0.0027**	**0.00001**	**0.0577**	

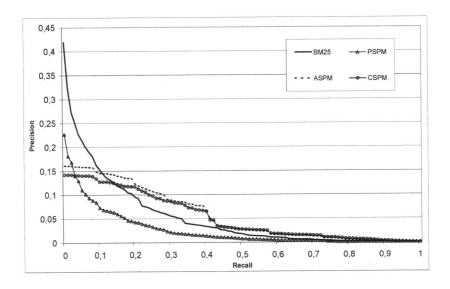

Fig. 1. MAiP of the three models evaluated on the 2007 collection

CSPM (method that only takes into account the closest tag). This is important since it shows that tags have somehow a long term dependency impact.

Our model is outperformed by the BM25 baseline at low recall levels (between P@0.00 and P@0.01), but it outperforms the BM25 baseline at other recall levels. In particular, we can see in table 3 that our model gives better results for P@0.90 and P@1.00. The precision of our model at P@1.00 is very low, but greater than zero. Each run submitted to INEX is composed by a ranked list of 1500 XML elements maximum. That means that our model is sometimes able to retrieve all the relevant elements among the first 1500 XML returned elements.

In order to better consider this fact in recall/precision curves, we have estimated $R[1500]$, the recall at 1500 elements, and we have used this score to normalize the recall-precision scores:

$$nMAiP = R[1500] * MAiP$$

These results show that our model outperforms the baseline (4.01% nMAiP versus 0.95% nMAiP), when exhaustivity (i.e. $R[1500]$) is considered (cf. table 4).

Table 4. R[1500] and nMAiP of the three models evaluated on the 2007 collection

Model	R[1500]	nMAiP
BM25 (baseline)	0.1778	0.0095
PSPM: all tags (weights product)	0.1255	0.0033
CSPM: closest tags only	**0.7026**	0.0372
ASPM: all tags (average weights)	0.6951	**0.0401**

4.4   Time Requirements

The learning step (resp. the indexing step using tags weights, and the querying step) took about 57 hours (resp. 55 hours and 3 hours), mainly due to XML Parsing. Each of these computations, parallelized on 18 PCs with 1.5Ghz-5Ghz processors and 512Mb-3Gb memory, took about 6 hours (resp. 5 hours and 12 minutes) of real time.

5   Conclusion

We proposed in this article a new way of integrating the XML structure in the classic probabilistic model. We consider both logical and formatting structure. The logical structure is used at indexing step to define elements that correspond to part of documents. These elements will be indexed and potentially returned to the user. The formatting structure is integrated in the document model itself. During a learning step a weight is computed for each formatting tag, based on the probability that this tag caracterizes relevant term. During the querying step, the relevance for an element is evaluated using the weights of the terms it contains, but each term weight is modified by the weights of the tags that label the term.

This model was evaluated on the INEX 2007 collection.

Our structured probabilistic model ASPM outperforms slightly a classical BM25 baseline. Moreover, experiments that takes into account the recall reached at 1500th rank ($R[1500]$) show that our model is better at high exhaustivity levels. We think that it is very important to integrate criteria like $R[1500]$ in evaluation measures. Indeed, in case of high exhaustivity is needed, it could be better to retrieve 69.51% of relevant information with 1500 elements (ASPM), than 17.78%, even with a better precision at low recall levels (BM25).

Beyond the fact that our structured probabilistic model ASPM outperforms a classical BM25 baseline, experiments with CSPM suggest that long term dependencies exist between tags and terms.

In a near future, we plan to analyze and take advantage of contextual information (*e.g.* long term-to-tag dependency, relationship of the tag in respect to other tags, etc.) and hope to obtain much better results. This can be done either from a practical point of view (*e.g.* using machine learning methods for modeling these relationships) or from theoretical point of view (*e.g.* adapting the aggregation between term and tag weight in the structured probabilistic model).

References

1. Baeza-Yates, R., Ribeiro-Neto, B.: Modern information retrieval. Addison-Wesley, Reading (1999)
2. Defude, B.: Etude et réalisation d'un système intelligent de recherche d'informations: Le prototype IOTA. PhD thesis, Institut National Polytechnique de Grenoble (January 1986)
3. Fourel, F.: Modélisation, indexation et recherche de documents structurés. PhD thesis, Université de Grenoble 1, France (1998)
4. Kamps, J., Pehcevski, J., Kazai, G., Lalmas, M., Robertson, S.: INEX 2007 evaluation measures. In: Fuhr, N., Lalmas, M., Trotman, A., Kamps, J. (eds.) INEX 2006. LNCS, vol. 4518, Springer, Heidelberg (2007)
5. Konopnicki, D., Schmueli, O.: W3qs: A query system for the world-wide web. In: 21th International Conference on Very Large Data Bases (VLDB 1995), September 1995, pp. 54–65 (1995)
6. Robertson, S.E., Sparck Jones, K.: Relevance weighting of search terms. Journal of the American Society for Information Sciences 27(3), 129–146 (1976)
7. Sauvagnat, K., Hlaoua, L., Boughanem, M.: XFIRM at INEX 2005: Ad-hoc and relevance feedback tracks. In: Fuhr, N., Lalmas, M., Malik, S., Kazai, G. (eds.) INEX 2005. LNCS, vol. 3977, pp. 88–103. Springer, Heidelberg (2006)
8. Swets, J.A.: Information retrieval systems. Science 141, 245–250 (1963)
9. Wilkinson, R.: Effective retrieval of structured documents. In: 17th ACM Conference on Research and Development in Information Retrieval (SIGIR 1994) (July 2007)

Phrase Detection in the Wikipedia

Miro Lehtonen[1] and Antoine Doucet[1,2]

[1] Department of Computer Science
P. O. Box 68 (Gustaf Hällströmin katu 2b)
FI–00014 University of Helsinki
Finland
{Miro.Lehtonen,Antoine.Doucet}@cs.helsinki.fi
[2] GREYC CNRS UMR 6072,
University of Caen Lower Normandy
F-14032 Caen Cedex
France
Antoine.Doucet@info.unicaen.fr

Abstract. The Wikipedia XML collection turned out to be rich of marked-up phrases as we carried out our INEX 2007 experiments. Assuming that a phrase occurs at the inline level of the markup, we were able to identify over 18 million phrase occurrences, most of which were either the anchor text of a hyperlink or a passage of text with added emphasis. As our IR system — EXTIRP — indexed the documents, the detected inline-level elements were duplicated in the markup with two direct consequences: 1) The frequency of the phrase terms increased, and 2) the word sequences changed. Because the markup was manipulated before computing word sequences for a phrase index, the actual multi-word phrases became easier to detect. The effect of duplicating the inline-level elements was tested by producing two run submissions in ways that were similar except for the duplication. According to the official INEX 2007 metric, the positive effect of duplicated phrases was clear.

1 Introduction

In previous years, our INEX-related experiments have included two dimensions to phrase detection, one at the markup level [1] and another in the term sequence analysis [2]. The methods have been tested on plain text corpora and scientific articles in XML format. The Wikipedia XML documents are the first collection of hypertext documents where our phrase detection methods are applied.

Regarding marked-up phrases, the nature of the markup in a hypertext document differs from that in a scientific article. The phrases that are marked in scientific texts are mostly meant to be displayed with a different typeface, e.g. italicised or underlined, whereas hypertext documents have similar XML structures for marking the anchor text related to a hyperlink. Both emphasised passages and anchors are important, but whether they can be treated equally is still an open question. The initial results support the idea that emphasised phrases and anchors are equal as long as they are marked with similar XML structures — inline-level elements.

N. Fuhr et al. (Eds.): INEX 2007, LNCS 4862, pp. 115–121, 2008.

This article is organised as follows. Section 2 describes our IR system as it was implemented in 2007. In Section 3, we go through the phrase detection process step by step, from the original XML fragment to an intermediate XML format and, further, to the vector representation. The observations of the inline elements in the actual test collections are summarised in Section 4. How we extracted multiword units this year is explained in Section 5. Our results are presented in Section 6, and finally, we draw conclusions and directions for future work in Section 7.

2   EXTIRP Baseline

The EXTIRP baseline without duplicated phrases is similar to our INEX 2006 submission [3] except for a few major bugs that have been fixed. The results are thus not comparable. First, EXTIRP scans through the document collection and selects disjoint fragments of XML to be indexed as atomic units. Typical fragments include XML elements marking sections, subsections, and paragraphs. In the Wikipedia, typical names for these elements are `article`, `section`, and `p`. The disjoint fragments are treated as traditional documents which are independent of each other. The pros include that the traditional IR methods apply, so we use the vector space model with a weighting scheme based on the tf*idf. The biggest of the cons is that the size of the indexed fragments is static, and if bigger or smaller answers are more appropriate for some query, the fragments have to be either divided further or combined into bigger fragments.

Second, two separate inverted indices are built for the fragments. A *word index* is created after punctuation and stopwords are removed and the remaining words are stemmed with the Porter algorithm [4]. The *phrase index* is based on Maximal Frequent Sequences (MFS) [5]. A sequence is said to be frequent if it occurs more often than a given sentence frequency threshold. It is said to be maximal if no other word can be inserted into the sequence without reducing the frequency below the threshold. This permits to obtain a compact set of document descriptors, that we use to build a phrase index of the collection. The frequency threshold is decided experimentally, because of the computational complexity of the algorithm. Although lower values for the threshold produce more MFSs, the computation itself would take too long to be practical. For the wikipedia collection, we used a frequency threshold of seven.

When processing the queries, we compute the cosine similarity between the document and the base term vectors which results in a `Word_RSV` value. In a similar fashion, each fragment vector gets a similarity score `MFS_RSV` for phrase similarity. These two scores are aggregated into a single RSV so that the aggregated RSV = α * `Word_RSV` + β * `MFS_RSV`, where α is the number of distinct query terms and β is the number of distinct query terms in the query phrases.

3   Phrase Detection and Duplication

The novelty in our system of 2007 was the analysis of the XML structure in order to locate marked-up phrases in the content. Table 1 shows an example of a passage in an XML fragment with two phrases marked up: "Britney Spears" and "I'm A Slave 4 U". Both of the phrases are presumably better descriptors of the passage than any of the other nouns, e.g. "success" or "single", which is also indicated by the corresponding document frequencies.

Table 1. A passage before phrase detection. The tag name `c` is an abbreviation of `collectionlink`. Link references are omitted.

```
He repeated the success by doing the same with
     <c>Britney Spears</c>' dance-pop single,
       &quot;<c>I'm A Slave 4 U</c>&quot;
```

3.1   Definition of Qualified Inline Elements

Because there are many XML elements in the document that have a similar structure but a very different function, we need a formal definition for the kind of elements that most likely contain a marked-up phrase. Therefore, we define a *Qualified inline element* as follows: An XML element is considered a qualified inline element when the corresponding element node in the document tree meets the following conditions:

(1) The text node siblings contain at least n characters after whitespace has been normalised.
(2) The text node descendants contain at least m characters after normalisation.
(3) The element has no element node descendants.
(4) The element content is separated from the text node siblings by word delimiters, e.g. whitespace or commas.

When the whitespace of a text node is normalised, all the leading and trailing whitespace characters are trimmed off. We set the parameters to a minimum of three (3) characters in at least one text node child and a minimum of five (5) characters in at least one text node sibling, so that $n = 5$ and $m = 3$. With this definition, we disqualify those inline-level elements that 1) only contain one or two characters, and 2) those that contain several sentences of text, and 3) those that contain other XML elements. Defining the lower bounds of n and m improves the quality of detected phrases in the qualified inline elements. However, regarding the effectiveness of IR, the benefit of setting the parameters is marginal: very short character strings are usually ignored, whereas several sentences of text rarely match any searched phrase.

3.2 Doubling "Britney Spears"

According to our hypothesis, whatever is emphasised in the document should also be emphasised in the index. Consequently, all the occurrences of qualified inline elements are duplicated before the text is indexed. An example of such duplication is shown in Table 2 which represents the intermediate XML format which is the basis for the eventual vector representation.

Table 2. The passage after phrase detection and duplication

```
        He repeated the success by doing the same with
<c>Britney Spears</c> <c>Britney Spears</c>' dance-pop single,
    &quot;<c>I'm A Slave 4 U</c> <c>I'm A Slave 4 U</c>&quot;
```

The example is representative of the most common appearances of "Britney Spears", which include the following:

```
freq.  appearance
-----  ----------
447    <collectionlink>Britney Spears</collectionlink>
 12    <emph2>Britney</emph2>
  5    <collectionlink>Britney</collectionlink>
```

After stemming and stopword removal, the corresponding word sequence becomes

```
britnei spear britnei spear
```

which is the input when extracting Maximal Frequent Sequences. Obviously, duplication has a dual effect of both increasing the term frequencies of the content that it concerns and changing the word sequence that phrases are extracted from. We believe the increase in term frequency is good because double "Britney Spears" is easier to spot than a single occurrence. The newly modified word sequence is also better as the MFS's that we extract also include the phrase `spear britnei` which, in addition to `britnei spear`, contributes to the score for phrase similarity (MFS_RSV). Duplicating phrases with more than two word units has a similar effect, as any word permutation within the phrase contributes to the MFS_RSV score.

4 Qualified Inline Elements in the Wikipedia XML

The most common elements that were duplicated are summarised in Table 3. The exhaustivity of an element type is the percentage of element occurrences duplicated out of all occurrences of that element.

Most of the qualified inline elements are links (83.8%) and only a minority mark emphasis (12.8%) in the Wikipedia XML collection, which is the opposite

Table 3. Distribution of the most frequent qualified inline elements by element type

XML Element	Count	Exhaustivity %	Percentage
collectionlink	12,971,384	76.2	69.1
unknownlink	2,372,870	60.0	12.6
emph2	1,339,345	49.2	7.1
emph3	992,373	67.0	5.3
p	282,438	10.3	1.5
outsidelink	230,675	26.8	1.2
title	222,917	14.0	1.2
languagelink	114,828	14.5	0.6
emph5	57,443	70.8	0.3
wikipedialink	42,009	23.8	0.2
All links	15,734,890	68.9	83.8
All emphasis	2,406,372	55.3*	12.8
Total	18,784,132	35.7	100

*All element types marking emphasis might not be included in the figures.

of the collection of IEEE Computer Society[1] journals and transactions where the share of links is only 2.0% while 85.0% mark emphasis. The frequency of qualified inline elements is bigger in the Wikipedia in general, as well: 35.7% of all elements meet the requirements, whereas the corresponding figure is 6.6% in the IEEE collection.

5   MFS Extraction

In this section, we are comparing our runs from the point of view of the MFSs that were extracted. We conjecture that the phrase duplication process facilitates the extraction of the more useful sequences, hereby inducing better retrieval performance. We will try to confirm this by analysing the extracted sequence sets corresponding to our runs.

Statistics are summarized in Table 4. As discussed earlier, the frequency threshold was always seven occurrences. That is, a sequence was considered frequent if it occurred in at least seven minimal units of a same document cluster. In the first run (UHel-Run1), we split the XML fragments extracted from the document collection into 500 disjoint clusters, whereas for UHel-Run2, the number of clusters is 250. Given the constant frequency threshold, a lower number of clusters causes a slower extraction but naturally permits finding more MFS occurrences. This is because it is easier to find seven occurrences of an MFS in larger clusters, that is, when the number of disjoint clusters is smaller [6].

To give a first hint on the benefit of our phrase duplication technique, we are displaying the 10 most frequent phrases that were duplicated in Table 5.

[1] http://www.computer.org/

Table 4. Per run statistics of the extracted MFS sets (frequency threshold: 7)

Run	Clusters	Number of sequences (total freq)	Average length	Average Frequence
UHel-Run1	500	21,009,668	2.248	19.9
UHel-Run2	250	37,252,061	2.184	26.4

Table 5. The 10 most frequent phrases that were duplicated

Frequency	Phrase
37,474	Native American
37,328	population density
37,047	African American
36,046	married couples
35,926	per capita income
35,829	other races
35,807	poverty line
35,764	Pacific Islander
32,974	United States Census Bureau
26,572	United States

6   Results

We submitted two runs for the adhoc track task of Focused retrieval. The results are shown in Table 6. The assessments of 107 topics are included in the evaluation. The performance of our systems is relatively low compared with other evaluated systems, but the level seems typical of systems using tfidf-based weighting.

What we learn from these results is that our second run is undeniably better than the first run at all recall levels. The p values of the one-tailed t-test show that Run 2 is significantly better than Run 1 overall as well as at the lowest recall levels (0.00 and 0.01), given the threshold of 0.05. It is thus not only "Britney Spears" that is easier to find when doubled, but many other phrases that were

Table 6. Performance of submissions "UHel-Run1" and "UHel-Run2" measured with interpolated precision at four recall levels. A total of 79 submissions are included in the ranking.

	Run1		Run2			t-test	Best official
Recall level	Rank	Score	Rank	Score	Improvement	p	Score
MAiP	53	0.0912	45	0.1024	12.3%	0.0107	0.2238
0.00	66	0.3639	60	0.4068	11.8%	0.0454	0.6056
0.01	63	0.3319	58	0.3773	13.7%	0.0287	0.5271
0.05	58	0.2729	56	0.3000	9.9%	0.0783	0.4697
0.10	58	0.2273	54	0.2447	7.6%	0.1386	0.4234

topic titles. Although the EXTIRP baseline has a relatively low performance, it has been stable the past few years, and any improvement over its performance is hardly coincidence. We believe therefore that also other systems would benefit of the phrase extraction as we have done it.

7 Conclusion

Phrase detection in the Wikipedia XML documents was a success as it improved our results at all recall levels. Analysing the XML markup did not involve any information about the document type, such as element names or tag names, so the technique is applicable to any XML documents. It can also be adopted by different systems as it is not tied to any specific document model or weighting method.

Our future work starts with the exploration of other algorithms for phrase extraction than the Maximal Frequent Sequences as we expect the duplication of inline elements to improve phrase extraction regardless of the algorithm. Another area of future development concerns the term weighting and matching in our system. We are interested in the effect of the phrase detection in more advanced and better performing systems, so we plan to discard the tfidf-based weights and move on to new directions.

References

1. Lehtonen, M.: Preparing heterogeneous XML for full-text search. ACM Trans. Inf. Syst. 24, 455–474 (2006)
2. Doucet, A., Ahonen-Myka, H.: Probability and expected document frequency of discontinued word sequences, an efficient method for their exact computation. Traitement Automatique des Langues (TAL) 46, 13–37 (2006)
3. Lehtonen, M., Doucet, A.: Extirp: Baseline retrieval from wikipedia. In: Malik, S., Trotman, A., Lalmas, M., Fuhr, N. (eds.) INEX 2006. LNCS, vol. 4518, pp. 119–124. Springer, Heidelberg (2007)
4. Porter, M.F.: An algorithm for suffix stripping. Program 14, 130–137 (1980)
5. Ahonen-Myka, H.: Finding all frequent maximal sequences in text. In: Mladenic, D., Grobelnik, M. (eds.) Proceedings of the 16th International Conference on Machine Learning ICML 1999 Workshop on Machine Learning in Text Data Analysis, Ljubljana, Slovenia, J. Stefan Institute, pp. 11–17 (1999)
6. Doucet, A., Ahonen-Myka, H.: Fast extraction of discontiguous sequences in text: a new approach based on maximal frequent sequences. In: Proceedings of IS-LTC 2006, pp. 186–191 (2006)

Indian Statistical Institute at INEX 2007 Adhoc Track: VSM Approach

Sukomal Pal and Mandar Mitra

Information Retrieval Lab, CVPR Unit,
Indian Statistical Institute, Kolkata
India
{sukomal_r,mandar}@isical.ac.in

Abstract. This paper describes the work that we did at Indian Statistical Institute towards XML retrieval for INEX 2007. As a continuation of our INEX 2006 work, we applied the Vector Space Model and enhanced our text retrieval system (SMART) to retrieve XML elements against the INEX Adhoc queries. Like last year, we considered Content-Only(CO) queries and submitted two runs for the FOCUSED sub-task. The baseline run does retrieval at the document level; for the second run, we submitted our first attempt at element level retrieval. This run uses a very naive approach and performs poorly, but the relative performance of the baseline run was fairly encouraging. After the official submissions, we conducted a few more experiments involving both document-level and element-level retrieval. These additional runs yield some improvements in retrieval effectiveness. We report the results of those runs in this paper. Though our document-level runs are promising, the element-level runs are still far from satisfactory. Our next step will be to explore ways to improve element-level retrieval.

1 Introduction

Traditional Information Retrieval systems return whole documents in response to queries, but the challenge in XML retrieval is to return the most relevant parts of XML documents which meet the given information need. INEX 2007 [1] marks a paradigm shift as far as retrieval granularity is concerned. This year, arbitrary passages are also permitted as retrievable units, besides the usual XML elements. A retrieved passage can be a sequence of textual content either from within an element or spanning a range of elements. INEX 2007 also classified the adhoc retrieval task into three sub-tasks: a) the FOCUSED task which asks systems to return a ranked list of elements or passages to the user; b) the RELEVANT in CONTEXT task which asks systems to return relevant elements or passages grouped by article; and c) the BEST in CONTEXT task which expects systems to return articles along with one best entry point to the user.

Each of the three subtasks can be based on two different query variants: Content-Only(CO) and Content-And-Structure(CAS) queries. In the CO task,

N. Fuhr et al. (Eds.): INEX 2007, LNCS 4862, pp. 122–128, 2008.

the user poses the query in free text and the retrieval system is supposed to return the most relevant elements/passages. A CAS query can provide explicit or implicit indications about what kind of element the user requires along with a textual query. Thus, a CAS query contains structural hints expressed in XPath-like [2] syntax, along with an *about()* predicate.

Our retrieval approach this year was based on the Vector Space Model which sees both the document and the query as bags of words, and uses their *tf-idf* based weight vectors to measure the inner product *similarity* between the document and the query. The documents are retrieved and ranked in decreasing order of the similarity value.

We used the SMART system for our experiments at INEX 2007 and submitted two runs for the *FOCUSED* sub-task of the Adhoc track considering CO queries only. We performed some additional experiments after the submission. In the following section, we describe our general approach for all these runs, and discuss results and further work in Section 3.

2 Approach

To extract the useful parts of the given documents, we manually shortlisted about thirty tags that were found to contain useful information: <p>, <ip1>, <it>, <st>, <fnm>, <snm>, <atl>, <ti>, <p1>, <h2a>,<h>, <wikipedialink>, <section>, <outsidelink>, <td>, <body>, etc. Documents were parsed using the libxml2 parser, and only the textual portions included within the short-listed tags (see above) were used for indexing. Similarly, for the topics, we considered only the *title* and *description* fields for indexing, and discarded the *inex-topic*, *castitle* and *narrative* tags. No structural information from either the queries or the documents was used.

The extracted portions of the documents and queries were indexed using single terms and a controlled vocabulary (or pre-defined set) of statistical phrases following Salton's blueprint for automatic indexing [3]. Stopwords were removed in two stages. First, we removed frequently occurring common words (like *know, find, information, want, articles, looking, searching, return, documents, relevant, section, retrieve, related, concerning,* etc.) from the INEX topic sets. Next, words listed in the standard stop-word list included within SMART were removed from both documents and queries. Words were stemmed using a variation of the Lovins stemmer implemented within SMART. Frequently occurring word bi-grams (loosely referred to as phrases) were also used as indexing units. We used the N-gram Statistics Package (NSP)[1] on the English Wikipedia text corpus and selected the 100,000 most frequent word bi-grams as the list of candidate phrases. Documents and queries were weighted using the *Lnu.ltn* [4] term-weighting formula. For each of 130 adhoc queries(414-543), we retrieved 1500 top-ranked XML documents or non-overlapping elements.

---

[1] http://www.d.umn.edu/~tpederse/nsp.html

2.1   Document-Level Run

For the baseline run, *VSMfb*, we retrieved whole documents only. We had intended to use blind feedback for this run, but ended up inadvertently submitting the results of simple, inner-product similarity based retrieval.

Later, we conducted the actual feedback run, in which we applied automatic query expansion following the steps given below for each query (for more details, please see [5]).

1. For each query, collect statistics about the co-occurrence of query terms within the set $\mathcal{S}$ of 1500 documents retrieved for the query by the baseline run. Let $df_\mathcal{S}(t)$ be the number of documents in $\mathcal{S}$ that contain term t.
2. Consider the 50 top-ranked documents retrieved by the baseline run. Break each document into overlapping 100-word windows.
3. Let $\{t_l, \ldots, t_m\}$ be the set of query terms (ordered by increasing $df_\mathcal{S}(t_i)$) present in a particular window. Calculate a similarity score *Sim* for the window using the following formula:

$$Sim = idf(t_1) + \sum_{i=2}^{m} idf(t_i) \times \min_{j=1}^{i-1}(1 - P(t_i|t_j))$$

where $P(t_i|t_j)$ is estimated based on the statistics collected in Step 1 and is given by

$$\frac{\#\ documents\ in\ \mathcal{S}\ containing\ words\ t_i\ and\ t_j}{\#\ documents\ in\ \mathcal{S}\ containing\ word\ t_j}$$

This formula is intended to reward windows that contain multiple matching query words. Also, while the first or "most rare" matching term contributes its full idf (inverse document frequency) to *Sim*, the contribution of any subsequent match is deprecated depending on how strongly this match was predicted by a previous match — if a matching term is highly correlated to a previous match, then the contribution of the new match is correspondingly down-weighted.
4. Calculate the maximum *Sim* value over all windows generated from a document. Assign to the document a new similarity equal to this maximum.
5. Rerank the top 50 documents based on the new similarity values.
6. Assuming the new set of top 20 documents to be relevant and all other documents to be non-relevant, use Rocchio relevance feedback to expand the query. The expansion parameters are given below:

$$number\ of\ words = 20$$
$$number\ of\ phrases = 5$$
$$Rocchio\ \alpha = 4$$
$$Rocchio\ \beta = 4$$
$$Rocchio\ \gamma = 2.$$

Finally, for each topic, 1500 documents were retrieved using the expanded query. This unofficial run is named *VSMfeedback*.

2.2   Element-Level Run

This year, we also attempted element-level retrieval for the first time. Since SMART does not natively support the construction of inverted indices at the element level, we adopted a 2-pass strategy. In the first pass, we retrieved 1500 documents for each query using query expansion.

In the second pass, these documents were parsed using the libxml2 parser, and leaf nodes having textual content were identified. Figure 1 shows a fragment of a file from the wikipedia collection. The leaf nodes that have textual content are enclosed in rectangles in the figure. The total set of such leaf-level textual elements obtained from the 1500 top-ranked documents were then indexed and compared to the query as before to obtain the final list of 1500 retrieved elements.

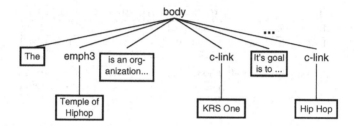

Fig. 1. Parse tree for a fragment of a wikipedia document

Since we considered only the leaf nodes as retrievable elements for *VSMfbElement*, the retrieved elements for the official run are automatically non-overlapping. However, as is clear from Figure 1, permitting only leaf-level textual elements to be retrieved has an obvious disadvantage: nodes such as <p> or <body> are very often not considered retrievable elements, because of the occurrence of nodes like <emph3> and <collectionlink> under the <p> or <body> node. It is not surprising, therefore, that the *VSMfbElement* run performs poorly (see the next section for details).

Post-submission, we incorporated within SMART the capability to retrieve elements at intermediate (i.e. non-leaf) levels of the XML tree. Retrieval results obtained using this capability are labelled *VSM-fdbk-elt* in the tables below. The *VSM-fdbk-elt* run is similar to the *VSMfbElement* run described earlier, and uses a 2-pass strategy. The difference is that, in the new run, the query is compared to all elements that contain text, instead of only the leaf-level textual nodes. In order to avoid any overlap in the final list of retrieved elements, the nodes for a document are sorted in decreasing order of similarity, and all nodes that have an overlap with a higher-ranked node are eliminated.

3   Results

The results for the official and unofficial runs (according to the updated evaluation script and relevance judgments for 107 topics) are shown in Tables 1 and 2 respectively.

Table 1. Official results for Element Retrieval (FOCUSED task, CO queries)

Run Id	P@0.00		P@0.01		P@0.05		P@0.10		MAiP	
	Score	Overall rank	Score	Overall rank	Score	Overall rank	Score	Overall rank	Score	Overall rank
VSMfb	0.4680	49	0.4524	39	0.3963	20	0.3797	14	0.1991	4
VSMfbElement	0.2406	76	0.1820	73	0.0990	74	0.0548	74	0.0159	76
BEST run	0.6056		0.5271		0.4697		0.4234		0.2238	

Table 2. Unofficial results for Element Retrieval (FOCUSED task, CO queries)

Run Id	P@0.00	P@0.01	P@0.05	P@0.10	MAiP
VSMfeedback	0.4839	0.4682	0.4236	0.3957	0.2116
VSM-fdbk-elt-slope0.2	0.4724	0.4171	0.3143	0.2497	0.0787
VSM-fdbk-elt-slope0.3	0.4873	0.4318	0.3358	0.2620	0.0803
VSM-fdbk-elt-slope0.4	0.5032	0.4558	0.3379	0.2374	0.0742
BEST run	0.6056	0.5271	0.4697	0.4234	0.2238

The first official run (VSMfb) fared quite well except at the early recall points. Since this run returns only whole documents, it compares unfavourably with other runs when evaluated using precision-oriented measures such as $P@0.00$ or $P@0.01$, but looks respectable in terms of $P@0.10$. Figure 2(a) (the red line) shows that it is among the top 4 runs for the later recall points, and it ranks 4th among 79 runs according to the MAiP measure. When blind feedback is used (the *VSMfeedback* run, represented by the cyan line in Figure 2(b)), results improve at all recall points, and the MAiP obtained is within about 5% of the best reported MAiP figure.

The element-level run *VSMfbElt* proved to be a damp squib. In hindsight, this is not surprising since during submission, our system did not consider elements at intermediate (non-leaf) levels. Leaf nodes are very often too small to contain any meaningful information; further, it is usually difficult to reliably rank such small pieces of text. After we implemented element retrieval at the non-leaf level (the *VSM-fdbk-elt* runs), results improved significantly.

However, this run was also far from satisfactory, especially at the early recall points. This suggests that the system is retrieving larger, less focused pieces of text than it should[2]. Increasing the degree of document length normalization could be one way to address this problem. The term-weighting scheme that we use – pivoted document length normalization [6] – gives us an easy way to test this hypothesis. Under this term-weighting scheme, the document length normalization factor is given by

$$(1 - slope) * pivot + slope * length$$

---

[2] Paradoxically, our document-level runs do significantly better than our element-level runs, an observation that is corroborated by other groups.

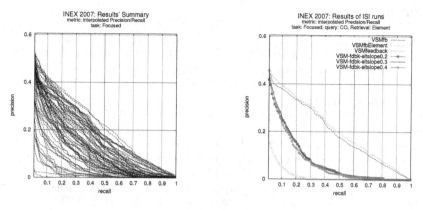

Fig. 2. Comparison of (a) all runs at INEX 07, and (b) ISI runs at INEX 07

Following [7], we had set the slope and pivot parameters to 0.20 and 80 respectively for our runs. Increasing the slope value is expected to promote elements that are shorter than the pivot length, while pushing longer elements to lower ranks. Accordingly, we experimented with two more *slope* values, viz. 0.3 and 0.4. Table 1 suggests that our intuition is correct: increasing the degree of normalization for long documents seems to improve early precision. However, the $P@0.05$ and $P@0.10$ figures reach a point of diminishing returns as the slope is increased. Also, the MAiP figures for these runs are rather dismal. More experiments are needed in order to understand the effect of normalization on precision at various recall points. We also need to explore ways to improve precision across all recall points.

4 Conclusion

This was our second year at INEX. Our main objective this year was to incorporate element-level retrieval within SMART. We started with retrieval only at the leaf level, and extended it to enable retrieval of elements at any level within the XML tree. We experimented with a 2-pass strategy where document level retrieval is done at the first pass, and the documents so retrieved are fed to the second pass for retrieval at a finer granularity. Except when considering precision at the early recall points, our baseline run was among the best, and improved further with blind feedback. However, our element-level runs were disappointing. Our intuition that short elements are victimised during retrieval seems to be validated by post-submission runs, but the effect of document length normalization on XML retrieval needs more careful study. We also hope to investigate ways to incorporate element/tag information into the term-weighting scheme. We hope these will be exciting exercises which we plan to continue in the coming years.

References

1. Fuhr, N., Lalmas, M., Trotman, A., Kamps, J. (eds.): Focused access to XML documents: 6th International Workshop of the Initiative for the Evaluation of XML Retrieval (INEX 2007), http://inex.is.informatik.uni-duisburg.de/2007
2. W3C: XPath-XML Path Language(XPath) Version 1.0), http://www.w3.org/TR/xpath
3. Salton, G.: A Blueprint for Automatic Indexing. ACM SIGIR Forum 16(2), 22–38 (Fall 1981)
4. Buckley, C., Singhal, A., Mitra, M.: Using Query Zoning and Correlation within SMART: TREC5. In: Voorhees, E., Harman, D. (eds.) Proc. Fifth Text Retrieval Conference (TREC-5), pp. 500–238. NIST Special Publication (1997)
5. Mitra, M., Singhal, A., Buckley, C.: Improving automatic query expansion. In: SIGIR 1998, Melbourne, Australia, pp. 206–214. ACM, New York (1998)
6. Singhal, A., Buckley, C., Mitra, M.: Pivoted document length normalization. In: SIGIR 1996: Proceedings of the 19th annual international ACM SIGIR conference on Research and development in information retrieval, pp. 21–29. ACM Press, New York (1996)
7. Singhal, A.: Term Weighting Revisited. PhD thesis, Cornell University (1996)

A Fast Retrieval Algorithm for Large-Scale XML Data

Hiroki Tanioka

Innovative Technology R&D, JustSystems Corporation,
108-4 Hiraishi-Wakamatsu Kawauchi-cho Tokushima-shi Tokushima, Japan
hiroki.tanioka@justsystems.com

Abstract. This paper proposes a novel approach for retrieving large-scale XML data using the vector space model. The vector space model is commonly used in the information retrieval community. Last year, for the Evaluation of XML Retrieval (INEX) 2006 Adhoc Track, we developed a system using fragment elements. The system made it possible to search over XML elements for queries with varying constraints on XML elements to be included in the search, without the need for reindexing the collection, supporting more flexible queries. However the system took significant time to unitize the fragment elements. To solve the problem, our new system is composed of an inverted-file list and a relative inverted-path list on the INEX 2007 Adhoc Track corpus.

1 Introduction

There are two approaches for XML information retrieval (IR): one based on database models, the other based on information retrieval models. Our system is based on the vector space model[9] from information retrieval. In the field of information retrieval, the retrieval unit returned by IR systems is typically a whole document or a document fragment, such as a paragraph in passage retrieval. Traditional IR systems based on the vector space model compute a postings file as term vectors for each retrieval unit, and calculate similarities between the units and the query. Specifically, the postings file maps each XML node from words, and the query consists of some words.

Our system uses keywords (multi-word terms, single words) as the query and separates XML [1] documents into two parts: content information (the keywords) and structural information. XML nodes correspond to retrieval units, and nodes that include query terms can be quickly retrieved using an inverted-file list. For very large XML documents, all XML nodes are indexed to each term directly included in the node itself, but not the node's children or more distantly related nodes. During the retrieval phase, the score of a retrieved node is calculated by merging the scores from its descendant nodes. To merge scores while identifying parent-child relationships, the system employs a relative inverted-path list (*RIP* list) that uses nested labels with offsets to save the structural information.

1.1 Related Works

The XML indexing strategy of dividing structural and content information has already been published in IR-CADG index[13]. A score merging method called the Bottom-UP Scheme (BUS) was proposed in [12]. In recent years, SIRIUS[15] achieved high

N. Fuhr et al. (Eds.): INEX 2007, LNCS 4862, pp. 129–137, 2008.
© Springer-Verlag Berlin Heidelberg 2008

precision using a combination of content and structural information, and GPX [16] used an index for some types of queries based on the BUS method.

However, GPX averaged 7.2 seconds per topic and required more than 30 seconds for certain query types. Unfortunately, this level of performance cannot be considered practical. Meanwhile, Hatano[17] proposed a means for eliminating nodes that the user (or the system) wants to exclude from the search. That system can handle larger XML documents, but it must reindex the collection to take into account any new node exclusion criteria specified by the user. For these reasons, It is preferred that the XML indexing and retrieving methods do not require reindexing and still keep the index size small.

In INEX 2006[4], our system was based on the vector space model and employed BUS[14]. However it took an average of 66.2 seconds per topic for retrieval. To improve performance for INEX 2007, our new system employs the RIP list. In retesting on the INEX 2007 Adhoc Track, our new system achieved an average query time of 3.94 seconds (with a worst case time of 9.95 seconds), while maintaining high precision.

The rest of this article is divided into three sections. In section 2, we describe the architecture of our indexing and retrieval system for XML documents. In section 3, we describe experimental results. And in section 4, we discuss results and future work.

2   XML Information Retrieval

In XML retrieval, the goal is to retrieve not only XML documents but also XML nodes in response to queries. In the database approach to XML retrieval, the system narrows the number of the retrieved nodes by a constraint condition such as XPath[2] and XQuery[3]. After that, the system performs a keyword search. Current research[17] indicates that such systems have low precision and require considerable time to complete the search operation. This is because keyword-based search systems must merge all query-term hits (as they cannot in principal prioritize hits via term scores) and they cannot easily rank results (as they have do not have precise document-scoring functions).

In the information retrieval community, this problem is partially addressed using the technique of passage retrieval[8], which returns a bounded segment of a document, such as a chapter, section, or paragraph. Alternatively, Evans[10] proposes that the boundary of the segments can be determined automatically, a technique know as sub-document retrieval. Research results have shown the effectiveness of both of these methods in many applications[11].

The ideal XML retrieval algorithm would compute scores at each hierarchical level of the document where the nodes at each level would be of uniform size. However, XML nodes have signficant variation in size. Thus we need to develop a node scoring model that normalizes for node size. In addition to size, XML nodes have structural information that should be included in the scoring model.

2.1   Inverted-File List

Our system starts with an inverted-file index that contains document information. In the system, terms and XML nodes become unique numerical identifiers, term IDs and node IDs, respectively. Then term IDs and node IDs are indexed as shown below,

Term-ID: {Node-ID, Term-Freq}

```
<? xml version="1.0" ?>
<article>
  <bdy>
    <sec>
      <p>I am XML.</p>
      <p></p>
      <p>First, Text is here. Here issues XML.</p>
      <p>
        <image>
          <title><p></p></title>
        </image>
      </p>
    </sec>
  </bdy>
</article>
```

"I"   → 0: {{3, 1}}
"am"   → 1: {{3, 1}}
"xml"   → 2: {{3, 1}, {5, 1}}
"first" → 3: {{5, 1}}
"text"   → 4: {{5, 1}}
"is"   → 5: {{5, 1}}
"here"   → 6: {{5, 2}}
"issue" → 7: {{5, 1}}

Fig. 1. XML document   **Fig. 2.** Inverted list

where the Term-Freq is the frequency of a term ID in a node indicated by a Node-ID. The inverted-file index for Figure 1 is as shown in Figure 2.

The XML Wikipedia collection in the INEX 2007 Adhoc Track has 52,562,497 nodes and 13,903,331 unique terms. The same collection was used for INEX 2006. If both a node ID and a term ID are stored as 4 byte integers, an inverted-file list takes about 1.78 GB on the hard disk.

2.2 Relative Inverted-Path List

There are a variety of methods for indexing and labeling structural information [18]. Our system performs a depth-first traversal of the XML DOM-tree and stores the node labels in this order. The resulting list is unique and contains all structural information, even though its size is relatively small. The list also preserves all distances between a node and its parent node, as follows,

$$\text{Node-ID: \{Distance\}}$$

where the distance is a relative distance between a Node-ID and its parent node's Node-ID. We call this data structure a *relative inverted-path* list (*RIP* list). Figure 3 shows that the RIP list allows rapid access from every Node-ID to its parent Node-ID.

The system merges the number of terms contained in every node, the scores of matching nodes, and the numbers of query terms contained in each node. Figure 4 shows the merging process for the number of terms contained in every node. Merging is a fast one-pass operation. In our system, the maximum number of nodes contained in a node is declared to be a 2-byte integer, allowing a node to contain up to 65,536 descendant nodes. Therefore the RIP list occupies about 105 MB in memory.

Let the number of nodes be n and the average number of descendant nodes contained within a node be m. To find a pair consisting of a node and its parent node, the number of nodes that must be checked is a function of the other nodes in the XML document. The average time complexity of the brute force merge algorithm is $O(m \cdot n)$. In contrast, the average time complexity of the merge algorithm using the RIP list is $O(n)$.

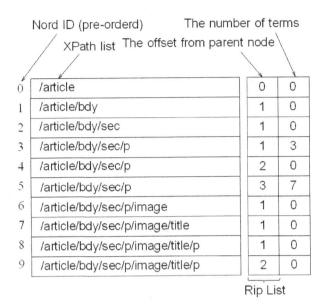

Fig. 3. Relative inverted-path list

2.3   Information Retrieval Model

Our system uses a TF-IDF scoring function for retrieval. TF-IDF is additive, therefore a node score can be easily calculated by merging the scores of its descendant nodes. The TF-IDF score L_j of the jth node is composed of the term frequency tf_i of the ith term in the query, the number of nodes f_i including the ith term, and the number of all the nodes n in the XML collection.

$$L_j = \sum_{i=1}^{t} \log(tf_i \cdot \frac{n}{f_i}) \tag{1}$$

However, if the node score is the sum of the scores of its descendants, there is the problem that the root node always has the highest score in the document. Therefore, the score R_j of the jth node is composed of the number T_j of terms contained in the jth node, the score L_k of the kth descendant of the jth node, and the number T_k of terms contained in the kth node.

$$R_j = \sum_{k \, children of \, j} D(k, T_k, T_j) \cdot L_k \tag{2}$$

$$T_j = \sum_{k \, children of \, j} T_k \tag{3}$$

The coefficient function $D(k, T_k, T_j)$ means a decaying function, which is as shown in the following equation,

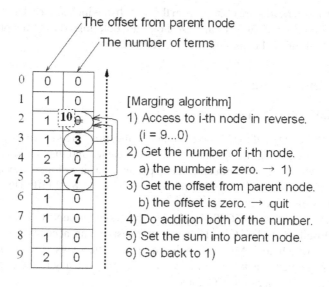

The offset from parent node
The number of terms

0	0	0
1	1	0
2	1	10
3	1	3
4	2	0
5	3	7
6	1	0
7	1	0
8	1	0
9	2	0

[Marging algorithm]
1) Access to i-th node in reverse.
   (i = 9...0)
2) Get the number of i-th node.
   a) the number is zero. $\rightarrow$ 1)
3) Get the offset from parent node.
   b) the offset is zero. $\rightarrow$ quit
4) Do addition both of the number.
5) Set the sum into parent node.
6) Go back to 1)

Fig. 4. Merging using a relative inverted-path list

$$D(k, T_k, T_j) = \begin{cases} 0 & T_k > T_1, \\ 0 & T_j > T_2, \\ 1/(\log d_k + 1) & \text{otherwise,} \end{cases}$$

where $T_1(= 100)$ is a threshold for the number of terms contained in the node to merge, $T_2(= 2,000)$ is a threshold for the number of terms contained in the merged node. According to the above coefficient function, a score L_k decays as a function of the distance d_k of kth node to jth node.

Then s_j is the occurrence of terms included in both the query and jth node,

$$s_j = count(\delta_j), \qquad \delta_j = \bigcup_{k\ childrenof\ j} \gamma_k,$$

where α is the set of terms included in the query, and β_j is the set of terms included in the jth node. The conjunction, $\gamma_j = \alpha \cap \beta_j$, is the set of query terms included in the jth node.

Then S_j is one of the heuristic scores we call a leveling score. If a node contains all terms in the query, the leveling score is the highest.

$$S_j = \frac{Q}{q} \cdot s_j, \qquad (4)$$

where $Q(= 500)$ is a constant number, and q is the number of terms in the query.

After that, the score RSV_j of jth node is composed of the TF-IDF score R_j, the leveling score S_j, and the logarithm of the number of terms T_j.

$$RSV_j = \frac{R_j + S_j}{\log T_j} \qquad (5)$$

Then the retrieved results are chosen from the node list, which is sorted in descending order of *RSV* scores. In addtion, all the thresholds are determined from the preliminary experimental result of the Focused task of the INEX 2006.

3   Experimental Results

3.1   INEX 2007 Adhoc Track

To index the XML 2007 Adhoc Track document collection, the system first parses all the structures of each XML document with an XML parser and then parses all the text nodes of each XML document with an English parser. The size of the index containing both content information and structure information is about 8.32 GB. Thereafter, the system uses the same index in every experiment.

Our experiment targets the CO Task only[5]. The system accepts CO queries, which are terms enclosed in <title> tags. However, the system can accept CAS queries as enhanced terms, which are XML tags enclosed in <castitle> tags. The enhanced term means the term frequency is twice as many as the term frequency in the query. Hence the system treats CAS queries as CO queries.

For each query set, there is the Focused task, the Relevant in Context task, and the Best in Context task in the INEX 2007 Adhoc Track. The Focused task is the only remaining official task from INEX 2006[6]. Hence the system parameters are tuned for the Focused task on topics from INEX 2006.

The system is installed on the same PC used on INEX 2006 for comparison. The PC has 2GHz CPU, 2GB RAM, and 300GB SATA HDD, and the system is implemented in Java 1.4.2_06. The time it takes to parse and load the 659,388 files on the PC is about 8.17 hours excluding file-copying time. The database size is about 3.18 GB on HDD.

3.2   Experimental Results

Table 1 compares the average retrieval time on INEX 2007 with the average retrieval time on INEX 2006. Under the same experimental conditions, the new system reduces retrieval time by a factor of sixteen. The system completes the search in an average of 3.92 seconds per topic, and in no worse than 9.95 seconds for any particular topic.

Figure 5 shows the frequency distribution of retrieval time and indicates the system has a relatively stable retrieval time on the INEX 2007 Adhoc Track. We submitted 18 runs, including three CO runs and three CAS runs over 130 topics for each of the three tasks.

Table 2, Table 3 and Table 4 show results for the three Adhoc tasks using the main evaluation measure [7] for INEX 2007.

Table 1. Average retrieving time

	Average retrieving time [s]
INEX 2006	66.2
INEX 2007	3.94

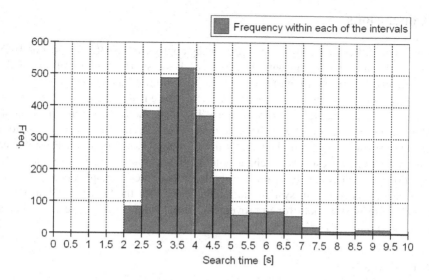

Fig. 5. Frequency distribution of search time

Table 2. Top 10 of 29 participants on Focused

Affiliation	iP[0.01]	Rank
Dalian	0.5271	1/79
Ecoles des M	0.5164	4/79
Waterloo	0.5108	7/79
CMU	0.5083	9/79
Max-Planck	0.5066	10/79
Tampere	0.4998	11/79
Doshisha	0.4975	12/79
Queensland	0.4924	15/79
RMIT	0.4834	18/79
Robert G	0.4758	22/79
. . . .		
JustSystems	**0.4632**	**34/79**

Table 3. Top 10 of 20 participants on Relevant in Context

Affiliation	MAgP	Rank
Dalian	0.1552	1/66
Queensland	0.1489	6/66
RMIT	0.1358	10/66
Amsterdam	0.1323	11/66
Robert G	0.1302	12/66
Cirquid	0.1233	15/66
INRIA	0.1147	20/66
JustSystems	**0.1107**	**23/66**
Ecoles des M	0.1081	25/66
Max-Planck	0.1077	28/66
. . . .		
. . . .		

Table 4. Top 10 of 21 participants on Best in Context

Affiliation	MAgP	Rank
RMIT	0.1919	1/71
Queensland	0.1831	2/71
Waterloo	0.1817	3/71
Dalian	0.1759	7/71
Robert G	0.1742	8/71
Amsterdam	0.1731	9/71
JustSystems	**0.1661**	**16/71**
INRIA	0.1633	17/71
Max-Planck	0.1350	30/71
Cirquid	0.1339	31/71
. . . .		
. . . .		

∗Metric: interpolated Precision / Recall (invalid submissions are discarded.), as interpolated precision at 0.01 recall.

∗Metric: generalized Precision / Recall (invalid submissions are discarded.).

∗Metric: generalized Precision / Recall (invalid submissions are discarded.).

The Focused task uses interpolated precision at 0.01 recall, while the Relevant in Context task and the Best in Context task use mean average generalized precision. Our system received a score of 0.4632 on the Focused task (14th of 29), 0.1107 on Relevant in Context task (8th of 20), and 0.1661 on the Best in Context task (7th of 21).

4   Conclusions

This paper describes a new high-speed processing approach to XML IR based on BUS using the RIP list. The system with the RIP list takes a shorter time to retrieve XML nodes than ever, using a variety of scoring functions with the same index.

One factor that contributes to high-speed performance is that the index of structural information is small enough to fit in memory. The other factor is that the merging algorithm has $O(n)$ time complexity, because the cost of searching for each parent node is vanishingly low. In addition, the structural information stored on the RIP list can be partially reindexed in response to changes to XML documents. Therefore, the XML information retrieval system has came a step closer to a practical application.

In terms of precision, our system did relatively well for the Focused task at INEX 2007, but there is still room for improvement. In the future, we will research better scoring algorithms for XML retrieval.

Acknowledgement

This paper was reported on the advice of David A. Evans of *JSERI (JustSystems Evans Research, Inc.)* The author would like to thank David and other *JSERI* members who did much comment on documentation.

References

1. Extensible Markup Language (XML) 1.1, 2nd edn., http://www.w3.org/TR/xml11/
2. XML Path Language (XPath) Version 1.0, http://www.w3.org/TR/xpath
3. XQuery 1.0: An XML Query Language, http://www.w3.org/TR/xquery/
4. INitiative for the Evaluation of XML Retrieval (INEX),
   http://inex.is.informatik.uni-duisburg.de/
5. Clarke, C., Kamps, J., Lalmas, M.: INEX 2006, Retrieval Task and Result Submission Specification (2006), http://inex.is.informatik.uni-duisburg.de/2006/inex06/pdf/INEX06_Tasks_v1.pdf
6. Kazai, G., Lalmas, M.: INEX 2005, Evaluation Metrics. In: Fuhr, N., Lalmas, M., Malik, S., Kazai, G. (eds.) INEX 2005. LNCS, vol. 3977, pp. 16–29. Springer, Heidelberg (2006), http://www.dcs.qmul.ac.uk/7Emounia/CV/Papers/inex-2005-metrics.pdf
7. Pehcevski, J., Kamps, J., Kazai, G., Lalmas, M., Ogilvie, P., Piwowarski, B., Robertson, S.: INEX 2007 Evaluation Measures (Draft) (2007),
   http://inex.is.informatik.uni-duisburg.de/
   2007/inex07/pdf/inex07-measures.pdf
8. Baeza-Yates, R., Ribeiro-Neto, B.: Modern Information Retrieval. ACM Press Series, pp. 1–69, 141–162. Addison-Wesley, Reading (1999)
9. Salton, G., Wong, A., Yang, C.S.: A vector space model for automatic indexing. Communications of the ACM 18, 613–620 (1975)
10. Evans, D.A., Lefferts, R.G.: Design and evaluation of the CLARIT-TREC-2 system. In: TREC, pp. 137–150 (1994)
11. Amer-Yahia, S., Lalmas, M.: XML search: languages, INEX and scoring. In: SIGMOD Rec., vol. 35(4), pp. 16–23. ACM Press, New York (2006)

12. Shin, D., Jang, H., Jin, H.: BUS: an effective indexing and retrieval scheme in structured documents. In: DL 1998: Proceedings of the third ACM conference on Digital libraries, pp. 235–243 (1998)
13. Weigel, F., Meuss, H., Schulz, K.U., Bry, F.: Content and structure in indexing and ranking XML. In: WebDB 2004 Proceedings of the 7th International Workshop on the Web and Databases, pp. 67–72 (2004)
14. Tanioka, H.: A Method of Preferential Unification of Plural Retrieved Elements for XML Retrieval Task. In: Fuhr, N., Lalmas, M., Trotman, A. (eds.) INEX 2006. LNCS, vol. 4518, pp. 45–56. Springer, Heidelberg (2007)
15. Eugen, P., Ménier, G., Marteau, P.-F.: SIRIUS XML IR System at INEX 2006: Approximate Matching of Structure and Textual Content. In: Fuhr, N., Lalmas, M., Trotman, A. (eds.) INEX 2006. LNCS, vol. 4518, pp. 185–199. Springer, Heidelberg (2007)
16. Geva, S.: GPX - Gardens Point XML IR at INEX 2006. In: Fuhr, N., Lalmas, M., Trotman, A. (eds.) INEX 2006. LNCS, vol. 4518, pp. 137–150. Springer, Heidelberg (2007)
17. Hatano, K., Kikutani, H., Yoshikawa, M., Uemura, S.: Determining the Retrieval Targets for XML Fragment Retrieval Systems Based on Statistical Information. The IEICE transactions on information and systems J89-D(3), 422–431 (2006)
18. Shimizu, T., Onizuka, M., Eda, T., Yoshikawa, M.: A Survey in Management and Stream Processing of XML Data. The IEICE transactions on information and systems J99-D, 159–184 (2007)

LIG at INEX 2007 Ad Hoc Track:
Using Collectionlinks as Context

Delphine Verbyst[1] and Philippe Mulhem[2]

[1] LIG - Université Joseph Fourier, Grenoble, France
`Delphine.Verbyst@imag.fr`
[2] LIG - CNRS, Grenoble, France
`Philippe.Mulhem@imag.fr`

Abstract. We present in this paper the work[1] of the Information Retrieval Modeling Group (MRIM) of the Computer Science Laboratory of Grenoble (LIG) at the INEX 2007 Ad Hoc Track. We study here the impact of non structural relations between structured document elements (doxels) on structured documents retrieval. We use existing links between doxels of the collection, encoded with the *collectionlink* tag, to integrate link and content aspects. We characterize the relation induced by the *collectionlink* tags with relative exhaustivity and specificity scores. As a consequence, the matching process is based on doxels content and these features. Results of experiments on the test collection are presented. Runs using non structural links overperform a baseline without such links.

1 Introduction

This paper describes the approach used for the Ad Hoc Track of the INEX 2007 competition. Our goal here is to show that the use of non structural links can increase the quality of the results provided by an information retrieval system on XML documents. We consider that handling links between documents in a smart way may help an information retrieval system, not only to provide better results, but also to organize the results in a way to overcome the usual simple list of documents. For INEX 2007, we only show that our approach impacts in a positive way the quality of the results provided.

The use of non structural links, such as Web links or similarity links has been studied in the past. Well known algorithms such as Pagerank [1] or HITS [3] do not integrate in a seamless way the links in the matching process. Savoy, in [6], showed that the use of non structural links may provide good results, without qualifying the strength of the inter-relations. In [7], Smucker and Allan show that similarity links may help navigation in the result space. We want, with the work described here, to go further in this direction.

In the following, the non structural relations between doxels will be referred to as the *context* of the doxels. Our assumption is that document parts are

---

[1] This work is supported by Orange France Telecom.

N. Fuhr et al. (Eds.): INEX 2007, LNCS 4862, pp. 138–147, 2008.

not only relevant because of their content, but also because they are related to other document parts that answer the query. In some way, we revisit the *Cluster Hypothesis* of van Rijsbergen [8], by considering that the relevance of each document is impacted by the values of related documents.

In our proposal, we first build inter-relations between doxels, and then characterize these relations using relative exhaustivity and specificity (see section 3.2) at indexing time. These elements are used later on by the matching process.

The nine officially submitted runs by the LIG for the Ad Hoc track integrate such non structural links. For each of the three tasks (Focused, Relevant in Context, Best in Context) a baseline without using such links was submitted. Taking into account the non structural links outperforms consistently this baseline.

The remaining of this paper is organized as follows: we describe the links that were used in our experiments in part 2, the doxel space is described in detail in section 3, in which we propose a document model using the context. Section 4 introduces our *matching in context* process. Results of the INEX 2007 Ad Hoc track are presented in Section 5.

2   Choice of Collectionlinks

The idea of considering neighbours was first proposed in [9], in order to facilitate the exploration of the result space by selecting the relevant doxels, and by indicating potential good neighbours to access from one doxel. For this task, the 4 Nearest Neighbours were computed.

The INEX 2007 collection contains several links between documents, like *unknownlinks*, *languagelinks* and *outsidelinks* for instance. We only considered existing relations between doxels with the *collectionlink* tag, because these links denote links inside the collection. We use the *xlink : href* attribute that indicates the target (file name) of the link. We notice that the targets of such links are only whole documents, and not documents parts; this aspect may negatively impact our expectations compared to our model that supports documents parts as targets. The table 1 shows these relations, with a first document D_1 (file 288042.xml) about "Croquembouche" and a second document D_2 (file 1502304.xml) about "Choux pastry". The third *collectionlink* tag in D_1 links D_1 to D_2; the source of this link is underlined in D_1 in the table and the target is the whole document D_2 which is also underlined. Overall, there are 17 013 512 collectionlinks in the INEX 2007 collection. We applied the following restrictions:

- for each leaf doxel d: the 4 first collectionlinks of d,
- for non-leaf doxels d': the union of 4 first collection links of its leaf doxels direct or indirect components.

With the restrictions above, we only take into account 12 352 989 collectionlinks.

Table 1. An example of collectionlinks from the INEX2007 corpus

Document D_1: file 288042.xml

```
<article>
<name id="288042">Croquembouche</name>
...
<body>A
<emph3>croquembouche</emph3>is a
<collectionlink ... xlink:href="10581.xml">French</collectionlink>
<collectionlink ... xlink:href="57572.xml">cake</collectionlink>
consisting of a conical heap of cream-filled

<collectionlink ... xlink:href=1502304.xml'>choux</collectionlink>

buns bound together with a brittle
<collectionlink ... xlink:href="64085.xml">caramel</collectionlink>
sauce, and usually decorated with ribbons or spun sugar.
...
</body>
</article>
```

Document D_2: file 1502304.xml

```
<article>
<name id="1502304">Choux pastry</name>
...
<body>
<emph3>Choux pastry</emph3>
<emph2>(pte  choux)</emph2>is a form of light
<collectionlink ... xlink:href="67062.xml">pastry</collectionlink>
used to make
<collectionlink ... xlink:href="697505.xml">profiterole</collectionlink>
s or
<collectionlink ... xlink:href="1980219.xml">eclair</collectionlink>
s. Its
<collectionlink ... xlink:href="198059.xml">raising agent</collectionlink>
is the high water content, which boils during cooking, puffing
out the pastry.
...
</body>
</article>
```

3   Doxel Space

3.1   Doxel Content

The representation of the content of doxel d_i is a vector generated from a usual vector space model using the whole content of the doxel: $d_i = (w_{i,1}, ..., w_{i,k})$. Such a representation has proved to give good results for structured document retrieval [2]. The weighting scheme retained is a simple $tf.idf$, with idf based on the whole corpus and with the following normalizations: the tf is normalized by the max of the tf of each doxel, and the idf is log-based, according to the document collection frequency. To avoid an unmanageable quantity of doxels, we kept only doxels having the following tags: article, p, collectionlink, title, section, item. The reason for using only these elements was because, except for the collectionlinks, we assume that the text content for these doxels are not too small. The overall number of doxels considered by us here is 29 291 417.

3.2 Doxel Context

Consider the two structured documents D_1 and D_2 linked as shown in table 1: they share *apriori* information. If a user looks for all the information about "croquembouche", the system should indicate if the link from D_1 to D_2 is relevant for the query. If the user only wants to have general informations about "croquembouche", D_1 is highly relevant, D_2 is less relevant, and moreover, the system should indicate that the link between D_1 and D_2 is not interesting for this query result. To characterize the relations between doxels, we propose to define relative exhaustivity and relative specificity between doxels. These features are inspired from the definitions of specificity and exhaustivity proposed at INEX 2005 [4]. Consider a non-compositional relation from the doxels d_1 to the doxel d_2:

- The relative specificity of this relation, noted $Spe(d_1, d_2)$, denotes the extent to which d_2 focuses on the topics of d_1. For instance, if d_2 deals only with elements from d_1, then $Spe(d_1, d_2)$ should be close to 1.
- The relative exhaustivity of this relation, noted $Exh(d_1, d_2)$, denotes the extent to which d_2 deals with all the topics of d_1. For instance, if d_2 discusses all the elements of d_1, then $Exh(d_1, d_2)$ should be close to 1.

The values of these features are in $[0, 1]$. We could think that these features behave in an opposite way: when $Spe(d_1, d_2)$ is high, then $Exh(d_1, d_2)$ is low, and vice versa. But $Spe(d_1, d_2)$ and $Exh(d_1, d_2)$ could be high both if d_1 and d_2 are encapsulated and deal with the same subject.

We propose to describe relative specificity and relative exhaustivity between two doxels d_1 and d_2 as extensions of the overlap function [5] of their index: these values reflect the amount of overlap between the source and target of the relation. We define relative specificity and relative exhaustivity in formulas (1) and (2) on the basis of the non normalized doxel vectors $w_{1,i}$ and $w_{2,i}$ (respectively for d_1 and d_2).

$$Exh(d_1, d_2) = \frac{\sum_i w_{1,i} \cdot w_{2,i}}{\sum_i w^2_{\oplus 1/w_{2,i}}} \tag{1}$$

$$Spe(d_1, d_2) = \frac{\sum_i w_{1,i} \cdot w_{2,i}}{\sum_i w^2_{\oplus 2/w_{1,i}}} \tag{2}$$

where: $w_{\oplus m/w_{n,i}} = \begin{cases} w_{m,i} & \text{if } w_{n,i} \leq 1 \\ \sqrt{w_{m,i} \cdot w_{n,i}} & \text{otherwise.} \end{cases}$

$w_{\oplus m/n,i}$ ensures that the scores are in $[0,1]$.

4   Model of Matching in Context

We assume that the matching process should return doxels relevant to the user's information needs, regarding both content, structure aspects, and considering also the context of each relevant doxel.

We define the matching function as a linear combination of a standard matching result without context and a matching result based on relative specificity and exhaustivity. The relevant status value $RSV(d, q)$ for a given doxel d and a given query q is thus given by:

$$RSV(d,q) = \alpha * RSV_{content}(d,q) + (1 - \alpha) * RSV_{context}(d,q), \qquad (3)$$

where $\alpha \in [0, 1]$ is experimentally fixed, $RSV_{content}(d, q)$ is the score without considering the set of neighbours $\mathcal{V}_d$ of d (i.e. cosine similarity) and

$$RSV_{context}(d,q) = \sum_{d' \in \mathcal{V}_d} \frac{\beta * Exh(d,d') + (1 - \beta) * Spe(d,d')}{|\mathcal{V}_d|} RSV_{content}(d',q)$$

$$(4)$$

where $\beta \in [0, 1]$ is used to privilege exhaustivity or specificity.

The matching in context model computes scores with both content and context dimensions to complete our model. Using a linear combination makes sense, as a doxel may be relevant *per se* without any other relevant context but a relevant context may increase the relevance of a doxel.

5   Experiments and Results

The INEX 2007 Adhoc track consists of three retrieval tasks: the Focused Task, the Relevant In Context Task, and the Best In Context Task. We submitted 3 runs for each of these tasks. For all these runs, we used only the *title* of the INEX 2007 queries as input for our system: we removed the words prefixed by a '-' character, and we did not consider the indicators for phrase search. The vocabulary used for the official runs is quite small (39 000 terms), but was assumed large enough to prove the validity of our proposal.

First of all, we have experimented our system with INEX 2006 collection to fix α and β parameters of formulas (3) and (4). The best results were achieved with a higher value for the exhaustivity than for the specificity. As a consequence, we decide to fix $\alpha = 0.75$ and $\beta = 0.75$ for our expected best results.

5.1 Focused Task

The INEX 2007 Focused Task is dedicated to find the most focused results that satisfy an information need, without returning "overlapping" elements. In our focused task, we experiment with two different rankings.

For the first run, the "default" one, namely $LIG\_075075\_FOC\_FOC$ with $\lambda = 0.75$ and $\beta = 0.75$, we rank the result based on matching in context proposed in section 4; overlap is removed by applying a post-processing.

For the second run, we choose to use the results of the Relevant In Context Task to produce our Focused Task results : relevant doxels are ranked by article, and we decide to score the doxels with the score of each corresponding article and list them according to their position in the document, and removing overlapping doxels. This run is called $LIG\_075075\_FOC\_RIC$, and we set $\lambda = 0.75$ and $\beta = 0.75$.

The last run, namely $LIG\_1000\_FOC\_RIC$ is our baseline. It is similar to the second run with $\lambda = 1.0$ and $\beta = 0.0$, i.e. it considers only the contents of the doxels.

Table 2. Focused Task for INEX2007 Ad Hoc

Run	precision at 0.0 recall	precision at 0.01 recall	precision at 0.05 recall	precision at 0.10 recall
$LIG\_075075\_FOC\_FOC$ $MAiP = 0.0158$	0.3107	0.1421	0.0655	0.0492
$LIG\_1000\_FOC\_RIC$ $MAiP = 0.0580$	0.3540	0.3192	0.2119	0.1734
$LIG\_075075\_FOC\_RIC$ $MAiP = 0.0647(+11.6\%)$	0.3475 (-1.8%)	0.3144 (-1.5%)	0.2480 (+17.0%)	0.2126 (+22.6%)

We present our results for the focused task in Table 2 showing precision values at given percentages of recall, and in Figure 1 showing the generalized precision/recall curve. These results show that runs based on Relevant In Context approach outperforms the "default" Focused Task run, $LIG\_075075\_FOC\_FOC$: after checking the code, we found a bug that leads to incorrect paths for the doxels, and this bug impacts in a lesser extent the second run. The first column of the Table 2 shows that, considering the Mean Average Interpolated Precision, the $LIG\_075075\_FOC\_RIC$ run outperforms the $LIG\_1000\_FOC\_RIC$ run by +11.6%, proving that the collectionlinks are usefull. Moreover, in Table 2 and in Figure 1, we see that for the results between 0.05 recall and 0.25 recall, the $LIG\_075075\_FOC\_RIC$ performs much better than the $LIG\_1000\_FOC\_RIC$. Our best run is ranked 60 on 79 runs.

5.2 Relevant in Context Task

For the Relevant In Context Task, we take "default" focused results and re-ordered the first 1500 doxels such that results from the same document are

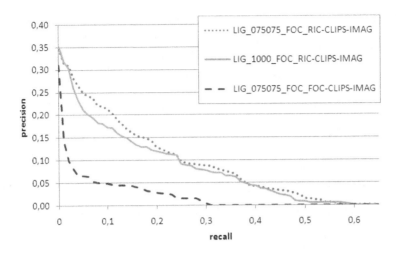

Fig. 1. Interpolated Precision/Recall - Focused Task

clustered together. It considers the article as the most natural unit and scores the article with the score of its doxel having the highest RSV.

We submitted three runs:

- LIG_1000_RIC : a baseline run which doesn't take into account the inner collectionlinks to score doxels. We set $\lambda = 1.0$ and $\beta = 0.0$;
- LIG_075075_RIC : a retrieval approach based on the collectionlinks use. We set $\lambda = 0.75$ and $\beta = 0.75$;
- LIG_00075_RIC : an approach that consider the RSV of a doxel only considering its context: we set $\lambda = 0.0$ and $\beta = 0.75$.

Table 3. Relevant In Context Task for INEX2007 Ad Hoc

Run	gP[5]	gP[10]	gP[25]	gP[50]
LIG_1000_RIC	0.0926	0.0826	0.0599	0.0448
$MAgP = 0.0329$				
LIG_075075_RIC	0.1031	0.0957	0.0731	0.0542
$MAgP = 0.0424\ (+28.9\%)$	(+11.3%)	(+15.9%)	(+22.0%)	(+21.0%)
LIG_00075_RIC	0.0779	0.0581	0.0401	0.0291
$MAgP = 0.0174\ (-47.1\%)$	(-15.9%)	(-29.7%)	(-33.1%)	(-35.0%)

For the relevant in context task, our results in terms of non-interpolated generalized precision at early ranks $gP[r], r \in \{5, 10, 25, 50\}$ and non-interpolated Mean Average Generalized Precision $MAgP$ are presented in Table 3. Figure 2 shows the generalized precision/recall curve. This shows that using collectionlinks and the doxels content (LIG_075075_RIC) improves the baseline by a ratio greater than 11%. The LIG_00075_RIC gives bad results, showing that the context of the doxels only is not relevant. In Figure 2, we see that the LIG_075075_RIC run is also above the default run. Our best run is ranked 56 on 66 runs.

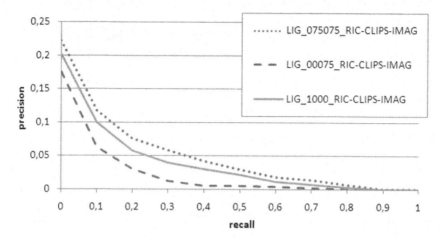

Fig. 2. Generalized Precision/Recall - Relevant In Context task

5.3   Best in Context Task

For the Best In Context Task, we examine whether the most focused doxel in a relevant document is the best entry point for starting to read relevant articles. We take "normal" focused results and the first 1500 doxels belonging to different files. For this task, we submitted three runs:

- LIG_1000_BIC : the baseline run which doesn't take into account collectionlinks: we set $\lambda = 1.0$ and $\beta = 0.0$;
- LIG_075075_BIC : the retrieval approach based on the use of collectionlinks. We set $\lambda = 0.75$ and $\beta = 0.75$;
- LIG_00075_BIC : the approach that uses only the context of doxels to compute their RSV: we set $\lambda = 0.0$ and $\beta = 0.75$.

Table 4. Best In Context Task for INEX2007 Ad Hoc

Run	gP[5]	gP[10]	gP[25]	gP[50]
LIG_1000_BIC $MAgP = 0.0630$	0.1194	0.1176	0.1035	0.0910
LIG_075075_BIC $MAgP = 0.0761\ (+20.8\%)$	0.1373 $(+15.0\%)$	0.1261 $(+7.2\%)$	0.1151 $(+11.2\%)$	0.0957 $(+5.2\%)$
LIG_00075_BIC $MAgP = 0.0639\ (+1.4\%)$	0.1303 $(+9.1\%)$	0.1107 (-5.9%)	0.0977 (-5.6%)	0.0819 (-0.1%)

For the best in context task, our results are presented in Table 4 and Figure 3 with the same measures as the Relevant In Context Task results. Our best run is ranked 54 on 71. Conclusions are the same: using collectionlinks and content improves the baseline by a mean average of more than 20%, and the LIG_00075_BIC run is consistently below the baseline. There is one result however, the LIG_00075_BIC run outperforms the baseline at $gP[5]$ by more than

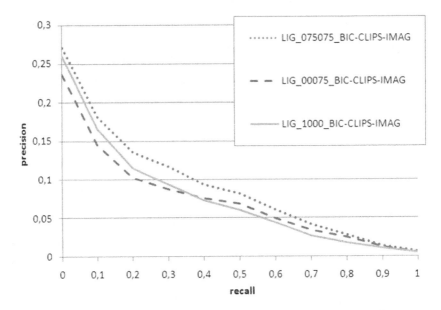

Fig. 3. Generalized Precision/Recall - Best In Context task

9% and in Figure 3 we see than the baseline and the LIG_00075_BIC are quite close to each others. This means that the *a priori* links are really meaningful.

6   Summary and Conclusion

We proposed a way to integrate the content of the doxels as well as their context (collectionlinks in INEX 2007 documents). We have submitted runs implementing our theoretical proposals for the different Ad Hoc tasks. For each of the tasks, we showed that combining content and context produce better results than considering content only and context only of the doxels, which is a first step in validating our proposal. According to the official evaluation of INEX 2007, our best runs are ranked in the last third of participants systems, for the Content-Only runs. However, we plan to improve our baseline to obtain better results in the following directions:

- As mentioned earlier, the size of the vocabulary used is too small, leading to query terms out of our vocabulary.
- When submitting our runs for our first participation at INEX competition we found some bugs related to the identifiers of the doxels, so the results were negatively impacted.
- We are working on the integration of negative terms in the query, in a way to get better results.

Since the submission of our official runs, we integrated a larger vocabulary (about 200 000 terms) and corrected our bugs, which led to an increase of 24%

for the MAiP, when using the official evaluation tool released in december 2007 and the version 2.0 of the assessments.

References

1. Brin, S., Page, L.: The anatomy of a large-scale hypertextual Web search engine. Computer Networks and ISDN Systems 30(1–7), 107–117 (1998)
2. Fang Huang, D.H., Watt, S., Clark, M.: Robert Gordon University at INEX 2006: Adhoc Track. In: INEX 2006 Workshop Pre-Proceeding, pp. 70–79 (2006)
3. Kleinberg, J.M.: Authoritative sources in a hyperlinked environment. J. ACM 46(5), 604–632 (1999)
4. Piwowarski, B., Lalmas, M.: Interface pour l'evaluation de systemes de recherche sur des documents XML. In: Premiere COnference en Recherche d'Information et Applications (CORIA 2004), Toulouse, France, Hermes (2004)
5. Salton, G., McGill, M.J.: Introduction to Modern Information Retrieval, ch. 6, p. 203. McGraw-Hill, Inc., New York (1986)
6. Savoy, J.: An extended vector-processing scheme for searching information in hypertext systems. Inf. Process. Manage. 32(2), 155–170 (1996)
7. Smucker, M.D., Allan, J.: Using similarity links as shortcuts to relevant web pages. In: SIGIR 2007: Proceedings of the 30th annual international ACM SIGIR conference on Research and development in information retrieval, pp. 863–864. ACM Press, New York (2007)
8. van Rijsbergen, C.: Information retrieval, 2nd edn., ch. 3. Butterworths (1979)
9. Verbyst, D., Mulhem, P.: Doxels in context for retrieval: from structure to neighbours. In: SAC 2008: Proceedings of the 2008 ACM symposium on Applied computing. ACM Press, New York (2008)

Overview of the INEX 2007 Book Search Track (BookSearch'07)

Gabriella Kazai[1] and Antoine Doucet[2]

[1] Microsoft Research Cambridge, United Kingdom
gabkaz@microsoft.com
[2] University of Caen, France
doucet@info.unicaen.fr

Abstract. This paper provides an overview of the newly launched Book Search Track at INEX 2007 (BookSearch'07), its participants, tasks, book corpus, test topics and relevance assessments, as well as some results.

1 Introduction

Libraries around the world and commercial companies like Amazon, Google and Microsoft are digitizing hundreds of thousands of books in an effort to enable online access to these collections. The Open Content Alliance (OCA)[1], a library initiative formed after Google announced its library book digitization project, has brought library digitization efforts into the public eye, even though libraries have been digitizing books for decades before that. However, unlike most library digitization projects of the past, which centered around preservation and involved the careful and individual selection of materials to be digitized, the recent mass-digitization efforts aim at the conversion of materials on an industrial scale with minimum human intervention [2].

The increasing availability of the full-text of digitized books on the Web and in digital libraries, both enables and prompts research into techniques that facilitate storage, access, presentation and use of the digitized content. Indeed, the unprecedented scale of the digitization efforts, the unique characteristics of the digitized material as well as the unexplored possibilities of user interactions make full-text book search an exciting area of research today.

Motivated by this need, the book search track was launched in 2007 as part of the INEX initiative. INEX was chosen as a suitable forum due to its roots in the evaluation of structured document retrieval (SDR) approaches and since searching for information in a collection of books can be seen as one of the natural application areas of SDR. For example, in focused retrieval a clear benefit to users is to gain direct access to parts of books (of potentially hundreds of pages) relevant to the information need.

The ultimate goal of the INEX book search track is to investigate book-specific relevance ranking strategies, UI issues and user behaviour, exploiting

---

[1] http://www.opencontentalliance.org/

N. Fuhr et al. (Eds.): INEX 2007, LNCS 4862, pp. 148–161, 2008.

special features, such as back of book indexes provided by authors, and linking to associated metadata like catalogue information from libraries. However, searching over large collections of digitzed books comes with many new challenges that need to be addressed first. For example, proper infrastructure has to be developed to allow for the scalable storage, indexing and retrieval of the digitized content. In addition, the setting up of a new track requires identifying suitable usage scenarios and tasks, establishing an evaluation framework complete with relevance criteria, judgement procedures and evaluation metrics, as well as the development of a support system infrastructure. In its first year, the track set to explore these issues with the aim to investigate the requirements for such an infrastructure.

This paper reports on the outcome of the BookSearch'07 track. It provides an overview of its participants, tasks, book corpus, test topics and relevance assessments, as well as some results and findings. Since, at the time of writing, the relevance assessments for one of the tasks (Page in Context) were still outstanding, the results for this task are not reported here.

This paper is organised as follows. Section 2 gives a brief summary of the participating organisations. In Section 3, we briefly describe the retrieval tasks defined at BookSearch'07. Section 4 details the book corpus, test topics, and relevance assessments. Section 5 presents the results of the evaluation. Finally, we close with a summary and plans for BookSearch'08.

2   Participating Organisations

In response to the call for participation, issued in April 2007, 27 organisations registered for BookSearch'07. Throughout the year, however, a number of groups droped out and only about a third remained active by the end of the year. Most groups reported difficulties due insufficient resources, including lack of space to store the dataset or scalable approach to process it, as well as lack of time or human resources required to tackle the various tasks.

The 27 groups along with details of their participation are summarized in Table 1. As it can be seen, only 10 groups remained active throughout. 16 groups downloaded the book corpus, 7 groups contributed search topics, and only 2 groups managed to submit runs.

3   Retrieval Tasks

The track defined four tasks: 1) Book Retrieval, 2) Page in Context retrieval, 3) Classification and 4) User intent taxonomy building. A summary of these are given in the following sections. Further details and the various submission DTDs are available in the track's Tasks and Submission Guidelines [7].

3.1   Book Retrieval Task

The goal of this task was to investigate the impact of book specific features on the effectiveness of book search systems, where the unit of retrieval is the

Table 1. Participating groups at BookSearch'07 (In the Status column, A stands for Active, C for Cancelled, and P for Passive; In the Runs column, BR stands for Book Retrieval task, and PiC for Page in Context task)

ID	Organisation	Status A/C/P	Corpus download	Topics created	Runs submitted
1	University of Kaiserslautern	C	Y	-	-
2	University of California, Berkeley	A	Y	-	4 BR
4	University of Granada	C	Y	-	-
5	Lexiclone Inc	P	-	-	-
9	Queensland University of Technology	A	Y	-	-
10	University of Otago	C	-	-	-
12	University of Strathclyde	C	-	-	-
14	Wuhan University, China	P	-	-	-
19	Indian Statistical Institute	C	Y	-	-
20	LAMSADE	P	-	-	-
22	Doshisha University	A	Y	1	-
23	Kyungpook National University	A	Y	1	-
25	Max-Planck-Institut für Informatik	P	Y	-	-
26	Dalian University of Technology	A	Y	5	-
28	University of Helsinki	A	Y	2	-
32	RMIT University	P	-	-	-
33	Information Engineering Lab, CSIRO	P	-	-	-
36	University of Amsterdam	A	Y	3	-
37	University of Waterloo	C	Y	-	-
40	Carnegie Mellon University	P	Y	-	-
42	LIP6	P	-	-	-
53	Ecoles des Mines de Saint-Etienne	P	-	-	-
54	Microsoft Research, Cambridge	A	Y	13	-
55	University of Tampere	A	Y	5	-
61	Hong Kong Uni. of Sci. and Tech.	P	-	-	-
68	University of Salford, UK	P	-	-	-
92	Cairo Microsoft Innovation Center	A	Y	-	6 BR, 7 PiC
	Total (27 organisations)	10/6/11	16	30	10 BR, 7 PiC

(complete) book. Users are thus assumed to be searching for (whole) books relevant to their information need that they may want to, e.g., purchase or borrow from a library.   Participants of this task were invited to submit pairs of runs with the following condition: one of the runs would be generated using generic IR techniques, while the other run would extend this technique by exploiting book-specific features (e.g. back-of-book index, citation statistics, library catalogue information, etc.) or specifically tuned algorithms. In both cases, the ranked list of books could contain a maximum of 1000 books estimated relevant to the given topic, ranked in order of estimated relevance to the query.

Participants were permitted to submit up to 3 pairs of runs.

3.2   Page in Context Task

This task was set up similarly to the ad hoc track's Relevant in Context task
[3], but here the task is applied to a collection of digitized books with shallow
structural markup. Accordingly, based on the assumption of a focused informational request, the task of a book search system was to return the user a ranked
list of books estimated relevant to the request, and then present within each
book, a ranking of relevant non-overlapping XML elements, passages, or book
pages. The difference from the Relevant in Context task is that book search
systems were required to rank the relevant elements/passages inside the books
(instead of returning sets of elements/passages). The challenge for existing INEX
participants was to test the scalability of their XML IR approaches and the
adaptability of their search engines to the new domain. This task, however, is,
and has proven to be, rather ambitious for most of the participants. For example, the Wikipedia corpus used in the ad hoc track experiments totals only
about 1GB, whereas the size of the BookSearch'07 corpus is around 210GB (see
Section 4).

Participants were allowed to submit up to 10 runs. One automatic (title-only)
and one manual run were compulsory. Additional manual runs were encouraged in order to help the construction of a reliable test collection. Each run
could contain for each test topic a maximum of 1000 books estimated relevant
to the topic, ordered by decreasing value of relevance. For each book, participants were asked to provide a ranked list of non-overlapping XML elements,
passages, or book page results that were estimated relevant to the query, ordered by decreasing value of relevance. A minimum of 1 element/passage/page
result per book was required. A submission could only contain one type of
result, i.e., only book pages or only passages; alas, result types could not be
mixed.

3.3   Classification Task

In this task, systems were tested on their ability to assign the correct classification labels from the Library of Congress (LoC) classification scheme to the books
of the test corpus based only on information available from the full text of the
books. The distributed corpus of about 42,000 books (see Section 4) served as the
training corpus, where classification labels were available for 20,692 books out
of the 39,176 that had an associated MARC record. The test corpus contained
2 sets of 1,000 books.

Participants were allowed to submit up to three runs per test set. Each run
was required to contain all 1,000 books of the given test set. For each book,
systems needed to return a ranked list of classification labels, with a minimum
of one label.

The list of Library of Congress classification headings extracted from the
MARC records of the 20,692 books was made available by organisers on the
INEX web site.

3.4   User Intent Taxonomy Task

User intent is a critical component in the understanding of users' search behaviour. It defines what kinds of search tasks users engage in. In traditional information retrieval, a user's intent is assumed to be informational in nature: It is driven by the user's need for information in order to complete a task at hand. Observations of Web use resulted in further two categories: navigational and transactional [1]. It is clear that these can also be applied to the book domain. However, it is possible that there are additional classes of user intent which are specific to books. It may also be the case that user tasks and user behaviour in the book domain will have specific traits and characteristics that may, for example, depend on genre. What are the possible classes of user intent and user tasks and what properties they have is a research question that this task was set to explore.

The goal of this task was to derive a taxonomy of user intent with its associated properties and search tasks. The use of examples of (actual or hypothetic) information needs demonstrating each class of intent and task was encouraged. Such an investigation could extend to include both research and design questions and possible answers regarding how a given user behaviour might be supported by a search system and its user interface. For example, a user hoping to buy a book is likely to be more interested in a price comparison feature, while an informational query will more likely benefit from a "find related books" feature.

Examples of questions that could be explored included: How is user intent dependent on book genre? What book specific features best support the different types of intent and tasks? How could intent be extracted from query logs? How should one design experiments to allow for the identification of user intent from system logs? What data would enable the prediction of intent in order to aid users? What user behaviour follows from them?

Participation in this task involved the submission of a research or opinion paper detailing the proposed taxonomy. Participants could choose to report findings from the analysis of collected user log data or provide recommendations for the design of user studies to help elicit such data.

4   Test Collection

4.1   Book Corpus

The corpus was provided by Microsoft Live Search Books and the Internet Archive (for non-commercial purposes only). It consists of 42,049 digitized out-of-copyright books, and totals around 210Gb in size. The collection contains books from a wide range of genre and includes reference works as well as poetry books. Most of the corpus is made up of history books (mostly American history), biographies, literary studies, religious texts and teachings. There are also encyclopedias, essays, proceedings and novels.

The OCR content of the books is stored in djvu.xml format. 39,176 of the 42,049 books also have associated metadata files (*.mrc), which contain publication (author, title, etc.) and classification information in MAchine-Readable Cataloging (MARC) record format.

The basic XML structure of a book (djvu.xml) is as follows:

```
<DjVuXML>
 <BODY>
  <OBJECT data="file..." [...]>
   <PARAM name="PAGE" value="[...]">
   [...]
   <REGION>
    <PARAGRAPH>
     <LINE>
      <WORD coords="[...]"> Moby </WORD>
      <WORD coords="[...]"> Dick </WORD>
      <WORD coords="[...]"> Herman </WORD>
      <WORD coords="[...]"> Melville </WORD>
      [...]
     </LINE>
     [...]
    </PARAGRAPH>
   </REGION>
   [...]
  </OBJECT>
  [...]
 </BODY>
</DjVuXML>
```

An <OBJECT> element corresponds to a page in a digitized book. A page counter is embedded in the @value attribute of the <PARAM> element which has the @name="PAGE" attribute. The actual page numbers (as printed inside the book) can be found (not always) in the header or the footer part of a page. Note, however, that headers/footers are not explicitly recognised in the OCR, i.e., the first paragraph on a page could be a header and the last one or more paragraphs could be part of a footer. Depending on the book, headers may include chapter titles and page numbers (although due to OCR error, the page number is not always present).

Inside a page, each paragraph is marked up. It should be noted that an actual paragraph that starts on one page and ends on the next would be marked up as two separate paragraphs within two page elements.

Each paragraph element consists of line elements, within which each word is marked up. Coordinates that correspond to the four points of a rectangle surrounding a word are given as attributes of word elements, and could be used to enable text highlighting.

No further structural markup is currently available, although some books have rich logical structure, including chapters, sections, table of contents, bibliography, back-of-book index, and so on.

4.2   Topics

The test topics in BookSearch'07 are representations of users' informational needs, i.e, where the user is assumed to search for information on a given subject. For this year, all topics were limited to deal with content only aspects (i.e., no structural query conditions). The structure of books, however, could still be used by search engines to improve their ranking of books or book parts estimated relevant to a query.

Two sets of topics were used: 1) a set of 250 queries extracted from the query log of Live Search Books was used for the Book Retrieval task; 2) a set of 30 topics was created by the participating organisations for the Page in Context task. The next sections detail the topic format, the topic creation process for the Page in Context task, and provide a summary of the collected topics.

Topic Format. Topics are made up of three parts, each of which describe the same information need, but for different purposes and at different level of detail:

<title> represents the search query that is to be used by systems. It serves as a short summary of the content of the user's information need.

<description> is a natural language definition of the information need.

<narrative> is a detailed explanation of the information need and a description of what makes an element/passage relevant or irrelevant. The narrative must be a clear and precise description of the information need in order to unambiguously determine whether or not a given text fragment in a book fulfills the need. The narrative is taken as the only true and accurate interpretation of the user's needs. Relevance assessments are made on compliance to the narrative alone.

Precise recording of the narrative is also important for scientific repeatability. To aid this, the narrative should explain not only what information is being sought, but also the context and motivation of the information need, i.e., why the information is being sought and what work-task it might help to solve. The narrative thus has the following two parts:

<task> is a description of the task for which information is sought, specifying the context, background and motivation for the information need.

<infneed> is a detailed explanation of what information is sought and what is considered relevant or irrelevant.

An example topic is given in Figure 1.

4.3   Topic Creation and Collected Topics

Topics for the Book Retrieval Task. 250 queries were extracted from the query logs of Live Search Books for which the test corpus contained at least one relevant book. No additional background or context information was available for these queries. Therefore these topics only have topic titles; and both the description and the narrative fields are left empty.

```
<title> Octavius Antony Cleopatra conflict </title>
<description> I am looking for information on the conflict between
    Octavius, Antony and Cleopatra. </description>
<narrative>
  <task> I am writing an essay on the relationship of Antony and Cleopatra
      and currently working on a chapter that explores the conflict between
      Octavius (the brother of Antony's wife, Octavia) and the lovers.
  </task>
  <infneed> Of interest is any information that details what motivated the
      conflict, how it developed and evolved through events such as the
      ceremony known as the Donations of Alexandria, Octavious' propaganda
      campaign in Rome against Antony, Antony's divorce from Octavia, and
      the battle of Actium in 31BC. Any information on the actions and
      emotions of the lovers during this period is relevant. Any
      non-documentary or non-biographical information, such as theatre plays
      (e.g., Shakespeare's play) or their critics are not relevant.
  </infneed>
</narrative>
```

Fig. 1. Example topic (not part of the BookSearch'07 test set)

On average, a query contained 2.188 words, the longest query being 6 words
long. The distribution of queries by length (in number of words) is shown in
Figure 2.

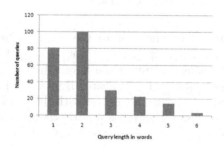

Fig. 2. Distribution of queries by length (in number of words)

Topics for the Page in Context Task. Participants were asked to submit
candidate topics for which at least 3 but no more than 20 relevant books were
found during the collection exploration stage [8]. Participants were provided with
a collection exploration tool to assist them in their topic creation. A screenshot
of the tool is given in Figure 3.

This tool gave participants the means to search and explore the book corpus.
This was achieved by building an interface to the search service provided by
Live Search Books[2]. The tool took advantage of the fact that all books in the
BookSearch'07 collection are indexed by Live Search Books. It worked by first

---

[2] http://books.live.com, or http://search.live.com/books#q=\&mkt=en-US

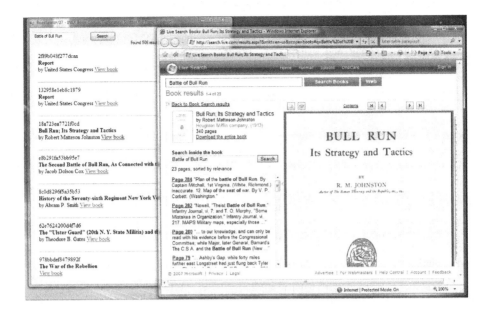

Fig. 3. Screenshot of the system used to assist topic creation at BookSearch'07. The window on the left shows the main search window of the system. It allowed participants to enter a query and view the ranked list of books. The screen on the right is the book viewer window of the Live Search Books service, which allowed participants to explore the contents of a book for relevant information.

sending the query entered by the user to the the Live Search Books search engine, and filtering the result list so that only books of the BookSearch'07 corpus were shown. Clicking on any of the books in the ranked list, took the user directly to the book viewer of Live Search Books (see Figure 3). Using this tool, participants could familiarize themselves with the collection and determine whether a candidate topic met the necessary criteria to be considered as a test topic: creation of topics with too few or too many relevant answers had to be aborted as these were deemed unsuitable for testing system performance [8].

A total of 30 topics were created, complete with topic title, description and narrative (as described in Section 4.2). Table 1 shows the number of topics each participating group contributed.

Participants were allowed to create topics based around the queries used for the Book Retrieval task. A query thus served as a starting point for participants to build up an information need during the collection exploration phase. Based on the similarity between the topic title of the created topic and the original query, we can distinguish the following categories: full, partial and no match. 10 topics belong in the full match category, meaning that the created topic title is exactly the same as the original query. 9 topics have partial matches, where participants refined the focus of the query, usually narrowing its scope. The remaining 11 topics were created from scratch.

4.4   Relevance Assessments

Relevance Assessments for the Book Retrieval Task. The relevance assessments for the 250 queries used for this task were collected by Live Search Books from human judges. Judges were presented with a query and a set of books to judge. Assessment were made along a four point scale: Excellent, Good, Fair, and Non-relevant.

In total, 3,918 relevant books are contained in the assessment set. These include 1061 Excellent, 1,655 Good, and 1,202 Fair judgments. The average number of relevant books per query is 15.672, and the maximum is 41. The distribution of number of relevant books per topic is shown in Figure 4.

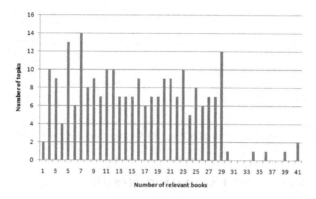

Fig. 4. Distribution of relevant books per query

Relevance Assessments for the Page in Context Task. Relevance judgments for the Page in Context task are to be provided by the participants of the track. At the time of writing, the relevance assessment system was still being finalised and thus the collection of judgments has not yet started.

The assessment system was implemented by adapting XRai [9] used at the ad hoc track and integrating it with the book viewer of Live Search Books. Figure 5 shows a screenshot of the assessment system. For a given topic, the list of books to be assessed forms the assessment pool. On accessing a book, the book is opened in the Live Search Books viewer while the list of pages inside the book that are to be judged are displayed as hyperlinks on the left hand side of the browser window. Clicking on a page link displays the corresponding page in the book viewer. To ease the assessment process, judges can assess groups of pages using the "select all", "select none", and "select range" options. The latter offers a simple syntax for selecting a number of pages with little user effort (e.g., "18-36; 49" selects all pages between 18 and 36 (inclusive), and page 49). For each book, assessors will be required to judge all pages in the assessment pool and will be encouraged to explore additional pages in the book. The location of additional, possibly relevant pages is supported through the relevance bar feature of the Live Search Books scrollbar, which highlights pages where the query terms occur.

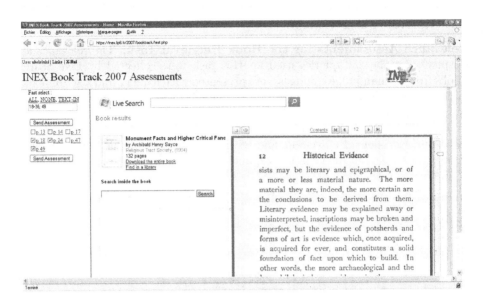

Fig. 5. Screenshot of the Page in Context assessment system

5 Submissions and Evaluation Results

As shown in Table 1, only two groups submitted retrieval runs to the Book Retrieval task and only one group to the Page in Context task. The University of California, Berkeley has also participated in the Classification task, but only evaluated its results unofficially. Furthermore, experiments on Book Retrieval were also conducted by participants at Microsoft Research Cambridge, the results of which were published in [12]. No submissions were received for the User Intent Taxonomy task. This was a bit of a surprise as our intent with this task was to attempt to open up the track allowing participation without a fully working book search engine.

To evaluate runs submitted to the Book Retrieval task, we adopted the Normalized Discounted Cumulated Gain measure of [6], using the discount function of $1/\log_b(i)$, where $b = 2$, and where the discounted cumulated gain for a given ranking is calculated as:

$$DCG[i] = \begin{cases} \sum_i G[i] & \text{if } i \leq b, \\ DCG[i-1] + \frac{\sum_i G[i]}{\log_b(i)} & \text{if } i > b. \end{cases} \quad (1)$$

The normalized DCG scores were obtained by dividing the DCG vector of the system's ranking by the DCG vector of the ideal ranking. The gain associated with a relevant book was 3 for a book rated Excellent, 2 for a book judged Good, and 1 for a Fairly relevant book. Irrelevant and unjudged books gave 0 gain.

Table 2. Performance scores for the Book Retrieval task (In the Query column, A stands for Automatic, and M is for Manual; in the Method column, N stands for Non-book-specific retrieval approach, and B for Book-specific approach). Paired runs are indicated by * (where the information was available).

ID	RunID	Query A/M	Method N/B	NDCG @1	NDCG @5	NDCG @10	NDCG @25	NDCG @100	NDCG @1000
2	BERK_T2_OBJ **	A	N	0.351	0.316	0.316	0.349	0.427	0.478
2	BERK_T2_OBJ2 *	A	N	0.351	0.316	0.316	0.349	0.427	0.478
2	BERK_MARC_T2FB **	A	B	0.446	0.349	0.334	0.343	0.394	0.439
2	BERK_T2_CC_MARC *	A	B	0.453	0.375	0.359	0.371	0.422	0.462
92	Indri-F-C-A	A	N	0.521	0.477	0.479	0.503	0.562	0.604
92	Indri-NF-C-A	A	N	0.527	0.490	0.490	0.514	0.573	0.613
92	Indri-NF-PC-A	A	N	0.319	0.319	0.331	0.359	0.425	0.488
92	Indri-F-TOC-A	A	B	0.241	0.218	0.206	0.219	0.262	0.305
92	Indri-NF-TOC-A	A	B	0.257	0.233	0.225	0.235	0.275	0.316
92	Indri-NF-H-A	A	B	0.511	0.421	0.398	0.399	0.440	0.478

Table 2 shows the NDCG scores, reported at various rank cutoffs. From the University of California, Berkeley results, it appears that performance at top ranks (up to rank 25) is improved using book-specific features or ranking methods. However, the results of Cairo Microsoft Innovation Center show that superior performance is achieved by simply applying traditional document retrieval techniques. Overall, we can observe a large variation in the performance scores, from 0.206 to 0.613. Interestingly, the runs `Indri-F-TOC-A` and `Indri-NF-TOC-A`, which only relied on pages that contained the table of contents or the back-of-book index performed the worst of all runs. On the other hand, the best performance was produced by the simplest strategy in `Indri-NF-C-A`, using a standard document retrieval framework. These findings highlight the need for further study in order to better understand the utility of book-specific retrieval features and suggest that there is plenty of room for further development. For details on the approaches of the two groups, please refer to [10] and [11].

6   Summary of BookSearch'07 and Plans for BookSearch'08

The Book Search track in 2007 focused on investigating infrastructure issues that come with the setting up of a new track. A range of tasks were defined: some of them extending established focused retrieval tasks studied at INEX into the book domain and some novel, book-specific tasks. The tasks were designed with the aim to provide new challenges for participants with existing search engines, as well as to attract new groups with an interest in digitized books. Although most of these tasks proved to be rather ambitious, they represent a significant step in the shaping of a research agenda for the future of book search. The level of interest (27 registered groups) suggests that book search is an area that is set to grow considerably in the coming years, especially as more and more groups will be able to muster the necessary resources to tackle the range of challenges.

The Book Search track in 2008 (BookSearch'08)[3] will aim to look beyond the topic of search and extend to issues that touch a wider research community. BookSearch'08 will aim to bring together researchers in Information Retrieval, Digital Libraries, Human Computer Interaction, and eBooks with the goal to work on a common research agenda around digitized book collections. Towards this goal, the track will investigate the following topics:

- Users' interactions with e-books and collections of digitized books
- IR techniques for searching full texts of digitized books
- Digital library services to increase accessibility of digitized books

We plan to propose five tasks for 2008 and invite participants in the setting up process. The tasks are 1) Structure extraction from digitized books, 2) Creation of virtual bookshelves, 3) Supporting active reading, and the 4) Book retrieval, and 5) Page in Context tasks from BookSearch'07. The different tasks will make use of different sets of digitized books, ranging from a collection of 100 books to 50,000.

The track is set to start in mid April 2008.

Acknowledgements

We thank Steven Robertson, Nick Craswell and Natasa Milic-Frayling for their valuable comments on aspects of the organisation of this track.

References

1. Broder, A.: A taxonomy of web search. SIGIR Forum 36(2), 3–10 (2002)
2. Coyle, K.: Mass digitization of books. Journal of Academic Librarianship 32(6), 641–645 (2006)
3. Fuhr, N., Kamps, J., Lalmas, M., Malik, S., Trotman, A.: Overview of the INEX, ad hoc track. In: Fuhr, N., et al. (eds.) [4]
4. Fuhr, N., Kamps, J., Lalmas, M., Trotman, A.: Focused access to XML documents, 6th International Workshop of the Initiative for the Evaluation of XML Retrieval (INEX), Revised Selected Papers. LNCS. Springer, Heidelberg (2008)
5. Fuhr, N., Lalmas, M., Trotman, A. (eds.): Pre-proceedings of the 6th International Workshop of the Initiative for the Evaluation of XML Retrieval (INEX (2007), http://inex.is.informatik.uni-duisburg.de/2007/inex07/pdf/2007-preproceedings.pdf
6. Järvelin, K., Kekäläinen, J.: Cumulated gain-based evaluation of IR techniques. ACM Transactions on Information Systems (ACM TOIS) 20(4), 422–446 (2002)
7. Kazai, G.: Tasks and submission guidelines for the INEX, book search track. In: Fuhr, N., et al. (eds.) [5], pp. 473–480
8. Kazai, G.: Topic development guidelines for the INEX, book search track. In: Fuhr, N., et al. (eds.) [5], 464–472

---

[3] http://www.inex.otago.ac.nz/tracks/books/books.asp

9. Lalmas, M., Piwowarski, B.: INEX 2007 relevance assessment guide. In: Fuhr, N., et al. (eds.). [4]
10. Larson, R.R.: Logistic regression and EVIs for XML books and the heterogeneous track. In: Fuhr, N., et al. (eds.) [4]
11. Magdy, W., Darwish, K.: CMIC at INEX,: Book search track. In: Fuhr, N., et al. (eds.) [4]
12. Wu, H., Kazai, G., Taylor, M.: Book search experiments: Investigating IR methods for the indexing and retrieval of books. In: Macdonald, C., Ounis, I., Plachouras, V., Ruthven, I., White, R.W. (eds.) ECIR 2008. LNCS, vol. 4956, pp. 234–245. Springer, Heidelberg (2008)

Logistic Regression and EVIs for XML Books and the Heterogeneous Track

Ray R. Larson

School of Information
University of California, Berkeley
Berkeley, California, USA, 94720-4600
ray@ischool.berkeley.edu

Abstract. For this year's INEX UC Berkeley focused on the Book track and the Heterogeneous track, For these runs we used the TREC2 logistic regression probabilistic model with blind feedback as well as Entry Vocabulary Indexes (EVIs) for the Books Collection MARC data. For the full text records of the book track we encountered a number of interesting problems in setting up the database, and ended up using page-level indexing of the full collection.

As (once again) the only group to actually submit runs for the Het track, we are guaranteed both the highest, and lowest, effectiveness scores for each task. However, because it was again deemed pointless to conduct the actual relevance assessments on the submissions of a single system, we do not know the exact values of these results.

1 Introduction

In this paper we will first discuss the algorithms and fusion operators used in our official INEX 2007 Book Track and Heterogenous (Het) track runs. Then we will look at how these algorithms and operators were used in the various submissions for these tracks, and finally we will discuss problems in implementation, and directions for future research.

2 The Retrieval Algorithms and Fusion Operators

This section describes the probabilistic retrieval algorithms used for both the Adhoc and Het track in INEX this year. These are the same algorithms that we have used in previous years for INEX, and also include the addition of a blind relevance feedback method used in combination with the TREC2 algorithm. In addition we will discuss the methods used to combine the results of searches of different XML components in the collections. The algorithms and combination methods are implemented as part of the Cheshire II XML/SGML search engine [19,17,15] which also supports a number of other algorithms for distributed search and operators for merging result lists from ranked or Boolean sub-queries.

N. Fuhr et al. (Eds.): INEX 2007, LNCS 4862, pp. 162–174, 2008.
© Springer-Verlag Berlin Heidelberg 2008

2.1   TREC2 Logistic Regression Algorithm

Once again the principle algorithm used for our INEX runs is based on the *Logistic Regression* (LR) algorithm originally developed at Berkeley by Cooper, et al. [8]. The version that we used for all tasks this year was the Cheshire II implementation of the "TREC2" [6,5] that provided good Thorough retrieval performance in the INEX 2005 evaluation [19]. As originally formulated, the LR model of probabilistic IR attempts to estimate the probability of relevance for each document based on a set of statistics about a document collection and a set of queries in combination with a set of weighting coefficients for those statistics. The statistics to be used and the values of the coefficients are obtained from regression analysis of a sample of a collection (or similar test collection) for some set of queries where relevance and non-relevance has been determined. More formally, given a particular query and a particular document in a collection $P(R \mid Q, D)$ is calculated and the documents or components are presented to the user ranked in order of decreasing values of that probability. To avoid invalid probability values, the usual calculation of $P(R \mid Q, D)$ uses the "log odds" of relevance given a set of S statistics derived from the query and database, such that:

$$\log O(R|C,Q) = \log \frac{p(R|C,Q)}{1 - p(R|C,Q)} = \log \frac{p(R|C,Q)}{p(\overline{R}|C,Q)}$$

$$= c_0 + c_1 * \frac{1}{\sqrt{|Q_c| + 1}} \sum_{i=1}^{|Q_c|} \frac{qtf_i}{ql + 35}$$

$$+ c_2 * \frac{1}{\sqrt{|Q_c| + 1}} \sum_{i=1}^{|Q_c|} \log \frac{tf_i}{cl + 80}$$

$$- c_3 * \frac{1}{\sqrt{|Q_c| + 1}} \sum_{i=1}^{|Q_c|} \log \frac{ctf_i}{N_t}$$

$$+ c_4 * |Q_c|$$

where C denotes a document component and Q a query, R is a relevance variable, and

$p(R|C,Q)$ is the probability that document component C is relevant to query Q,

$p(\overline{R}|C,Q)$ the probability that document component C is not relevant to query Q, (which is 1.0 - $p(R|C,Q)$)

$|Q_c|$ is the number of matching terms between a document component and a query,

qtf_i is the within-query frequency of the ith matching term,

tf_i is the within-document frequency of the ith matching term,

ctf_i is the occurrence frequency in a collection of the ith matching term,

ql is query length (i.e., number of terms in a query like $|Q|$ for non-feedback situations),

cl is component length (i.e., number of terms in a component), and
N_t is collection length (i.e., number of terms in a test collection).
c_k are the k coefficients obtained though the regression analysis.

Assuming that stopwords are removed during index creation, then ql, cl, and N_t are the query length, document length, and collection length, respectively. If the query terms are re-weighted (in feedback, for example), then qtf_i is no longer the original term frequency, but the new weight, and ql is the sum of the new weight values for the query terms. Note that, unlike the document and collection lengths, query length is the "optimized" relative frequency without first taking the log over the matching terms.

The coefficients were determined by fitting the logistic regression model specified in $\log O(R|C, Q)$ to TREC training data using a statistical software package. The coefficients, c_k, used for our official runs are the same as those described by Chen[3]. These were: $c_0 = -3.51$, $c_1 = 37.4$, $c_2 = 0.330$, $c_3 = 0.1937$ and $c_4 = 0.0929$. Further details on the TREC2 version of the Logistic Regression algorithm may be found in Cooper et al. [6].

2.2 Blind Relevance Feedback

It is well known that blind (also called pseudo) relevance feedback can substantially improve retrieval effectiveness in tasks such as TREC and CLEF. (See for example the papers of the groups who participated in the Ad Hoc tasks in TREC-7 (Voorhees and Harman 1998)[23] and TREC-8 (Voorhees and Harman 1999)[24].)

Blind relevance feedback is typically performed in two stages. First, an initial search using the original queries is performed, after which a number of terms are selected from the top-ranked documents (which are presumed to be relevant). The selected terms are weighted and then merged with the initial query to formulate a new query. Finally the reweighted and expanded query is run against the same collection to produce a final ranked list of documents. It was a simple extension to adapt these document-level algorithms to document components for INEX.

The TREC2 algorithm has been been combined with a blind feedback method developed by Aitao Chen for cross-language retrieval in CLEF. Chen[4] presents a technique for incorporating blind relevance feedback into the logistic regression-based document ranking framework. Several factors are important in using blind relevance feedback. These are: determining the number of top ranked documents that will be presumed relevant and from which new terms will be extracted, how to rank the selected terms and determining the number of terms that should be selected, how to assign weights to the selected terms. Many techniques have been used for deciding the number of terms to be selected, the number of top-ranked documents from which to extract terms, and ranking the terms. Harman [12] provides a survey of relevance feedback techniques that have been used.

Lacking comparable data from previous years, we adopted some rather arbitrary parameters for these options for INEX 2007. We used top 10 ranked components from the initial search of each component type, and enhanced and reweighted the query terms using term relevance weights derived from well-known Robertson and Sparck Jones[22] relevance weights, as described by Chen and Gey[5]. The top 10 terms that occurred in the (presumed) relevant top 10 documents, that were not already in the query were added for the feedback search.

2.3   TREC3 Logistic Regression Algorithm

In addition to the TREC2 algorithm described above, we also used the TREC3 algorithm in some of our Het track runs. This algorithm has be used repeatedly in our INEX work, and described many times, but we include it below for ease on comparison. The full equation describing the "TREC3" LR algorithm used in these experiments is:

$$\log O(R \mid Q, C) =$$

$$b_0 + \left(b_1 \cdot \left(\frac{1}{|Q_c|} \sum_{j=1}^{|Q_c|} \log qtf_j \right) \right)$$

$$+ \left(b_2 \cdot \sqrt{|Q|} \right)$$

$$+ \left(b_3 \cdot \left(\frac{1}{|Q_c|} \sum_{j=1}^{|Q_c|} \log tf_j \right) \right) \tag{1}$$

$$+ \left(b_4 \cdot \sqrt{cl} \right)$$

$$+ \left(b_5 \cdot \left(\frac{1}{|Q_c|} \sum_{j=1}^{|Q_c|} \log \frac{N - n_{t_j}}{n_{t_j}} \right) \right)$$

$$+ \left(b_6 \cdot \log |Q_d| \right)$$

Where:

Q is a query containing terms T,
$|Q|$ is the total number of terms in Q,
$|Q_c|$ is the number of terms in Q that also occur in the document component,
tf_j is the frequency of the jth term in a specific document component,
qtf_j is the frequency of the jth term in Q,
n_{t_j} is the number of components (of a given type) containing the jth term,
cl is the document component length measured in bytes.
N is the number of components of a given type in the collection.
b_i are the coefficients obtained though the regression analysis.

This equation, used in estimating the probability of relevance for some of the Het runs in this research, is essentially the same as that used in [7]. The b_i coefficients in the original version of this algorithm were estimated using relevance judgements and statistics from the TREC/TIPSTER test collection. In INEX 2005 we did not use the original or "Base" version, but instead used a version where the coeffients for each of the major document components were estimated separately and combined through component fusion. This year, lacking relevance data from Wikipedia for training, we used the base version again. The coefficients for the Base version were $b_0 = -3.70$, $b_1 = 1.269$, $b_2 = -0.310$, $b_3 = 0.679$, $b_4 = -0.0674$, $b_5 = 0.223$ and $b_6 = 2.01$.

2.4   CORI Collection Ranking Algorithm

The resource selection task in the Heterogeneous track is basically the same as the collection selection task in distributed IR. For this task we drew on our previously experiments with distributed search and collection ranking [15,16], where we used the above "TREC3" algorithm for collection selection and compared it with other reported distributed search results.

The collection selection task attempts to discover which distributed databases are likely to be the best places for the user to begin a search. This problem, distributed information retrieval, has been an area of active research interest for many years. Distributed IR presents three central research problems:

1. How to select appropriate databases or collections for search from a large number of distributed databases;
2. How to perform parallel or sequential distributed search over the selected databases, possibly using different query structures or search formulations, in a networked environment where not all resources are always available; and
3. How to merge results from the different search engines and collections, with differing record contents and structures (sometimes referred to as the collection fusion problem).

Each of these research problems presents a number of challenges that must be addressed to provide effective and efficient solutions to the overall problem of distributed information retrieval. Some of the best known work in this area has been Gravano, et al's work on GlOSS [11,10] and Callan, et al's [2,25,1] application of inference networks to distributed IR. One of the best performing collection selection algorithms developed by Callan was the "CORI" algorithm. This algorithm was adapted for the Cheshire II system, and used for some of our Resource Selection runs for the Het track this year. The CORI algorithm defines a belief value for each query term using a form of tfidf ranking for each term and collection:

$$T = \frac{df}{df + 50 + 150 \cdot cw/\overline{cw}}$$

$$I = \frac{\log(\frac{|DB|+0.5}{cf})}{\log(|DB| + 1.0)}$$

$$p(r_k|db_i) = 0.4 + 0.6 \cdot T \cdot I$$

Where:

df is the number of documents containing terms r_k,
cf is the number of databases or collections containing r_k,
$|DB|$ is the number of databases or collections being ranked,
cw is the number of terms in database or collection db_i,
$\overline{cw}$ is the average cw of the collections being ranked, and
$p(r_k|db_i)$ is the belief value in collection db_i due to observing term r_k

These belief values are summed over all of the query terms to provide the collection ranking value.

2.5 Result Combination Operators

As we have also reported previously, the Cheshire II system used in this evaluation provides a number of operators to combine the intermediate results of a search from different components or indexes. With these operators we have available an entire spectrum of combination methods ranging from strict Boolean operations to fuzzy Boolean and normalized score combinations for probabilistic and Boolean results. These operators are the means available for performing fusion operations between the results for different retrieval algorithms and the search results from different different components of a document. For Hetergeneous search we used a variant of the combination operators, where MINMAX normalization across the probability of relevance for each entry in results from each sub-collection was calculated and the final result ranking was based on these normalized scores.

In addition, for the Adhoc Thorough runs we used a merge/reweighting operator based on the "Pivot" method described by Mass and Mandelbrod[20] to combine the results for each type of document component considered. In our case the new probability of relevance for a component is a weighted combination of the initial estimate probability of relevance for the component and the probability of relevance for the entire article for the same query terms. Formally this is:

$$P(R \mid Q, C_{new}) = (X * P(R \mid Q, C_{comp})) + ((1 - X) * P(R \mid Q, C_{art})) \quad (2)$$

Where X is a pivot value between 0 and 1, and $P(R \mid Q, C_{new})$, $P(R \mid Q, C_{comp})$ and $P(R \mid Q, C_{art})$ are the new weight, the original component weight, and article weight for a given query. Although we found that a pivot value of 0.54 was most effective for INEX04 data and measures, we adopted the "neutral" pivot value of 0.5 for all of our 2007 adhoc runs, given the uncertainties of how this approach would fare with the new database.

3 Database and Indexing Issues

Because we were using the same databases for the Heterogeneous track as in 2007 we refer the reader to our INEX 2006 paper[18] where the indexing issues and approaches were discussed. We focus in this section on the Books database and the issues with it (as well as how the MARC data included with the Books database was converted and made searchable as XML, and how the EVIs are created).

All of the submitted runs for this year's Book track and Heterogeneous track used the Cheshire II system for indexing and retrieval. For the Book Track the "Classification Clustering" feature of the system was used to generate the EVIs used in query expansion. The original approach for Classification Clustering was in searching was described in [13] and [14]. Although the method has experienced considerable changes in implementation, the basic approach is still the same: topic-rich elements extracted from individual records in the database (such as titles, classification codes, or subject headings) are merged based on a normalized version of a particular organizing element (usually the classification or subject headings), and each such *classification cluster* is treated as a single "document" containing the combined topic-rich elements of all the individual documents that have the same values of the organizing element. The EVI creation and search approach taken for this research is described in detail in the following section.

3.1   Book Track: MARC and Entry Vocabulary Indexes

The earliest versions of Entry Vocabulary Indexes were developed to facilitate automatic classification of MARC library catalog records, and first used in searching in [14]. Given the MARC data included with almost all of the documents for the Book track (MARC data was available for 42013 of the 42080 books in the collection) it seemed an interesting experiment to test how well EVIs "library catalog" searching would work with the books collection in addition to the full XML search approaches. It also seemed interesting to combine these two approaches.

The early work used a simple frequency-based probabilistic model in searching, but a primary feature was that the "Classification clusters", were treated as documents and the terms associated with top-ranked clusters were combined with the original query, in a method similar to "blind feedback", to provide an enhanced second stage of search.

Our later work with EVIs used a maximum likelihood weighting for each term (word or phrase) in each classification. This was the approach described in [9] and used for Cross-language Domain-Specific retrieval for CLEF 2005. One limitation of that approach is that the EVI can produce maximum likelihood estimates for only a single term at a time, and alternative approaches needed to be explored for combining terms (see [21] for the various approaches).

In place of the simpler probabilistic model used in the early research, we use the same logistic regression based algorithm that is used for text retrieval. In effect, we just search the "Classification Clusters" as if they were documents using the TREC2 algorithm as described above, then take some number of the top-ranked terms and use those to expand the query for submission to the normal document collection. Alternatively, because of the one-to-one match of books and MARC records in this collection, MARC searches or classification cluster two-stage searches can be considered a form of document search.

Table 1. MARC Indexes for INEX Book Track 2007

Name	Description	Contents	Vector?
names	All Personal and Corporate names	//FLD[1670]00, //FLD[1678]10, //FLD[1670]11	No
pauthor	Personal Author Names	//FLD[170]00	No
title	Book Titles	//FLD130, //FLD245, //FLD240, //FLD730, //FLD740, //FLD440, //FLD490, //FLD830	No
subject	All Subject Headings	//FLD6..	No
topic	Topical Elements	//FLD6.., //FLD245, //FLD240, //FLD4.., //FLD8.., //FLD130, //FLD730, //FLD740, //FLD500, //FLD501, //FLD502 //FLD505, //FLD520, //FLD590	Yes
lcclass	Library of Congress Classification	//FLD050, //FLD950	No
doctype	Material Type Code	//USMARC@MATERIAL	No
localnum	ID Number	//FLD001	No
ISBN	ISBN	//FLD020	No
publisher	Publisher	//FLD260/b	No
place	Place of Publication	//FLD260/a	No
date	Date of Publication	//FLD008	No
lang	Language of Publication	//FLD008	No

Two separate EVIs were built for the MARC data extracted from the Books database. The first uses the library classification code (MARC field 050) as the organizing basis and takes the searchable terms from all titles and subject headings in the MARC record (E.g., MARC fields 245, 440, 490, 830, 740, 600, 610, 620, 630, 640, 650). The second uses the topical subject fields (MARC field 650) with the same searchable fields.

The indexes used in the MARC data are shown in Table 1. Note that the tags represented in the "Contents" column of the table are from Cheshire's MARC to XML conversion, and are represented as regular expressions (i.e., square brackets indicate a choice of a single character).

3.2   Indexing the Books XML Database

All indexing in the Cheshire II system is controlled by an XML/SGML Configuration file which describes the database to be created. This configuration file is subsequently used in search processing to control the mapping of search command index names (or Z39.50 numeric attributes representing particular types of bibliographic data) to the physical index files used and also to associated component indexes with particular components and documents.

Because the structure of the Books database was derived from the OCR of the original paper books, it is primarily focused on the page organization and layout and not on the more common structuring elements such as "chapters" or "sections". Because this emphasis on page layout goes all the way down to the individual word and its position on the page, there is a very large amount of markup for page with content. The entire document in XML form is typically multiple megabytes in size. Given the nature of the XML/SGML parser used in the Cheshire II system, each document was taking several minutes for parsing and indexing due to the large internal represention of the parsed document taking up all available RAM space and a large portion of swap space on the available indexing machine. After indexing was run for a full 24 hours, and only 54 items had been indexed, a different approach was taken. Instead of parsing the entire document, we treated each page representation (tagged as "object" in the XML markup) as if it were a separate document. Thus the 42,080 full books were treated as a collection of 14407042 page-sized documents (i.e., there were an average of 342 pages per book).

As noted above the Cheshire system permits parts of the document subtree to be treated as separate documents with their own separate indexes. Thus, paragraph-level components were extracted from the page-sized documents. Because unique object (page) level indentifiers are included in each object, and these identifiers are simple extensions of the document (book) level identifier, we were able to use the page-level identifier to determine where in a given book-level document a particular page or paragraph occurs, and generate an appropriate XPath for it.

Indexes were created to allow searching of full page (object) contents, and component indexes for the full content of each of individual paragraphs on a page. Because of the physical layout based structure used by the Books collection, paragraphs split across pages are marked up (and therefore indexed) as two paragraphs. Indexes were also created to permit searching by object id, allowing search for specific individual pages, or ranges of pages.

4   INEX 2007 Book Track and Heterogeneous Runs

4.1   Book Track Runs

Berkeley submitted 4 runs for the Book Retrieval Task of the Book Track. According to the Book Search Task Guide (v.5) available on the INEX Web site:

> The goal of this task is to investigate the impact of book specific features on the effectiveness of book retrieval systems, where the unit of retrieval is the (complete) book. Users are thus assumed to be searching for (whole) books relevant to their information need that they can, e.g., borrow from a Library or purchase from a retailer, etc.
>
> Participants of this task are invited to submit pairs of runs, where one run should be the result of applying generic IR techniques to return a ranked list of books to the user in response to a query. The other run should be generated using the same techniques (where possible) but with the use of additional book- specific features (e.g. back-of-book index, citation statistics, in or out of print, etc.) or specifically tuned methods.

In both cases, a result list should contain a maximum of 1000 books estimated relevant to the given topic, ranked in order of estimated relevance to the query.

The test queries used for this task have been extracted from the query log of a commercial search engine, and relevance judgements have been collected on a four point scale: Excellent, Good, Fair, and Not- relevant. The evaluation (subject to change) will be based on the measure of Normalised Discounted Cumulated Gain at various cut-off values.

Table 2 shows the submitted runs and their NDCG official values. The submitted pairs were "MARC_T2FB" with "T2_OBJ" and "T2_CC_MARC" with "T2_OBJ2". As Table 2 shows, the results for "T2_OBJ" and "T2_OBJ2" are identical.

The MARC_T2FB run searched the full content of the MARC data using the TREC2 LR algithm with blind feedback. Each MARC record retrieved in the searches was mapped to its corresponding XML book id for the submitted results.

The T2_CC_MARC run performed a two-stage search: The first stage searched the topic title in the EVI created from MARC data for matching Library of Congress classes of works (using the TREC2 LR algorithm). The second stage took the top-ranked class from the EVI and added that class number to the topic title, which was then searched in the MARC data using the TREC2 LR algorithm with blind feedback. Each MARC record retrieved in the searches was mapped to its corresponding XML book id for the submitted results.

Both the T2_OBJ and T2_OBJ2 runs directly searched the page-level XML records for each document. Because the results could have multiple page hits from each book, only the top-ranked page for each book was retained. Each page hit was mapped to the appropriate book id in the submitted results.

Table 2. Book Track Runs: Official Results

Run-ID	Type	ndcg@1	ndcg@5	ndcg@10	ndcg@25	ndcg@100	ndcg@1000
MARC_T2FB	book-spec.	0.446	0.349	0.334	0.342	0.394	0.439
T2_CC_MARC	book-spec.	0.453	0.375	0.358	0.371	0.422	0.462
T2_OBJ	non-spec.	0.351	0.316	0.316	0.349	0.427	0.477
T2_OBJ2	non-spec.	0.351	0.316	0.316	0.349	0.427	0.477

As the averaged Normalised Discounted Cumulated Gain measures for the submitted runs show, both the direct MARC search and the EVI-based MARC search were more effective than the page-level search of the full XML. The EVI also performed better than the direct MARC-only search. This is a very encouraging finding, and appears to confirm the results of early work with "Classification Clustering".

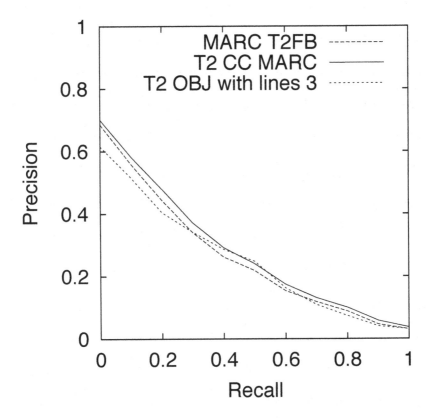

Fig. 1. Berkeley Runs for the Book Retrieval Task

4.2   Heterogeneous Runs

Three runs were submitted for the Resource Selection task, and 2 for the Content-Only task. The Resource selection runs used the TREC2, TREC3, and CORI algorithms, respectively, with no blind feedback. The two Content-Only runs used the TREC2 and TREC3 algorithms, also with no blind feedback.

Since Berkeley was the only group to submit Het track runs, it was decided not to go to the effort of evaluation with such a limited pool, so we do not have any figures on the actual or relative performance of these different techniques for the Heterogeneous track.

5   Conclusions and Future Directions

This year our participation focused on the Book Search Task and Heterogeneous track. The Heterogeneous track has been retired, and will not be offered again next year, so it is doubtful that any actual results will ever be available. The Book Search Track, however, is just beginning and still has tasks underway (including the book classification task which was started in February 2008). We have found the Book Search data and tasks to be challenging and interesting in their implications for XML search.

Figure 1 shows the results of further analysis using the Books search task QRELS (converted to binary relevance) and the standard trec_eval program to generate conventional Recall/Precision curves. Again, the relative ranking for the three approaches, as indicated by NDGC measures in Table 2, is borne out in the Precision Recall curves. The calculated mean average precision for T2_CC_MARC was 0.2633, for MARC_T2FB was 0.2438 and for T2_OBJ was 0.2358. These differences are not dramatic and may not be statistically significant - we have not yet conducted that analysis since the data became available only in the past week. However, we do find it interesting that the human-derived metadata from the MARC records appears to provide better results than the full-text available in the page-level search. Of course, there are some problems with full-text from OCR sources (misspelled words, etc.). However, we see many potential future experiments that attempt to leverage both the full XML data and the human-derived MARC metadata.

References

1. Callan, J.: Distributed information retrieval. In: Croft, W.B. (ed.) Advances in Information Retrieval: Recent research from the Center for Intelligent Information Retrieval, ch. 5, pp. 127–150. Kluwer, Boston (2000)
2. Callan, J.P., Lu, Z., Croft, W.B.: Searching Distributed Collections with Inference Networks. In: Fox, E.A., Ingwersen, P., Fidel, R. (eds.) Proceedings of the 18th Annual International ACM SIGIR Conference on Research and Development in Information Retrieval, Seattle, Washington, pp. 21–28. ACM Press, New York (1995)
3. Chen, A.: Multilingual information retrieval using english and chinese queries. In: Peters, C., Braschler, M., Gonzalo, J., Kluck, M. (eds.) CLEF 2001. LNCS, vol. 2406, pp. 44–58. Springer, Heidelberg (2002)
4. Chen, A.: Cross-Language Retrieval Experiments at CLEF 2002. In: Peters, C., Braschler, M., Gonzalo, J. (eds.) CLEF 2002. LNCS, vol. 2785, pp. 28–48. Springer, Heidelberg (2003)
5. Chen, A., Gey, F.C.: Multilingual information retrieval using machine translation, relevance feedback and decompounding. Information Retrieval 7, 149–182 (2004)
6. Cooper, W.S., Chen, A., Gey, F.C.: Full Text Retrieval based on Probabilistic Equations with Coefficients fitted by Logistic Regression. In: Text REtrieval Conference (TREC-2), pp. 57–66 (1994)
7. Cooper, W.S., Gey, F.C., Chen, A.: Full text retrieval based on a probabilistic equation with coefficients fitted by logistic regression. In: Harman, D.K. (ed.) The Second Text Retrieval Conference (TREC-2) (NIST Special Publication 500-215), Gaithersburg, MD, pp. 57–66. National Institute of Standards and Technology (1994)
8. Cooper, W.S., Gey, F.C., Dabney, D.P.: Probabilistic retrieval based on staged logistic regression. In: 15th Annual International ACM SIGIR Conference on Research and Development in Information Retrieval, Copenhagen, Denmark, June 21-24, 1992, pp. 198–210. ACM Press, New York (1992)

9. Gey, F., Buckland, M., Chen, A., Larson, R.: Entry vocabulary – a technology to enhance digital search. In: Proceedings of HLT 2001, First International Conference on Human Language Technology, San Diego, March 2001, pp. 91–95 (2001)

10. Gravano, L., García-Molina, H.: Generalizing GlOSS to vector-space databases and broker hierarchies. In: International Conference on Very Large Databases, VLDB, pp. 78–89 (1995)

11. Gravano, L., García-Molina, H., Tomasic, A.: GlOSS: text-source discovery over the Internet. ACM Transactions on Database Systems 24(2), 229–264 (1999)

12. Harman, D.: Relevance feedback and other query modification techniques. In: Frakes, W., Baeza-Yates, R. (eds.) Information Retrieval: Data Structures & Algorithms, pp. 241–263. Prentice-Hall, Englewood Cliffs (1992)

13. Larson, R.R.: Classification clustering, probabilistic information retrieval, and the online catalog. Library Quarterly 61(2), 133–173 (1991)

14. Larson, R.R.: Evaluation of advanced retrieval techniques in an experimental online catalog. Journal of the American Society for Information Science 43(1), 34–53 (1992)

15. Larson, R.R.: A logistic regression approach to distributed IR. In: SIGIR 2002: Proceedings of the 25th Annual International ACM SIGIR Conference on Research and Development in Information Retrieval, Tampere, Finland, August 11-15, 2002, pp. 399–400. ACM Press, New York (2002)

16. Larson, R.R.: Distributed IR for digital libraries. In: Koch, T., Sølvberg, I.T. (eds.) ECDL 2003. LNCS, vol. 2769, pp. 487–498. Springer, Heidelberg (2003)

17. Larson, R.R.: A fusion approach to XML structured document retrieval. Information Retrieval 8, 601–629 (2005)

18. Larson, R.R.: Probabilistic retrieval approaches for thorough and heterogeneous xml retrieval. In: Fuhr, N., Lalmas, M., Trotman, A. (eds.) INEX 2006. LNCS, vol. 4518, pp. 318–330. Springer, Heidelberg (2007)

19. Larson, R.R.: Probabilistic retrieval, component fusion and blind feedback for XML retrieval. In: Fuhr, N., Lalmas, M., Malik, S., Kazai, G. (eds.) INEX 2005. LNCS, vol. 3977, pp. 225–239. Springer, Heidelberg (2006)

20. Mass, Y., Mandelbrod, M.: Component ranking and automatic query refinement for xml retrieval. In: Fuhr, N., Lalmas, M., Malik, S., Szlávik, Z. (eds.) INEX 2004. LNCS, vol. 3493, pp. 73–84. Springer, Heidelberg (2005)

21. Petras, V., Gey, F., Larson, R.: Domain-specific CLIR of english, german and russian using fusion and subject metadata for query expansion. In: Peters, C., Gey, F.C., Gonzalo, J., Müller, H., Jones, G.J.F., Kluck, M., Magnini, B., de Rijke, M., Giampiccolo, D. (eds.) CLEF 2005. LNCS, vol. 4022, pp. 226–237. Springer, Heidelberg (2006)

22. Robertson, S.E., Jones, K.S.: Relevance weighting of search terms. Journal of the American Society for Information Science, 129–146, May–June (1976)

23. Voorhees, E., Harman, D. (eds.): The Seventh Text Retrieval Conference (TREC-7). NIST (1998)

24. Voorhees, E., Harman, D. (eds.): The Eighth Text Retrieval Conference (TREC-8). NIST (1999)

25. Xu, J., Callan, J.: Effective retrieval with distributed collections. In: Proceedings of the 21st International ACM SIGIR Conference on Research and Development in Information Retrieval, pp. 112–120 (1998)

CMIC at INEX 2007: Book Search Track

Walid Magdy and Kareem Darwish

Cairo Microsoft Innovation Center,
Smart Village, Bldg. B115, Km. 28 Cairo/Alexandria Desert Rd.
Abou Rawash, Egypt
{wmagdy,kareemd}@microsoft.com

Abstract. With massive book digitization efforts underway, the need for effective retrieval of books and pages in books is an important problem. This paper describes our submissions to the INEX 2007 Book Search track. We explored using book specific features such as table of content and index pages and headers along with non-book specific features. Our results show that indexing the entire contents of books and headers provided the most effective retrieval strategy.

Keywords: book search, OCR retrieval.

1 Introduction

Since the advent of the printing press in the fifteenth century, the amount of printed text has grown overwhelmingly. Although a great deal of text is now generated in electronic character-coded formats (HTML, word processor files ... etc.), many documents, available only in print, remain important. This is due in part to the existence of large collections of legacy documents available only in print, and in part because printed text remains an important distribution channel that can effectively deliver information without the technical infrastructure that is required to deliver character-coded text. Printed documents can be browsed and indexed for retrieval relatively easily in limited quantities, but effective access to the contents of large collections requires some form of automation.

One such form of automation is to scan the documents (to produce document images) and subsequently perform OCR on the document images to convert them into text. Many recent initiatives, such as the Million Book Project, have focused on digitizing and OCR'ing large repositories of legacy books (Thoma and Ford, 2002; Simske and Lin, 2004; Barret et al. 2004)[1]. Such initiatives have been successful in digitizing millions of books in a variety of languages[2,3,4,5,6]. Typically, the OCR process

---

[1] Project Gutenberg website, http://www.gutenberg.org
[2] Google Print website, http://books.google.com
[3] Live Search Books, http://search.live.com/results.aspx?&scope=books
[4] Internet Archive website, http://www.archive.org
[5] Amazon "Search inside Book" announcement,
   http://www.amazon.com/exec/obidos/tg/feature/-/507108/104-7825001-2871961
[6] Open Content Alliance website, http://www.opencontentalliance.org

N. Fuhr et al. (Eds.): INEX 2007, LNCS 4862, pp. 175–182, 2008.
© Springer-Verlag Berlin Heidelberg 2008

introduces errors in the text representation of the document images. The error level is affected by the quality of paper, printing, and scanning. The introduced errors adversely affect retrieval effectiveness of OCR'ed documents. This paper describes the runs we performed at the Cairo Microsoft Innovation Center (CMIC) for the 2007 INEX Book Search track and reports on our results. We participated in the book retrieval task, which involves finding books that are relevant to a particular subject and the page in context retrieval task, which involves finding specific pages in books that are relevant to a topic. We explored generalized retrieval approaches and specialized book-specific approaches that made use of the indices, tables of content pages, and other fields in books.

Section 2 provides an overview on previous work relating to the retrieval of OCR'ed documents; Section 3 describes the experimental setup including the data collection and the IR engine that we used, a description of the book retrieval task runs, and our approach to the page in context task; Section 4 reports and discusses the results; and Section 5 concludes the paper and provides possible future directions.

2 Background

Retrieval of OCR degraded text documents has been reported on for many languages, including English (Harding et al., 1997; Kantor and Voorhees, 1996; Taghva et al., 1994a; Taghva et al., 1995; Taghva et al., 1996b); Chinese (Tseng and Oard, 2001); and Arabic (Darwish and Oard, 2002).

For English, Doermann (1997) reports that retrieval effectiveness decrease significantly for OCR'ed documents with an error rate at some point between 5-20%. Taghva reported experiments which involved using English collections with documents ranging in number between 204 and 674 documents that were about 38 pages long on average (Taghva et al., 1994b; Taghva et al., 1995). The documents were scanned and OCR'd. His results show negligible decline in retrieval effectiveness due to OCR errors. Taghva's work was criticized for being done on very small collections of very long documents (Tseng and Oard, 2001). Small collections might not behave like larger ones, and thus they might not be reflective of real life applications in which retrieval from a large number of documents is required (Harman, 1992). Similar results for English were reported by Smith (1990) in which he reported no significant drop in retrieval effectiveness with the introduction of simulated OCR degradation in which characters were randomly replaced by a symbol indicating failure to recognize. These results contradict other studies in which retrieval effectiveness deteriorated dramatically with the increase in degradation. Hawking reported a significant drop in retrieval effectiveness at a 5% character error rate on the TREC-4 "confusion track" (Hawking, 1996). In the TREC-4 confusion track, approximately 50,000 English documents from the federal registry were degraded by applying random edit operations to random characters in the documents (Kantor and Voorhees, 1996). The contradiction might be due to the degradation method, the size of the collection, the size of the documents, or a combination of these factors. In general retrieval effectiveness is adversely affected by the increase in degradation and decrease in redundancy of search terms in the documents (Doermann, 1998).

Several studies reported the results of using n-grams. A study by Harding, Croft, and Weir (1997), compared the use of different length n-grams to words on four English collections, in which errors are artificially introduced. The documents were degraded iteratively using a model of OCR degradation until retrieval effectiveness of using words as index terms started to significantly deteriorate. The error rate in the documents was unknown. For n-grams, a combination of 2 and 3 grams and a combination of 2, 3, 4, and 5 grams were compared to words. Their results show that n-gram indexing consistently outperformed word indexing, and combining more n-grams was better than combining fewer. In another study by Tseng and Oard, they experimented with different combinations of n-grams on a Chinese collection of 8,438 document images and 30 Chinese queries (Tseng and Oard, 2001). Although ground-truth was not available for the image collection to determine the effect of degradation on retrieval effectiveness, the effectiveness of different index terms were compared. They experimented with unigrams, bigrams, and a combination of both. Chinese words were not segmented and bigrams crossed word boundaries. The results of the experiments show that a combination of unigrams and bigrams consistently and significantly outperform character bigrams, which in turn consistently and significantly outperforms character unigrams.

For Arabic, Darwish and Oard (2002) reported that character 3-gram and 4-grams were the best index terms for searching OCR degraded text. They conducted their experiments on a small collection of 2,730 scanned documents. In general, blind relevance feedback does not help for the retrieval of OCR degraded documents (Darwish and Emam, 2005; Lam-Adesina and Jones, 2006; Taghva et al., 1996a; Tseng and Oard, 2001).

3 Experimental Setup

3.1 Data Collection and Used Search Toolkit

The collection, provided by Microsoft Live Book Search and the Internet Archive, consists of 42,049 digitized out-of-copyright books, with books typically being printed before the 1930's. The actual number of books we used was 41,825, where 224 books were missed due to extraction errors or empty content books. The OCR content of the books was stored in djvu.xml format, which provides XML fields specifying pages, paragraphs, lines, and words along with their coordinates in the page. Associated with the books were other metadata such as names of authors, publisher, publication data, Library of Congress classification, etc.

For all submitted runs, Indri search toolkit was used for indexing and searching the collection of books. Indri was used with stop-word removal, but with no stemming, and several runs were performed twice while enabling or disabling blind relevance feedback. Indri combines inference network model with language modeling (Metzler and Croft, 2004).

3.2 Book Retrieval Task

This task aimed to help users identify books of interest based on a stated information need. There were 250 queries about general subjects: typically consisting of one word and commonly containing named entities. Two sample queries are: "Botany" and "Rigveda." Pairs of runs were required. For each pair, one run would apply generic IR techniques and the other would use additional book-specific features such as Table of Content (TOC) pages, index pages, and page headers. Each run was expected to return a ranked list of 1,000 books. We performed three pairs of runs.

The 3 runs using none book-specific features were as follows:

1. Each document was made up of the entire contents of each book. The books were subsequently indexed and searched using the provided queries. (*Run ID = BC "Book Content"*)

2. The run was identical to the first run, except that blind relevance feedback was used, where 30 terms where extracted from the top 25 retrieved books to expand the original query. (*Run ID = BC-FB "Book Content with Feedback"*)

3. Each document was a single page in each book. All the documents were subsequently indexed and searched using the provided queries. Using the top 5,000 results for a given query, the score of the book was the sum scores of the individual scores within the ranked list as follows:

$$Score_{book_i} = \sum_{\forall\, page_j \in book_i} 10^{score_{page_j}} \tag{1}$$

The reason for using 10 to the power of the score is that Indri scores are log values. Given the scores of the books, a new ranked list was produced. In essence, the books with the most pages mentioning a specific topic would typically ranked first. (*Run ID = PCS "Page Content Score"*)

The runs using book-specific features were as follows:

1. Each document was composed of all the headers in a book. The headers were assumed to be the first line in each page not composed entirely of digits. The documents were indexed and searched using the provided queries. The advantage of using headers is that they generally reflect the main topics in books and the titles of longer chapters are repeated more often, hence giving different weights to different titles. (*Run ID = BH "Book Headers"*)

2. Each document was composed of the TOC and index pages in a book. We deemed a page to be a TOC or index page if any of the following conditions are met:
   i. Presence of the key phrase "Table of Contents."
   ii. Presence of ordinary key words such as "contents", "page", or "index", with moderate number of lines that end with digits.
   iii. Absence of keywords indicating a TOC or index page, but the presence of a large number of lines that end with digits.
   iv. Presence of keywords such as contents, page, or index in a page that was immediately preceded by a page that was deemed as a TOC or index page.

In case we were not able to extract TOC and index pages, we used the first 3,000 characters from the OCR output and last ten pages of a book instead, as they are likely to contain TOC and index pages or the pages with the introduction and/or preface. The rational for using the first 3,000 characters instead of a fixed number of pages is that we found that many books typically contained many empty pages in the beginning. (*Run ID = TOC*)

3. Each document was identical to documents in the second run except that we used blind relevance feedback, where 20 terms were extracted from the top 25 retrieved documents to expand the queries. (*Run ID = TOC-FB*).

3.3 Book Page in Context Retrieval Task

In this task, each system was expected to return a ranked list of 1,000 books and for each book, a ranked list of relevant pages to a user's information need. For the 30 provided queries, we performed seven runs, six of which were automatic and one was manual. All the runs were identical to the run number "three" in the book search task in which no book-specific features were used to generate the ranked list of books. For each, we generated a ranked list of pages based on the score of each page. The differences between the seven runs were all due to the way the queries were formulated. The formulations used the title, description, and narrative fields as follows:

1. Title only
2. Title only with blind relevance feedback
3. Title and description
4. Title and description with blind relevance feedback
5. Title, description, and narrative
6. Title, description, and narrative with blind relevance feedback
7. Manually reformulated queries that were done with consultation of Wikipedia on the topics.

For runs with blind relevance feedback, the queries were expanded with 20 terms extracted from the top 25 retrieved documents. Unfortunately the relevance judgments for this task were not ready by the time we were writing in this paper, hence no results for this task are reported.

3.4 Evaluation Method

For evaluation, relevance judgments were received for book search task. The judging was done on a scale from 0 to 4, with 0 = bad, 1 = fair, 2 = good, 3 = excellent, and 4 = perfect. The figure of merit used is Normalized Discounted Cumulative Gain (NDCG), which is a metric that is becoming increasing popular for evaluating web search engines (Voorhees, 2001). The computation of NDCG was done as described by (Matveeva et al., 2006). NDCG attempts to compute the information gained by the user when he reads the top n results, with documents with higher scores portraying more information gain. NDCG was computed at using top n results, where n were selected to be 1, 5, 10, 25, 100, and 1000 top results.

Although NDCG scores were computed for book search runs, no relevance judgments were received for the page on context retrieval task, leading to no reporting for results for this task.

4 Results and Discussion

Table 1 and Figure 1 show the averages NDCG scores for the six runs in book retrieval task. Results show that indexing the whole books content achieves the best results for retrieval and blind relevance feedback consistently has negative impact on the retrieval effectiveness for both book specific and non-specific runs, which is consistent with previously reported studies. Summing page content scores for book retrieval was shown to be less effective than indexing the entire book. The same was true for TOC indexing, which achieved the worst results. This result seems peculiar especially that TOC pages generally contain all the topics in a book. Our intuition is that either our method for finding TOC pages was inadequate or although TOC pages generally listed all topics, topics were listed once and hence no term frequency information of available to properly weight the different topics. This requires further investigation. The result of the BH run, which involved indexing books by the headers, was the most impressive, achieving scores similar to those involving indexing the entire contents of books. The approach achieved 90% of the effectiveness of search using the entire contents of books, with 95% smaller index. We suspect that even though headers generally included all the entries in TOC pages the redundancy of headers provides more accurate term frequency estimates.

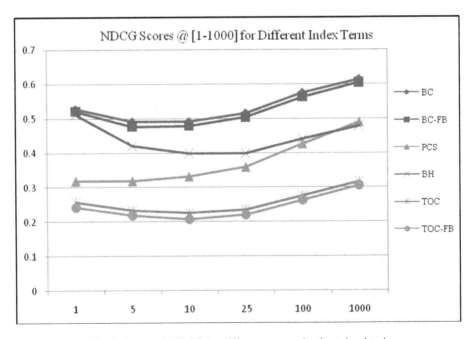

Fig. 1. Averaged NDCG for different runs on book retrieval task

Table 1. Averaged NDCG @ [1-1000] for different runs within book search task

Retrieval Type	Run ID	Averaged NDCG @					
		1	5	10	25	100	1000
non-book specific	BC	0.527	0.490	0.490	0.514	0.573	0.613
	BC-FB	0.521	0.477	0.479	0.503	0.562	0.604
	PCS	0.319	0.319	0.331	0.359	0.425	0.488
book-specific	BH	0.511	0.421	0.398	0.399	0.440	0.478
	TOC	0.257	0.233	0.225	0.235	0.275	0.316
	TOC-FB	0.241	0.218	0.206	0.219	0.262	0.305

5   Conclusion and Future Work

In our submitted runs we experimented with non-book-specific as well as book-specific features for the book search and page in context tasks. Unfortunately, no relevance judgments were received for page in context task, and hence no conclusion could be inferred for this task. For the book search task, indexing the whole book was the most effective method for indexing the book. Indexing using headers in a book only achieves 90% of the retrieval effectiveness of using the entire contents of books with an index that is 95% smaller. Contrary to our initial intuition, indexing TOC pages and indices only led to significant degradation in retrieval effectiveness. Lastly, using blind relevance feedback generally degraded retrieval effectiveness.

For future work, more experimentation is warranted to find better representations of book content to improve retrieval effectiveness. Also, we need to investigate better identification of TOC and index pages to determine conclusively if the poor results for indexing TOC and index pages were due to our identification technique. Lastly, we await the relevance judgments and results for the page in context task.

References

1. Barret, W., Hutchison, L., Quass, D., Nielson, H., Kennard, D.: Digital Mountain: From Granite Archive to Global Access. In: Proc. of International Workshop on Document Image Analysis for Libraries, Palo Alto, January 2004, pp. 104–121 (2004)
2. Croft, W.B., Harding, S., Taghva, K., Andborsak, J.: An evaluation of information retrieval accuracy with simulated OCR output. In: Proceedings of the 3rd Annual Symposium on Document Analysis and Information Retrieval, University of Nevada, Las Vegas, Nev, pp. 115–126 (1994)
3. Darwish, K., Emam, O.: The Effect of Blind Relevance Feedback on a New Arabic OCR Degraded Text Collection. In: International Conference on Machine Intelligence: Special Session on Arabic Document Image Analysis (2005)
4. Darwish, K., Oard, D.: Term Selection for Searching Printed Arabic. In: SIGIR 2002, pp. 261–268 (2002)

5. Doerman, D.: The Retrieval of Document Images: A Brief Survey. In: ICDAR, pp. 945–949 (1997)
6. Doermann, D.: The Indexing and Retrieval of Document Images: A Survey. Computer Vision and Image Understanding 70(3), 287–298 (1998)
7. Harding, S., Croft, W., Weir, C.: Probabilistic Retrieval of OCR-degraded Text Using N-Grams. In: European Conference on Digital Libraries, pp. 345–359 (1997)
8. Harman, D.: Overview of the First Text REtrieval Conference. In: Proceedings of the 16th annual international ACM SIGIR conference on Research and development in information retrieval, Pittsburgh, Pennsylvania, United States, pp. 36–47 (1992)
9. Hawking, D.: Document Retrieval in OCR-Scanned Text. In: Sixth Parallel Computing Workshop, paper P2-F (1996)
10. Kantor, P., Voorhees, E.: Report on the TREC-5 Confusion Track. TREC-5, p. 65 (1996)
11. Lam-Adesina, A.M., Jones, G.J.: Examining and improving the effectiveness of relevance feedback for retrieval of scanned text documents. Inf. Process. Manage. 42(3), 633–649 (2006)
12. Matveeva, I., Burges, C., Burkard, T., Laucius, A., Wong, L.: High accuracy retrieval with multiple nested rankers. In: SIGIR 2006 (2006)
13. Metzler, D., Croft, W.B.: Combining the Language Model and Inference Network Approaches to Retrieval. Information Processing and Management Special Issue on Bayesian Networks and Information Retrieval 40(5), 735–750 (2004)
14. Simske, S., Lin, X.: Creating Digital Libraries: Content Generation and Re-mastering. In: Proc. International Workshop on Document Image Analysis for Libraries, Palo Alto, January 2004, pp. 33–45 (2004)
15. Smith, S.: An Analysis of the Effects of Data Corruption on Text Retrieval Performance. Technical Report DR90-1, Thinking Machines Corp: Cambridge, MA (1990)
16. Taghva, K., Borsack, J., Condit, A.: An Expert System for Automatically Correcting OCR Output. In: Proc. IS&T/SPIE 1994 Intl. Symp. on Electronic Imaging Science and Technology, San Jose, CA, pp. 270–278 (1994a)
17. Taghva, K., Borasack, J., Condit, A., Gilbreth, J.: Results and Implications of the Noisy Data Projects. Technical Report 94-01, Information Science Research Institute, University of Nevada, Las Vegas (1994b)
18. Taghva, K., Borasack, J., Condit, A., Inaparthy, P.: Querying Short OCR'd Documents. Technical Report 94-10, Information Science Research Institute, University of Nevada, Las Vegas (1995)
19. Taghva, K., Borsack, J., Condit, A.: Evaluation of Model-Based Retrieval Effectiveness OCR Text. ACM Transactions on Information Systems 14(1), 64–93 (1996a)
20. Taghva, K., Borsack, J., Condit, A.: Effects of OCR errors on Ranking and Feedback using the Vector Space Model. Information Processing and Management 32(3), 317–327 (1996b)
21. Thoma, G., Ford, G.: Automated Data Entry System: Performance Issues. In: Proc. SPIE Conference on Document Recognition and Retrieval IX, San Jose, pp. 181–190 (2002)
22. Tseng, Y., Oard, D.: Document Image Retrieval Techniques for Chinese. In: Symposium on Document Image Understanding Technology, Columbia, MD, pp. 151–158 (2001)
23. Voorhees, E.: Evaluation by highly relevant documents. In: Proceedings of SIGIR, pp. 74–82 (2001)

Clustering XML Documents Using Closed Frequent Subtrees: A Structural Similarity Approach

Sangeetha Kutty, Tien Tran, Richi Nayak, and Yuefeng Li

Faculty of Information Technology,
Queensland University of Technology, Brisbane, Australia
{s.kutty,t4.tran,r.nayak,y2.li}@qut.edu.au

Abstract. This paper presents the experimental study conducted over the INEX 2007 Document Mining Challenge corpus employing a frequent subtree-based incremental clustering approach. Using the structural information of the XML documents, the closed frequent subtrees are generated. A matrix is then developed representing the closed frequent subtree distribution in documents. This matrix is used to progressively cluster the XML documents. In spite of the large number of documents in INEX 2007 Wikipedia dataset, the proposed frequent subtree-based incremental clustering approach was successful in clustering the documents.

Keywords: Clustering, XML document mining, Frequent Mining, Frequent subtrees, INEX, Structural mining.

1 Introduction

The rapid growth of XML (eXtensible Mark-up Language) after its standardization has marked its acceptance in a wide array of industries ranging from education to entertainment and business to government sectors. The major reason for its success can be attributed to its flexibility and self-describing nature in using structure to store its content. With the increasing number of XML documents there arise many issues concerning the efficient data management and retrieval. XML document clustering has been perceived as an effective solution to improve information retrieval, database indexing, data integration, improved query processing [8] and so on.

Clustering task on XML documents involves grouping XML documents based on their similarity without any prior knowledge on the taxonomy[10]. Clustering has been frequently applied to group text documents based on the similarity of its content. However, clustering XML documents presents a new challenge as it contains structural information with text data (or content). The structure of the XML documents is hierarchical in nature and it represents the relationship between the elements at various levels.

Clustering XML documents is a challenging task[10]. Majority of the existing algorithms utilize the tree-edit distance to compute the structural similarity between each pair of documents. This may lead to incorrect results as the calculated tree-edit distance can be large for very similar trees conforming to the same schema for different size

N. Fuhr et al. (Eds.): INEX 2007, LNCS 4862, pp. 183–194, 2008.
© Springer-Verlag Berlin Heidelberg 2008

trees [12]. A recent study showed that XML document clustering using tree summaries provide high accuracy for documents [3]. The structural summaries of the XML documents were extracted and used to compute the tree-edit distance. Due to the need of calculating the tree-edit distance between each pair of document structural summaries, this process becomes expensive for very large dataset such as INEX Wikipedia test collection that contains 48305 documents. This lays the ground to employ a clustering algorithm which does not utilise the expensive tree-edit distance computation.

In this paper, we propose CFSPC(Closed Frequent Structures-based Progressive Clustering) technique to cluster XML documents incrementally using the closed frequent subtrees. These closed frequent subtrees are called as the Pre-Cluster Form (PCF). Using the PCFs of the XML documents the global similarity between the XML documents is computed incrementally.

The assumption that we have made in this paper, based on the previous research [9] is that documents having a similar structure can be grouped together. For instance, the document from a publication domain will have a different structure than a document from a movie domain. Using this assumption we utilize only the hierarchical structure of the documents to group the XML documents. We have not included the content of the documents as it incurs a huge overhead in mining frequent trees and finding similarity between documents.

Rest of the paper is organized as follows: Section 2 provides the overview of the CFSPC method. Section 3 covers the pre-processing of XML documents for mining. Section 4 details the mining process which includes frequent mining and clustering. In Section 5, we present the experimental results and discussion. We conclude in Section 6 by presenting our future works in XML document mining.

2   The CFSPC (Closed Frequent Structures-Based Progressive Clustering) Method: Overview

As illustrated in Fig.1 CFSPC involves two major phases Pre-processing and Mining. The pre-processing phase involves extraction of the structure of a given XML document to obtain a *document tree*. Each *document tree* contains nodes which represent the tag names. The mining phase includes application of *frequent subtree mining* and

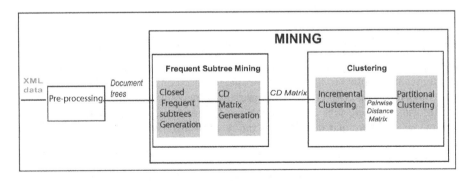

Fig. 1. The CFSPC Methodology

clustering. The frequent subtree mining determines the closed frequent subtrees from the *document trees* for a given support threshold. The closed frequent subtrees are condensed representations of the frequent subtrees. The distribution of the closed frequent subtrees in the corpus is modelled as a subtree-document matrix, $CD_{|CFS| \times |DT|}$, where *CFS* represents the closed frequent subtrees and *DT* represents the document trees in the given document tree collection. Each cell in the *CD* matrix represents the presence or absence of a given closed frequent subtree in the document tree. This matrix is used in calculating the similarity between documents.

As discussed earlier, the generation of distance matrix between each pair of documents is expensive for the INEX Wikipedia corpus due to its high dimension. Hence in the second phase of mining, the incremental clustering method is used to progressively cluster the documents in the corpus by comparing each document tree to the existing clusters. The similarity is measured by computing the Common SubTree coefficient (Ω) using the *CD* matrix based on the number of common closed frequent subtree between the document tree and existing clusters. Based on Ω, the document tree is grouped into an existing cluster with which it has the maximum Ω and greater than the user-defined cluster threshold otherwise the *document tree* is assigned to a new cluster.

As incremental clustering avoids the expensive pair-wise computation, it can cluster large data sets such as INEX 2007 Wikipedia dataset. However, this process results in undefined number of clusters according to the similarity measure used. In order to obtain the user-defined number of clusters, we utilize the pair-wise partitioning clustering algorithm [5]. The similarity between each pair of clusters is computed using Ω. Due to the reduced size of clusters, it is now computationally feasible to generate the pair-wise similarity matrix. This similarity matrix becomes the input to the partitional clustering algorithm. This algorithm generates the required number of clusters.

By combining the incremental and pair-wise clustering method, the CFSPC method is able to produce the clustering solution for the large data sets.

3   CFSPC Phase 1: Pre-processing

In the pre-processing phase, the XML document is decomposed into a tree structure with nodes representing only the tag names. The tag names are then mapped to unique integers for ease of computation. The semantic and syntactic meanings of tags are ignored. The Wikipedia documents conform to the same schema using the same tag set. Additionally previous research has shown that the semantic variations of tags do not provide any significant contribution in the clustering process [9, 10]. Other node information such as data types and constraints are also ignored.

There are several research works on clustering that use paths extracted from XML documents as a document representation and form the basis of calculating similarity between the documents[1, 10]. We have chosen to use the tree format to represent the XML documents. The tree format includes the sibling information of the nodes which is not included when an XML document is represented as a series of paths.

As shown in Fig. 2, the pre-processing of XML documents involves three subphases. They are namely:

1. Parsing
2. Representation
3. Duplicate branches removal.

3.1 Parsing

The XML data model is a graph structure comprising of atomic and complex objects. It can be modelled as a *tree*. Each XML document in INEX Wikipedia corpus is parsed and modelled as a rooted labeled ordered *document tree*. The document tree is *rooted* and *labeled* as there always exists a root node in the document tree and all the nodes are labeled using the tag names. The left-to-right *ordering* is preserved among the child nodes of a given parent in the document tree and therefore they are ordered.

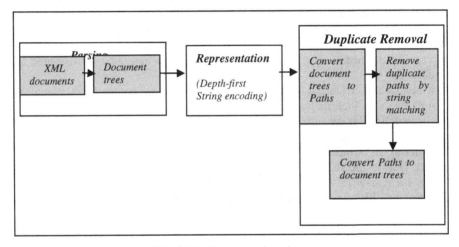

Fig. 2. The Pre-processing phase

3.2 Representation

The document trees need to be represented in a way that is suitable for mining in the next phase. A popular representation for trees, the depth-first string format[2], is used to represent the document trees. The *depth-first string encoding* traverses a tree in the *depth-first order*. It represents the *depth-first traversal* of a given document tree in a string like format where every node has a "–1" to represent backtracking and "#" to represent the end of the string encoding. For a document tree T with only one node r, the depth-first string of T is $S(T) = l,\#$ where l is the label of the root node r. For a document tree T with multiple nodes, where r is the root node and the children nodes of r are $r_1,...,r_k$ preserving left to right ordering, the depth-first string of T is $S(T)= l_r$ $l_{r_1}\text{-}1\ l_{r_2}\text{-}1...l_{r_k}\text{-}1\#$.

3.3 Duplicate Branches Removal

An analysis of the INEX Wikipedia dataset reveals that a large number of *document trees* contain duplicate branches. These duplicate branches are redundant information and hence they could cause additional overhead in the mining process. In order to remove the duplicate branches, the document tree is converted to a series of paths. The duplicate paths of the document trees are identified using string matching and removed. The remaining paths are combined together to create the document trees without having duplicate branches.

4 CFSPC Phase 2: Mining

The mining phase includes two phases namely incremental clustering and pair-wise clustering. We first explain the generation of closed frequent subtrees from document trees. We then explain the process of clustering with the use of closed frequent subtrees.

4.1 Incremental Clustering

Frequent Subtree mining is first applied on the XML documents to identify closed frequent subtrees for a given user-specified support threshold. Closed frequent subtrees are condensed representations of frequent subtrees without any information loss[7]. Frequent subtree mining on XML documents can be formally defined as follows:

Problem Definition for the Frequent Subtree Mining on XML Documents
Given a collection of XML documents $D = \{D_1, D_2, D_3,...,D_n\}$ modelled as document trees $DT = \{DT_1, DT_2, DT_3,...,DT_n\}$ where n represents the number of XML documents or document trees. There exists a subtree $DT' \subseteq DT_k$ that preserves the parent-child relationship among the nodes as that of the document tree DT_k.

Support(DT') (or frequency(DT')) is defined as the number of document trees in DT where DT' is a subtree. A subtree DT' is frequent if its support is not less than a user-defined minimum support threshold. In other words, DT' is a frequent subtree in the document trees in DT such that,

$$frequency\ (DT')/|DT| \geq min\_supp \tag{1}$$

where min_supp is the user-given support threshold and $|DT|$ is the number of document trees in the document tree dataset DT.

Due to the large number of frequent subtrees generated at lower support thresholds, recent researchers have focused on using condensed representation without any information loss [6]. The popular condensed representation is the closed frequent subtrees which is defined as follows.

Problem Definition for Closed Subtree
For a given document tree dataset, $DT = \{DT_1, DT_2, DT_3,...,DT_n\}$, if there exists two frequent subtrees DT' and DT'' ,the frequent subtree DT' is closed of DT'' iff for every $DT' \supseteq DT''$, $supp(DT') = supp(DT'')$ and there exists no superset for DT' having the same support as that of DT' . This property is called as *closure*.

In order to generate the closed frequent subtrees from the pre-processed document trees, the CMTreeMiner[2] is utilized. This algorithm adopts the apriori-based approach of generate-and-test to determine closed frequent subtrees. Having generated the closed frequent subtrees, their distribution in the corpus is modelled as a Boolean subtree-document matrix, $CD_{|CFS|\times|DT|}$, where *CFS* represents the closed frequent sub-trees and *DT* represents the document trees in the given document tree collection. Each cell in the *CD* matrix represents the presence or absence of a given closed frequent subtree {cfs_1, cfs_2,...,cfs_l} in the document tree {DT_1,DT_2,DT_3,...,DT_n}. Fig. 3 shows a $CD_{|CFS|\times|DT|}$ with closed frequent subtree {cfs_1, cfs_2, cfs_3} in the document trees $DT = \{DT_1, DT_2, DT_3, DT_4\}$.

This matrix is used to compute the similarity between the document trees for clustering. The column of *CD* matrix for each document tree is referred as Pre-cluster Form (PCF).

	DT_1	DT_2	DT_3	DT_4
cfs_1	1	0	1	1
cfs_2	0	1	0	1
cfs_3	1	1	1	0

Fig. 3. *CD* matrix

The computation of structural similarity between documents and clusters in the incremental clustering process is given below.

Structural Similarity Computation
Using *CD* matrix, we compute the structural similarity between

1. *two document trees*
2. *a tree and a cluster*

Tree-to-Tree Similarity
To begin with, there exists no cluster. Firstly, the two trees are used to compute the pair-wise similarity to form a cluster. It is measured by first finding the common closed frequent subtrees between the two document trees using the *CD* matrix.

Problem Definition for Tree-to-Tree Similarity
Let there be two document trees DT_x and DT_y and their pre-cluster forms (PCFs), d_x and d_y respectively in the given *CD* matrix. For a given *CD* matrix, let *CFS*= {cfs_1,..., cfs_n} be a set of closed frequent subtrees representing the rows and let *DT* = {DT_1,DT_2,DT_3,...,DT_n} be the document trees representing the columns, the PCF of a document tree DT_x is $d_x = \{x_1, x_2,..., x_n\}$ where $x_1...x_n \in \{0,1\}$ and $n = |CFS|$.

To compute the tree-to–tree similarity using the PCFs d_x and d_y in the *CD* matrix, the common closed frequent subtrees ($\zeta_i(d_x, d_y)$), between the two document trees DT_x and DT_y are computed for a given *i-th* closed frequent subtree using,

$$\zeta_i(d_x, d_y) = (d_x(i) \,\&\, d_y(i)=1) \,?\, 1 : 0 \tag{2}$$

Using the PCFs d_x and d_y in the CD matrix, the possible i-th closed frequent sub-trees ($\alpha_i(d_x, d_y)$) is calculated between the two document trees DT_x and DT_y using,

$$\alpha_i(d_x, d_y) = (d_x(i) \,|\, d_y(i)=1) \,?\, 1 : 0 \tag{3}$$

The degree of similarity ($\Omega_{dx,\,dy}$) between the two document trees using their PCFs, d_x and d_y is finally computed by combining the equations (2) and (3). The degree of similarity between the two document trees is the probability of the occurrence of a common closed frequent subtree in the possible closed frequent subtree space. It is defined as the ratio of sum of the common closed frequent subtrees over the total number of the possible closed frequent subtrees between a pair of document trees.

$$\Omega_{dx,\,dy} = \frac{\sum_{i=1}^{j} \zeta_i(d_x, d_y)}{\sum_{i=1}^{j} \alpha_i(d_x, d_y)} \quad \text{where } j = |CFS| \tag{4}$$

If the tree-to-tree similarity value ($\Omega_{dx,\,dy}$) between the PCFs, d_x and d_y of DT_x and DT_y respectively is higher than the user-defined minimum cluster threshold (μ), then, d_x and d_y are grouped into the same cluster otherwise they are assigned to two separate clusters. If they are grouped into the same cluster then the two PCFs are merged by union operation.

$$d_{clust}(i) = (d_x(i) \,|\, d_y(i)=1) \,?\, 1 : 0 \tag{5}$$

Tree to Cluster Similarity
Once a cluster is formed, the similarity between the incoming document tree and the existing cluster is computed using their PCFs given by d_x and d_{clust} respectively. It is computed using the Equation (2) given by

$$\zeta_i(d_x, d_{clust}) = (d_x(i) \,\&\, d_{clust}(i)=1) \,?\, 1 : 0 \tag{6}$$

Similar to Equation (3), the possible closed frequent subtrees between a document tree and a cluster is computed as follows,

$$\alpha_i(d_x, d_{clust}) = (d_x(i) \,|\, d_{clust}(i)=1) \,?\, 1 : 0 \tag{7}$$

Using equations (6) and (7), the degree of similarity between a document tree and a cluster is computed. The degree of similarity between the document tree and a cluster is the probability of the occurrence of a common closed frequent subtree in the possible closed frequent subtree space. It is defined as the ratio of the sum of common closed frequent subtrees over the total number of possible closed frequent subtrees between a document tree and its cluster.

$$\Omega_{dx, \text{clust}} = \sum_{i=1}^{j} \frac{\zeta_i(d_x, d_{clust})}{\alpha_i(d_x, d_{clust})} \quad \text{where } j = |CFS| \tag{8}$$

If the tree-to-cluster similarity value ($\Omega_{dx, \text{clust}}$) between PCFs d_x and d_{clust} of DT_x and DT_{clust} is higher than the user-defined minimum cluster threshold (μ), then, d_x and d_{clust} are grouped into the DT_{clust} cluster otherwise d_x is assigned to a separate cluster. In situations where d_x is grouped into the DT_{clust} cluster then the two clusters are merged by union operation.

$$d_{clust}(i) = (d_x(i)| d_{clust}(i)=1) ? 1 : 0 \tag{9}$$

CFSPC is a progressive clustering algorithm. The clusters are formed in an incremental fashion. The process starts without any cluster. When a new tree arrives, it is assigned to a new cluster. A cluster is represented as the PCF of the document tree if it has a single member. A cluster with multiple member document trees is represented by the union of their PCFs. When the next tree arrives, the similarity between the current document tree and the document tree in the cluster is computed using the tree to tree similarity method. If the similarity value is greater than the user-defined cluster threshold (μ) then the incoming document tree is grouped into the cluster otherwise it is assigned to a new cluster. If there exists new PCF information with respect to the closed frequent subtrees in the recently clustered document tree, then the additional information is merged with the clustering information.

The incremental clustering results in a large number of clusters. This is due to allowing the documents to form a separate cluster when an appropriate cluster is not found for them. In order to control the number of clusters, the clusters are further merged using pair-wise clustering.

4.2 Partitional Clustering

A similarity matrix is generated by computing the degree of similarity between each pair of PCFs representing the clusters using the following equations,

$$\alpha_i(d_{clust_1}, d_{clust_2}) = (d_{clust_1}(i)| d_{clust_2}(i) = 1) ? 1 : 0 \tag{10}$$

$$\Omega_{clust_1, clust_2} = \sum_{i=1}^{j} \frac{\zeta_i(d_{clust_1}, d_{clust_2})}{\alpha_i(d_{clust_1}, d_{clust_2})} \quad \text{where } j = |CFS| \tag{11}$$

where $\Omega_{clust_1, clust_2}$ is the cluster-to-cluster similarity value. The similarity matrix is fed to a partitional clustering algorithm such as the k-way clustering solution[5]. The k-way clustering algorithm groups the documents to the required number of clusters. The k-way clustering solution computes cluster by performing a sequence of $k-1$ repeated bisections. In this approach, the matrix is first clustered into two groups, and then one of these groups is chosen and bisected further. This process of bisection continues until the desired number of bisections is reached. During each step of bisection, the cluster is bisected so that the resulting 2-way clustering solution locally optimizes a particular criterion function [5].

5 Experiments and Discussion

We implemented the CFSPC algorithm using Microsoft Visual C++ 2005 and con-ducted experiments on the Wikipedia corpus from the INEX XML Mining Challenge 2007. The required numbers of clusters for INEX result submission were 21 and 10 clusters. The incremental clustering technique for a given clustering threshold often generates a large number of clusters. Hence, the k-way clustering algorithm option in CLUTO[5] is used to group the intermediate clusters to the required number of clus-ters (21 and 10 clusters).

We submitted 2 results, one with 21 clusters and the other with 10 clusters using the cluster threshold of 0.4. The following table summarizes the results based on Mi-cro F1 and Macro F1 measure evaluation metrics for 10 and 21 clusters with the clus-tering threshold of 0.4.

Table 1. Submitted clustering results for INEX Wikipedia XML Mining Track 2007

Clustering Threshold	Number of Clusters using incremental clustering	Number of Clus-ters	Micro F1	Macro F1
0.4	2396	21	0.251	0.251
		10	0.251	0.250

We conducted several more experiments with varying support threshold and clus-tering threshold. The experimental results for varying clustering threshold are shown in Table 2.

Table 2. Results from INEX Wikipedia XML Mining Track 2007 with varying clustering threshold

Clustering Threshold	Number of Clusters using incremental clustering	Number of Clusters	Micro F1	Macro F1
0.5	3735	21	0.252	0.248
		10	0.251	0.249
0.3	1682	21	0.253	0.249
		10	0.251	0.247
0.2	1217	21	0.252	0.261
		10	0.251	0.249
0.1	857	21	0.251	0.258
		10	0.251	0.263

As indicated in the Tables 1 and 2, the number of clusters using incremental clustering increases with the clustering threshold. The partitional clustering could provide the required number of clusters. It can be seen from the Table 2 that there is not much improvement in the Micro F1 average; however, there is an improvement for Macro F1 average for lower clustering threshold. The results on the Wikipedia dataset clearly indicates that there is not any significant improvement in performance for varying clustering threshold using structural only information in clustering.

To analyse whether the number of closed frequent subtrees is an influential factor in final clustering results, experiments are conducted with the higher support threshold than the previous set of experiments. We ran the experiments with varying clustering thresholds setting the 10% support threshold to generate the frequent trees.

Table 3. Results from INEX Wikipedia XML Mining Track 2007 for 10% Support threshold and various clustering threshold

Support Threshold	No. of Closed Frequent Subtrees	Clustering Threshold	No. of Clusters using Inc. Clustering	No. of Clusters from Part. clustering	Micro average (F1)	Macro average (F1)
10%	387	0.4	1118	21	0.253	0.269
				10	0.252	0.245
		0.5	1633	21	0.253	0.256
				10	0.251	0.247
		0.6	2510	21	0.252	0.248
				10	0.251	0.243

Also, we wanted to analyse whether the number of clusters plays a significant role. The above Table 3 summarizes the results on various numbers of clusters at 0.5 clustering threshold with 10% support threshold. The results from Table 3 show that the clustering performance does not vary much with the change of various parameters.

Table 4. Comparison of our approach against other structure-only approaches on INEX Wikipedia dataset

Approaches	Number of clusters	Micro F1	Macro F1
Hagenbuchner et.al[4]	10	0.251	0.257
	21	0.264	0.269
Hagenbuchner[4]	10	0.252	0.267
	21	0.258	0.252
Tien et. Al[10]	10	0.251	0.252
	21	0.251	0.253
Our approach	10	0.251	0.263
	21	0.253	0.269

Table 4 lists the comparison between our approach and other approaches using structure-only on INEX 2007 Wikipedia dataset. There were two other participants using structure-only and their results are presented in Table 4. It is evident from Table 4 that there is no significant difference between our approach and other approaches using only the structure of XML documents. Based on our experiments and the comparison with other approaches[4, 11] using structure-only in the INEX 2007 Document Mining challenge, it can be concluded that clustering using structural similarity between documents is not suitable for the INEX 2007 Wikipedia data set. As the INEX 2007 Wikipedia dataset is a homogeneous collection with most of the documents having only one schema and hence the structure of the XML document plays a less important role than the content.

6   Conclusions and Future Direction

In this paper, we have proposed and presented the results of our progressive clustering algorithm for mining only the structure of XML documents in INEX 2007 Wikipedia dataset. The main aim of this study is to explore and understand the importance of structure of the XML documents over the content of XML for clustering task. In order to cluster the XML documents, we have used a frequent subtree – document matrix generated from closed frequent subtrees. Using the matrix, we have computed the similarity between XML documents and incrementally clustered them based on their similarity values. From the experimental results, it is evident that the structure plays a minor role in determining the similarity between the INEX documents.

This is the first study conducted on INEX dataset using common subtrees and hence in the future, we will aim in devising efficient similarity computation techniques to effectively cluster the XML documents. Also, as a future work, we will be focusing on including the content of XML documents to provide more meaningful cluster.

References

1. Aggarwal, C.C., et al.: Xproj: a framework for projected structural clustering of xml documents. In: Proceedings of the 13th ACM SIGKDD international conference on Knowledge discovery and data mining, pp. 46–55. ACM, San Jose (2007)
2. Chi, Y., et al.: Frequent Subtree Mining- An Overview. In: Fundamenta Informaticae, pp. 161–198. IOS Press, Amsterdam (2005)
3. Dalamagas, T., et al.: A methodology for clustering XML documents by structure. Inf. Syst. 31(3), 187–228 (2006)
4. Hagenbuchner, M., et al.: Efficient clustering of structured documents using Graph Self-Organizing Maps. In: Pre-proceedings of the Sixth Workshop of Initiative for the Evaluation of XML Retrieval, Dagstuhl, Germany (2007)
5. Karypis, G.: CLUTO - Software for Clustering High-Dimensional Datasets Karypis Lab, May 25 (2007)
6. Kutty, S., Nayak, R., Li, Y.: PCITMiner- Prefix-based Closed Induced Tree Miner for finding closed induced frequent subtrees. In: Sixth Australasian Data Mining Conference (AusDM 2007), ACS, Gold Coast (2007)

7. Kutty, S., Nayak, R., Li, Y.: XML Data Mining: Process and Applications. In: Song, M., Wu, Y.-F. (eds.) Handbook of Research on Text and Web Mining Technologies. Idea Group Inc., USA (2008)
8. Nayak, R., Witt, R., Tonev, A.: Data Mining and XML Documents. In: International Conference on Internet Computing (2002)
9. Nayak, R.: Investigating Semantic Measures in XML Clustering. In: Proceedings of the 2006 IEEE/WIC/ACM International Conference on Web Intelligence, pp. 1042–1045. IEEE Computer Society Press, Los Alamitos (2006)
10. Tran, T., Nayak, R.: Evaluating the Performance of XML Document Clustering by Structure Only. In: Comparative Evaluation of XML Information Retrieval Systems, pp. 473–484 (2007)
11. Tran, T., Nayak, R.: Document Clustering using Incremental and Pairwise Approaches. In: Pre-proceedings of the Sixth Workshop of Initiative for the Evaluation of XML Retrieval, Dagstuhl, Germany (2007)
12. Xing, G., Xia, Z., Guo, J.: Clustering XML Documents Based on Structural Similarity. In: Advances in Databases: Concepts, Systems and Applications, pp. 905–911 (2007)

Probabilistic Methods for Structured Document Classification at INEX'07

Luis M. de Campos, Juan M. Fernández-Luna, Juan F. Huete,
and Alfonso E. Romero

Departamento de Ciencias de la Computación e Inteligencia Artificial
E.T.S.I. Informática y de Telecomunicación, Universidad de Granada,
18071 – Granada, Spain
{lci,jmfluna,jhg,aeromero}@decsai.ugr.es

Abstract. This paper exposes the results of our participation in the Document Mining track at INEX'07. We have focused on the task of classification of XML documents. Our approach to deal with structured document representations uses classification methods for plain text, applied to flattened versions of the documents, where some of their structural properties have been translated to plain text. We have explored several options to convert structured documents into flat documents, in combination with two probabilistic methods for text categorization. The main conclusion of our experiments is that taking advantage of document structure to improve classification results is a difficult task.

1 Introduction

This is the first year that members of the research group "Uncertainty Treatment in Artificial Intelligence" at the University of Granada submit runs to the Document Mining track of INEX. As we had previous experience in automatic classification, particularly in learning Bayesian network classifiers [1,3], we have limited our participation only to the task of text categorization.

The proposed methodology does not use text classification algorithms specifically designed to manage and exploit structured document representations. Instead, we use algorithms that apply to flat documents and do not take structure into consideration at all. What we want to test is whether these methods can be used, in combination with some simple techniques to transform document structure into a modified flat document representation having additional characteristics (new or transformed features, different frequencies,...), in order to improve the classification results obtained by the same methods but using purely flat document representations.

The rest of the paper is organized in the following way: in Section 2 we describe the probabilistic flat text classifiers we shall use. Section 3 gives details of the different approaches to map structured documents into flat ones. Section 4 is focused on the experimental results. Finally, Section 5 contains the concluding remarks and some proposals for future work.

N. Fuhr et al. (Eds.): INEX 2007, LNCS 4862, pp. 195–206, 2008.
© Springer-Verlag Berlin Heidelberg 2008

2   Probabilistic Methods for Flat Text Classification

In this section we are going to explain the two methods for non-structured (flat) text classification that we are going to use in combination with several methods for managing structured documents. One of them is the well-known naive Bayes classifier, whereas the other is a new method, based on a restricted type of Bayesian network.

The classical probabilistic approach to text classification may be stated as follows: We have a class variable C taking values in the set $\{c_1, c_2, \ldots, c_n\}$ and, given a document d_j to be classified, the posterior probability of each class, $p(c_i|d_j)$, is computed according to the Bayes formula:

$$p(c_i|d_j) = \frac{p(c_i)p(d_j|c_i)}{p(d_j)} \propto p(c_i)p(d_j|c_i) \qquad (1)$$

and the document is assigned to the class having the greatest posterior probability, i.e.

$$c^*(d_j) = \arg\max_{c_i}\{p(c_i)p(d_j|c_i)\}$$

Then the problem is how to estimate the probabilities $p(c_i)$ and $p(d_j|c_i)$.

2.1   The Naive Bayes Classifier

The naive Bayes classifier is the simplest probabilistic classification model that, despite its strong and often unrealistic assumptions, performs frequently surprisingly well. It assumes that all the attribute variables are conditionally independent of each other given the class variable. In fact, the naive Bayes classifier can be considered as a Bayesian network-based classifier, where the network structure contains only arcs from the class variable to the attribute variables, as shown in Figure 1. In the context of text classification, there exist two different models called naive Bayes, the multivariate Bernouilli naive Bayes model [4,5,9] and the multinomial naive Bayes model [6,7]. In this paper we are going to use the multinomial model.

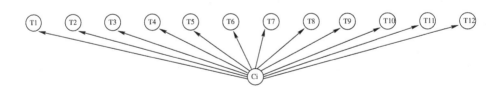

Fig. 1. The naive Bayes classifier

In this model a document is an ordered sequence of words or terms drawn from the same vocabulary, and the naive Bayes assumption here means that the occurrences of the terms in a document are conditionally independent given the class, and the *positions* of these terms in the document are also independent given the class. Thus, each document d_j is drawn from a multinomial distribution of words with as many independent trials as the length of d_j. Then,

$$p(d_j|c_i) = p(|d_j|) \frac{|d_j|!}{\prod_{t_k \in d_j} n_{jk}!} \prod_{t_k \in d_j} p(t_k|c_i)^{n_{jk}} \tag{2}$$

where t_k are the distinct words in d_j, n_{jk} is the number of times the word t_k appears in the document d_j and $|d_j| = \sum_{t_k \in d_j} n_{jk}$ is the number of words in d_j. As $p(|d_j|) \frac{|d_j|!}{\prod_{t_k \in d_j} n_{jk}!}$ does not depend on the class, we can omit it from the computations, so that we only need to calculate

$$p(d_j|c_i) \propto \prod_{t_k \in d_j} p(t_k|c_i)^{n_{jk}} \tag{3}$$

The estimation of the term probabilities given the class, $p(t_k|c_i)$, is usually carried out by means of the Laplace estimation:

$$p(t_k|c_i) = \frac{N_{ik} + 1}{N_{i\bullet} + M} \tag{4}$$

where N_{ik} is the number of times the term t_k appears in documents of class c_i, $N_{i\bullet}$ is the total number of words in documents of class c_i and M is the size of the vocabulary (i.e. the number of distinct words in the documents of the training set).

The estimation of the prior probabilities of the classes, $p(c_i)$, is usually done by maximum likelihood, i.e.:

$$p(c_i) = \frac{N_{i,doc}}{N_{doc}} \tag{5}$$

where N_{doc} is the number of documents in the training set and $N_{i,doc}$ is the number of documents in the training set which are assigned to class c_i.

In our case we have used this multinomial naive Bayes model but, instead of considering only one class variable C having n values, we decompose the problem using n binary class variables C_i taking their values in the sets $\{c_i, \bar{c}_i\}$. This is a quite common transformation in text classification [10], especially for multilabel problems, where a document may be associated to several classes. In this case we build n naive Bayes classifiers, each one giving a posterior probability $p_i(c_i|d_j)$ for each document. As in the Wikipedia XML Corpus each document may be assigned to only one class, we select the class $c^*(d_j)$ such that $c^*(d_j) = \arg\max_{c_i} \{p_i(c_i|d_j)\}$. Notice that in this case, as the term $p_i(d_j)$ in the expression $p_i(c_i|d_j) = p_i(d_j|c_i)p_i(c_i)/p_i(d_j)$ is not necessarily the same for all the class values, we need to compute it explicitly through

$$p_i(d_j) = p_i(d_j|c_i)p_i(c_i) + p_i(d_j|\bar{c}_i)(1 - p_i(c_i))$$

This means that we have also to compute $p_i(d_j|\bar{c}_i)$. This value is estimated using the corresponding counterparts of eqs. (3) and (4), where

$$p(t_k|\bar{c}_i) = \frac{N_{\bullet k} - N_{ik} + 1}{N - N_{i\bullet} + M} \qquad (6)$$

$N_{\bullet k}$ is the numbers of times that the term t_k appears in the training documents and N is the total number of words in the training documents.

2.2 The OR Gate Bayesian Network Classifier

The second classification method for flat documents that we are going to use is based on a Bayesian network with the following topology: Each term t_k appearing in the training documents (or a subset of these terms in the case of using some method for feature selection) is associated to a binary variable T_k taking its values in the set $\{t_k, \bar{t}_k\}$, which in turn is represented in the network by the corresponding node. There are also n binary variables C_i taking their values in the sets $\{c_i, \bar{c}_i\}$ (as in the previous binary version of the naive Bayes model) and the corresponding class nodes. The network structure is fixed, having an arc going from each term node T_k to the class node C_i if the term t_k appears in training documents which are of class c_i. In this way we have a network topology with two layers, where the term nodes are the "causes" and the class nodes are the "effects". An example of this network topology is displayed in Figure 2.

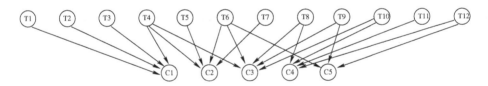

Fig. 2. The OR gate classifier

The quantitative information associated to this network are the conditional probabilities $p(C_i|pa(C_i))$, where $Pa(C_i)$ is the set of parents of node C_i in the network (i.e. the set of terms appearing in documents of class c_i) and $pa(C_i)$ is any configuration of the parent set (any assignment of values to the variables in this set). As the number of configurations is exponential with the size of the parent set, we use a canonical model to define these probabilities, which reduce the number of required numerical values from exponential to linear size. More precisely, we use a noisy OR Gate model [8].

The conditional probabilities in a noisy OR gate are defined in the following way:

$$p(c_i|pa(C_i)) = 1 - \prod_{T_k \in R(pa(C_i))} (1 - w(T_k, C_i)) \, , \, p(\bar{c}_i|pa(C_i)) = 1 - p(c_i|pa(C_i)) . \quad (7)$$

where $R(pa(C_i)) = \{T_k \in Pa(C_i) \,|\, t_k \in pa(C_i)\}$, i.e. $R(pa(C_i))$ is the subset of parents of C_i which are instantiated to its t_k value in the configuration $pa(C_i)$. $w(T_k, C_i)$ is a weight representing the probability that the occurrence of the "cause" T_k alone (T_k being instantiated to t_k and all the other parents T_h instantiated to $\bar{t}_h$) makes the "effect" true (i.e., forces class c_i to occur).

Once the weights $w(T_k, C_i)$ have been estimated, and given a document d_j to be classified, we instantiate in the network each of the variables T_k corresponding to the terms appearing in d_j to the value t_k (i.e. $p(t_k|d_j) = 1$ if $t_k \in d_j$), and all the other variables T_h (those associated to terms that do not appear in d_j) to the value $\bar{t}_h$ (i.e. $p(t_h|d_j) = 0 \;\forall t_h \notin d_j$). Then, we compute for each class node C_i the posterior probabilities $p(c_i|d_j)$. As in the case of the naive Bayes model, we assign to d_j the class having the greatest posterior probability.

The combination of network topology and numerical values represented by OR gates allows us to compute very efficiently and in an exact way the posterior probabilities:

$$p(c_i|d_j) = 1 - \prod_{T_k \in Pa(C_i)} (1 - w(T_k, C_i) \times p(t_k|d_j)) = 1 - \prod_{T_k \in Pa(C_i) \cap d_j} (1 - w(T_k, C_i))$$

(8)

In order to take into account the number of times a word t_k occurs in a document d_j, n_{jk}, we replicate each node T_k n_{jk} times, so that the posterior probabilities then become

$$p(c_i|d_j) = 1 - \prod_{T_k \in Pa(C_i) \cap d_j} (1 - w(T_k, C_i))^{n_{jk}} \,.$$

(9)

The estimation of the weights in the OR gates, $w(T_k, C_i)$, can be done in several ways. The simplest one is to estimate $w(T_k, C_i)$ as $p(c_i|t_k)$, the conditional probability of class c_i given that the term t_k is present. We can do it by maximum likelihood:

$$w(T_k, C_i) = \frac{N_{ik}}{N_{\bullet k}}$$

(10)

Another, more accurate way of estimating $w(T_k, C_i)$ is directly as $p(c_i|t_k, \bar{t}_h \,\forall T_h \in Pa(C_i), T_h \neq T_k)$. However, this probability cannot be reliably estimated, so that we are going to compute an approximation in the following way[1]:

$$p(c_i|t_k, \bar{t}_h \,\forall h \neq k) \approx p(c_i|t_k) \prod_{h \neq k} \frac{p(c_i|\bar{t}_h)}{p(c_i)}$$

(11)

The values of $p(c_i|t_k)$ and $p(c_i|\bar{t}_h)/p(c_i)$ in eq. (12) are also estimated using maximum likelihood. Then, the weights $w(T_k, C_i)$ are in this case:

$$w(T_k, C_i) = \frac{N_{ik}}{N_{\bullet k}} \times \prod_{h \neq k} \frac{(N_{i\bullet} - N_{ih})N}{(N - N_{\bullet h})N_{i\bullet}}$$

(12)

---

[1] This approximation results from assuming a conditional independence statement similar to that of the naive Bayes classifier, namely $p(t_k, \bar{t}_h \,\forall h \neq k|c_i) \approx p(t_k|c_i) \prod_{h \neq k} p(\bar{t}_h|c_i)$.

3   Document Representation

In this section we deal with the problem of document representation. As we have seen before, we are using flat-document classifiers for this track, so we need methods to translate structural properties to plain text document.

Because these methods are independent of the classifier used, it is possible to make all possible combinations of classifiers and transformation methods, wich gives us a large amount of categorization procedures.

We shall use the small XML document (the beginning of "Don Quijote") displayed in Figure 3 to illustrate the proposed transformations. Next we explain the different approaches to map structural documents into flat ones.

```
<book>
 <title>El ingenioso hidalgo Don Quijote de la Mancha</title>
 <author>Miguel de Cervantes Saavedra</author>
 <contents>
   <chapter>Uno</chapter>
    <text>En un lugar de La Mancha de cuyo nombre no quiero
    acordarme...</text>
 </contents>
</book>
```

Fig. 3. "Don Quijote", XML fragment used to illustrate the different transformations

3.1   Method 1: "Only Text"

This is the naive approach. It consists in removing all the structural marks from the XML file, obtaining a plain text file. Used with the previous example, we obtain the document displayed in Figure 4:

```
El ingenioso hidalgo Don Quijote de la Mancha Miguel de Cervantes
Saavedra Uno En un lugar de La Mancha de cuyo nombre no quiero
acordarme...
```

Fig. 4. "Don Quijote", with the "only text" approach

This method should be taken as a *baseline*, as we are losing all the structural information. We would like to improve its classification accuracy by using more advanced representations.

3.2   Method 2: "Adding"

This method adds structural features to the document, different from the textual features. That is to say, structural marks are introduced into the document as if they were "additional terms". We can consider structural marks in an atomic way, or in the context of the other marks where they are contained (i.e. using part of the path from the mark being considered to the root element, until a certain depth level). Using the previous example, the `text` mark can be considered standalone ("adding_1", with depth = 1), `contents_text` ("adding_2", depth = 2) or `book_contents_text` ("adding_0", maximum depth value, the complete path to the root mark).

We show in Figure 5 the transformed flat document of the example document using "adding" with depth = 2. Leading underscores are used to distinguish between textual terms and terms representing structural marks:

```
_book _book_title El ingenioso hidalgo Don Quijote de la Mancha
_book_author Miguel de Cervantes Saavedra
_book_contents _contents_chapter Uno _contents_text En un lugar
de La Mancha de cuyo nombre no quiero acordarme...
```

Fig. 5. "Don Quijote", with the "adding_2" approach

3.3 Method 3: "Tagging"

This approach is the same as the one described in [2], also named "tagging". It considers that two appearances of a term are different if it appears inside two different structural marks. To modelize this, terms are "tagged" with a representation of the structural mark they appear in. This can be easily simulated prepending a preffix to the term, representing its container. We can also experiment at different depth levels, as we did in the method "adding".

Data preprocesed with this method can be very sparse, and very large lexicon could be built from medium sized collections. For our example document this method, with depth = 1, obtains the flat document displayed in Figure 6.

```
title_El title_ingenioso title_hidalgo title_Don title_Quijote
title_de title_la title_Mancha author_Miguel author_de
author_Cervantes author_Saavedra chapter_Uno text_En text_un
text_lugar text_de text_La text_Mancha text_de text_cuyo
text_nombre text_no text_quiero text_acordarme...
```

Fig. 6. "Don Quijote", with the "tagging_1" approach

3.4 Method 4: "No Text"

This method tries to unveil the categorization power using only structural units, processed in the same way as in the "adding" method. Roughly speaking, it is equivalent to "adding" and then removing textual terms. In Figure 7 we can see the "notext_0" processing of the previous example.

```
_book _book_title _book_author _book_contents
_book_contents_chapter _book_contents_text
```

Fig. 7. "Don Quijote", with the "notext_0" approach

3.5   Method 5: "Text Replication"

The previous methods deal with a structured collection, having no previous knowledge about it. That is to say, they do not take into account the kind of mark, in order to select one action or another. This approach assigns an integer value to each mark, proportional to its informative content for categorization (the higher the value, the more informative). This value is used to *replicate* terms, multiplying their frequencies in a mark by that factor. Notice that only values for structural marks directly containing terms must be supplied.

In the previous example, suppose we assign the following set of replication values:

```
title: 1, author: 0, chapter: 0; text: 2
```

Notice that a value equal to 0 indicates that the terms in that mark will be removed. The resulting text is displayed in Figure 8.

```
El ingenioso hidalgo Don Quijote de la Mancha En En un un lugar
lugar de de La La Mancha Mancha de de cuyo cuyo nombre nombre no
no quiero quiero acordarme acordarme...
```

Fig. 8. "Don Quijote", with the "replication" approach, using values proposed before

This method is very flexible, and it generalizes several ones, as the "only text" approach (one may select 1 for all the replication values). The method consisting of just selecting text from certain marks can be simulated here using 1 and 0 replication values if the text within a given mark is to be considered or not, respectively.

The main drawback of "text replication" is that we need some experience with the collection, in order to build the table of replication values before processing the files.

4   Experimentation

Previous to the production runs, and in order to select the best combinations of classifiers and representations, we have carried out some experiments using only the training set, by means of cross-validation (dividing the training set into 5 parts). The selected evaluation measures are the microaverage and macroaverage breakeven point (BEP) (for *soft categorization*) and microaverage and macroaverage F1 (for *hard categorization*) [10]. In every case, the "only text" representation will be used as a baseline to compare results among different alternatives.

Table 1 displays the replication values used in the experiments with the "text replication" approach, for the different tags. Tags with unspecified replication values are always set to 1.

We have also carried out experiments with some feature/term selection methods. For the naive Bayes model we used a simple method that removes all the terms that appear in less that a specified number of documents. For the OR gate model we used a local selection method (different terms may be selected for different class values) based on computing the mutual information measure between each term and each class variable C_i.

Table 1. Replication values used in the experiments

Tag	id=2	id=3	id=4	id=5	id=8	id=11
conversionwarning	0	0	0	0	0	0
emph2	2	3	4	5	10	30
emph3	2	3	4	5	10	30
name	2	3	4	5	20	100
title	2	3	4	5	20	50
caption	2	3	4	5	10	10
collectionlink	2	3	4	5	10	10
languagelink	0	0	0	0	0	0
template	0	0	0	0	0	0

The results of this preliminary experimentation are displayed in Table 2. In this table, "OR Gate (ML)" means the OR gate classifier using eq. (10); "OR Gate (AP)" is the OR gate classifier using eq. (12); "$\geq i$ docs." means using term selection, where only terms that appear in more than or equal to i documents are selected; "MI" means local term selection using mutual information.

The best classifier for the four performance measures is the OR Gate classifier using the weights in eq. (12); it gets the best results with the "only text" approach, together with a very light term selection method. The simpler version of this OR Gate classifier (the one using maximum likelihood) obtains quite poor results, except if we use a much more aggresive term selection method based on mutual information.

It is a clear fact that the "replication" approach helps the naive Bayes classifier. One of the main drawbacks of this classifier are the generally bad results obtained in macro measures (this is probably due to the nature of the classifier, that benefits the classes with higher number of training examples). This drawback can be alleviated by using a replication approach with moderate replication values.

On the other hand, the "adding" and "tagging" methods do not seem to give good results in combination with any of these probabilistic classifiers. The runs with the "notext" approach were also really disappointing and they are not listed here.

4.1 Official Runs

Finally, we decided to submit to the Document Mining track the five runs described in Table 3. The evaluation measures of the official runs are the microaverage and macroaverage recall (which coincide in this case with the microaverage

Table 2. Results of the preliminary experimentation with the training set using 5-fold cross-validation

Classifier	Representation	Term Selec.	micro BEP	macro BEP	micro F1	macro F1
Naïve Bayes	Only text	None	0.76160	0.58608	0.78139	0.64324
Naïve Bayes	Only text	≥ 2 docs.	0.72269	0.67379	0.77576	0.69309
Naïve Bayes	Only text	≥ 3 docs.	0.69753	0.67467	0.76191	0.68856
Naïve Bayes	Adding_1	None	0.75829	0.56165	0.76668	0.58591
Naïve Bayes	Adding_1	≥ 3 docs.	0.68505	0.66215	0.74650	0.65390
Naïve Bayes	Adding_2	None	0.73885	0.55134	0.74413	0.54971
Naïve Bayes	Adding_2	≥ 3 docs.	0.66851	0.62747	0.71242	0.59286
Naïve Bayes	Adding_3	None	0.71756	0.53322	0.72571	0.51125
Naïve Bayes	Adding_3	≥ 3 docs.	0.64985	0.59896	0.68079	0.53859
Naïve Bayes	Tagging_1	None	0.72745	0.49530	0.72999	0.50925
Naïve Bayes	Tagging_1	≥ 3 docs.	0.65519	0.60254	0.71755	0.60594
Naïve Bayes	Replic. (id=2)	None	0.76005	0.64491	0.78233	0.66635
Naïve Bayes	Replic. (id=2)	≥ 2 docs.	0.71270	0.68386	0.61321	0.73780
Naïve Bayes	Replic. (id=2)	≥ 3 docs.	0.70916	0.68793	0.73270	0.65697
Naïve Bayes	Replic. (id=3)	None	0.75809	0.67327	0.77622	0.67101
Naïve Bayes	Replic. (id=4)	None	0.75921	0.69176	0.76968	0.67013
Naïve Bayes	Replic. (id=5)	None	0.75976	0.70045	0.76216	0.66412
Naïve Bayes	Replic. (id=8)	None	0.74406	0.69865	0.72728	0.61602
Naïve Bayes	Replic. (id=11)	None	0.72722	0.67965	0.71422	0.60451
OR Gate (ML)	Only text	None	0.37784	0.38222	0.59111	0.37818
OR Gate (ML)	Only text	MI	0.74014	0.72816	0.74003	0.68430
OR Gate (AP)	Only text	None	0.79160	0.76946	0.79160	0.74922
OR Gate (AP)	Only text	≥ 3 docs.	0.77916	0.78025	0.77916	0.73544
OR Gate (AP)	**Only text**	**≥ 2 docs.**	**0.79253**	**0.78135**	**0.79253**	**0.75300**
OR Gate (ML)	Adding_1	None	0.40503	0.43058	0.58777	0.39361
OR Gate (ML)	Adding_1	≥ 3 docs.	0.39141	0.41191	0.57809	0.36936
OR Gate (ML)	Adding_1	MI	0.69944	0.72460	0.69943	0.58835
OR Gate (ML)	Adding_2	None	0.40573	0.43335	0.58908	0.39841
OR Gate (ML)	Adding_2	≥ 3 docs.	0.39204	0.41490	0.57951	0.37346
OR Gate (ML)	Adding_2	MI	0.65642	0.70755	0.65642	0.52611
OR Gate (ML)	Notext_2	None	0.40507	0.42914	0.48818	0.38736
OR Gate (ML)	Tagging_1	None	0.37859	0.40726	0.57274	0.35418
OR Gate (ML)	Tagging_1	≥ 3 docs.	0.36871	0.38475	0.56030	0.32546
OR Gate (ML)	Tagging_1	MI	0.59754	0.67800	0.59754	0.39141
OR Gate (AP)	Tagging_1	None	0.73784	0.74066	0.73789	0.70121
OR Gate (ML)	Replic. (id=2)	MI	0.74434	0.73908	0.74432	0.66995
OR Gate (AP)	Replic. (id=2)	None	0.78042	0.76158	0.78042	0.73768
OR Gate (ML)	Replic. (id=3)	MI	0.74612	0.74275	0.74608	0.67249
OR Gate (AP)	Replic. (id=3)	None	0.78127	0.76095	0.78127	0.73756
OR Gate (ML)	Replic. (id=4)	MI	0.74815	0.74623	0.74813	0.67357
OR Gate (AP)	Replic. (id=4)	None	0.78059	0.75971	0.78059	0.73511
OR Gate (ML)	Replic. (id=5)	MI	0.74918	0.74643	0.74916	0.67498
OR Gate (AP)	Replic. (id=5)	None	0.77977	0.75833	0.77978	0.73245
OR Gate (ML)	Replic. (id=8)	MI	0.75059	0.75254	0.75059	0.66702
OR Gate (AP)	Replic. (id=8)	None	0.77270	0.74943	0.77270	0.72186
OR Gate (ML)	Replic. (id=11)	MI	0.72656	0.71326	0.72656	0.64101
OR Gate (AP)	Replic. (id=11)	None	0.73041	0.70260	0.73041	0.66733

and macroaverage F1 mesaures, because only one class may be assigned to each document), whose values are also displayed in Table 3.

Notice that the relative ordering among these classifiers is the same than in the previous table (OR Gate AP > NB > OR Gate ML), and the final evaluation measures are close to the previously presented estimators.

Table 3. Submitted runs

Classifier	Representation	Term Selec.	micro Recall	macro Recall
Naïve Bayes	Only text	None	0.77630	0.58536
Naïve Bayes	Replic. (id=2)	None	0.78107	0.63730
OR Gate (ML)	Replic. (id=5)	MI	0.75354	0.61298
OR Gate (ML)	Replic. (id=8)	MI	0.75097	0.61973
OR Gate (AP)	**Only text**	**≥ 2**	**0.78998**	**0.76054**

5   Concluding Remarks

Our participation in the XML Document Mining track of the INEX 2007 Workshop has been described in this work. This is the first year that we apply for this track but, despite the low number of participants in the Categorization approach, our participation was remarkable. The main relevant results presented here are the following:

- We have described a new approach for flat document classification, the so called "OR Gate classifier", with two different variants: ML estimation, and a more accurate approximation of the required conditional probabilities.
- We have shown different methods of representing structured documents as plain text ones. We must also recall that some of them, particularly the replication method, are new.
- According to the results, we found that we could improve categorization of structured documents using a multinomial naive Bayes classifier, which is widely known and is included in almost every text-mining software package, in combination with the replication method.

On the other hand, the present paper raises the following questions, which can be stated as future lines of research:

- How are the results of our models compared with a SVM (Support Vector Machine) using only the text of the documents?
- Can the naive Bayes classifier be improved more using a more sophisticated feature selection method?
- Having in mind that the replication approach is the one that has given the best results, what are the optimum replication parameters that can be used in Wikipedia? In other words, what marks are more informative and how much?

- Is there a way to make a representation of the structure of documents that could be used to improve the results of the OR Gate classifier (specially in its more promising AP version)?
- Do the "adding", "tagging" and "no text" approaches help other categorization methods, like, for instance, Rocchio or SVMs?

Managing structure in this problem has been revealed as a difficult task. Besides, it is not really clear if the structure can make a good improvement of categorization results. So, we hope to start answering the previous questions in future editions of this track.

Acknowledgments. This work has been jointly supported by the Spanish Consejería de Innovación, Ciencia y Empresa de la Junta de Andalucía, Ministerio de Educación y Ciencia and the research programme Consolider Ingenio 2010, under projects TIC-276, TIN2005-02516 and CSD2007-00018, respectively.

References

1. Acid, S., de Campos, L.M., Castellano, J.G.: Learning Bayesian network classifiers: searching in a space of acyclic partially directed graphs. Machine Learning 59(3), 213–235 (2005)
2. Bratko, A., Filipic, B.: Exploiting structural information for semi-structured document categorization. Information Processing and Management 42(3), 679–694 (2006)
3. de Campos, L.M., Huete, J.F.: A new approach for learning belief networks using independence criteria. International Journal of Approximate Reasoning 24(1), 11–37 (2000)
4. Koller, D., Sahami, M.: Hierarchically classifying documents using very few words. In: Proceedings of the Fourteenth International Conference on Machine Learning, pp. 170–178 (1997)
5. Larkey, L.S., Croft, W.B.: Combining classifiers in text categorization. In: Proceedings of the 19th International ACM SIGIR Conference on Research and Development in Information Retrieval, pp. 289–297 (1996)
6. Lewis, D., Gale, W.: A sequential algorithm for training text classifiers. In: Proceedings of the 17th International ACM SIGIR Conference on Research and Development in Information Retrieval, pp. 3–12 (1994)
7. McCallum, A., Nigam, K.: A Comparison of event models for Naive Bayes text classification. In: AAAI/ICML Workshop on Learning for Text Categorization, pp. 137–142. AAAI Press, Menlo Park (1998)
8. Pearl, J.: Probabilistic Reasoning in Intelligent Systems: Networks of Plausible Inference. Morgan and Kaufmann, San Mateo (1988)
9. Robertson, S.E., Sparck-Jones, K.: Relevance weighting of search terms. Journal of the American Society for Information Science 27, 129–146 (1976)
10. Sebastiani, F.: Machine Learning in automated text categorization. ACM Computing Surveys 34, 1–47 (2002)

Efficient Clustering of Structured Documents Using Graph Self-Organizing Maps

Markus Hagenbuchner[1], Ah Chung Tsoi[2], Alessandro Sperduti[3], and Milly Kc[1]

[1] University of Wollongong, Wollongong, Australia
{markus,millykc}@uow.edu.au
[2] Hong Kong Baptist University, Hong Kong
act@hkbu.edu.hk
[3] University of Padova, Padova, Italy
sperduti@math.unipd.it

Abstract. Graph Self-Organizing Maps (GraphSOMs) are a new concept in the processing of structured objects using machine learning methods. The GraphSOM is a generalization of the Self-Organizing Maps for Structured Domain (SOM-SD) which had been shown to be a capable unsupervised machine learning method for some types of graph structured information. An application of the SOM-SD to document mining tasks as part of an international competition: Initiative for the Evaluation of XML Retrieval (INEX), on the clustering of XML formatted documents was conducted, and the method subsequently won the competition in 2005 and 2006 respectively. This paper applies the GraphSOM to the clustering of a larger dataset in the INEX competition 2007. The results are compared with those obtained when utilizing the more traditional SOM-SD approach. Experimental results show that (1) the GraphSOM is computationally more efficient than the SOM-SD, (2) the performances of both approaches on the larger dataset in INEX 2007 are not competitive when compared with those obtained by other participants of the competition using other approaches, and, (3) different structural representation of the same dataset can influence the performance of the proposed GraphSOM technique.

1 Introduction

In general, structured objects can be described by graphs, e.g. acyclic directed graphs, cyclic graphs, un-directed graphs, etc. Graphs are generalizations of the more common vectorial representation as a graph can encode relationships among structural elements of objects, or provide contextual information concerning data points which may be described in vectorial form.

The machine learning community recognizes that any model which is capable of dealing with structured information can potentially be more powerful than approaches which are limited to the processing of vectorial information. This observation motivates us to develop machine learning methods which are capable of encoding structured information. A noteworthy result of such efforts is the Graph Neural Network (GNN) which is a supervised machine learning method

N. Fuhr et al. (Eds.): INEX 2007, LNCS 4862, pp. 207–221, 2008.

capable of learning from a set of graphs [1]. The GNN is probably one of the more powerful supervised machine learning methods devised since it is capable of processing arbitrary types of graphs, e.g. cyclic, un-directed, where (numeric) labels may be attached to nodes and links in the graph. In other words, a GNN can encode the topology of a given set of graph structures as well as the numerical information which may be attached to the nodes or links in the graph.

Supervised machine learning methods require the availability of target information for some of the data, and are typically applied to tasks requiring the categorization or approximation of information. Unsupervised machine learning methods have no such requirement on the target information, and are typically applied to tasks requiring the clustering or segmentation of information. Unsupervised machine learning techniques for graph structured information are often based on the well-known Self-Organizing Maps [2] and are called Self-Organizing Maps for Structures (SOM-SD) [3]. While a SOM-SD is restricted to the processing of bounded positional acyclic directed graphs, it is found that this is sufficient for many practical applications. An application of the SOM-SD to the clustering of XML structured scientific documents at an international competition on document mining: Initiative for the Evaluation of XML Retrieval (INEX) was conducted, and this technique won in the year 2005 [4].

The introduction of a contextual SOM-SD (CSOM-SD) extended the capabilities of the SOM-SD model to allow for the contextual processing of bounded positional directed graphs which may contain cycles [5]. The SOM-SD and CSOM-SD were again applied to document mining tasks at INEX 2006. Both approaches produced winning results albeit amongst a fairly small group of participants [6]. However, it was observed that the CSOM-SD has a nonlinear computational complexity; in most cases, this is close to quadratic. This would limit the application of the CSOM-SD technique to small datasets. In this paper we will use a modification of the CSOM-SD method which we called Graph Self-Organizing Map (*GraphSOM*) [7], which (1) has a linear computational complexity, and (2) allows the encoding of more general types of graphs which may be unbound, cyclic, undirected, and non-positional. This paper demonstrates the efficiency and capability of the GraphSOM technique. Comparisons are made with the existing machine learning method: SOM-SD [3].

A drawback of the SOM-SD is that it does not scale well with the size of a graph. In particular, the computational demand increases quadratically with the maximum outdegree of any node in the dataset. Moreover, the SOM-SD requires prior knowledge of the maximum outdegree, and hence, has limitations in problem domains where the maximum outdegree is not known a priori, or for which the outdegree cannot be fixed a priori. The GraphSOM addresses these shortcomings through a modification of the underlying learning procedures [7]. The effect is that the computational complexity is reduced to a linear one, and, as a side effect, allows the processing of much more general types of graphs which may feature loops, undirected links, and for which the maximum outdegree is not known a priori. A more detailed theoretical analysis of the computational complexity of the GraphSOM is presented in [7].

This paper is structured as follows: Section 2 introduces to the SOM-SD and GraphSOM, and offers some comparisons. The experimental setting and experimental findings are presented in Section 3. Conclusions are drawn in Section 4.

2   Self-Organizing Maps

This section gives an overview to how unsupervised learning of graph structured information is achieved when using Self-Organizing Map techniques[1]. Another unsupervised neural network method capable of learning from graphs is [9] which realizes an auto-associative memory for graph structures, and hence, is quite different to clustering methods discussed in this paper. An alternative approach is constituted by the use of standard clustering methods in conjunction with metrics explicitly defined on graphs or induced by kernels for graphs, such as in [10] where a version of SOM that uses a version of the edit distance for graphs is presented. We are not aware of papers where kernels for undirected and unbounded graphs are used within a traditional clustering method. Finally, MLSOM [11] is an improved self-organizing map for handling tree structured data and cannot deal with graphs.

Traditionally, Self-Organizing Maps (SOMs) are an extension of the Vector Quantization technique [2] in which prototype units are arranged on an n-dimensional lattice. Each element of the lattice is associated with one unit which has adjustable weights. SOMs are trained on vectorial inputs in an unsupervised fashion through a suitable adjustment of the associated weights of the best matching prototype unit and its neighbors. Training is repeated for a number of iterations. The result is a topology preserving mapping of possibly high-dimensional data onto a lower dimensional one, often 2-dimensional mapping space. In practice, SOMs have found a wide range of applications to problem domains requiring the clustering or projection onto lower dimensional space of unlabeled high dimensional vectors. Self-Organizing Maps (SOMs) are a classic concept in machine learning allowing the mapping of high-dimensional data onto a low-dimensional display space [2].

An extension to data which can be described by graphs was made with the introduction of the SOM-SD [3]. With SOM-SD it has become possible for the first time to have an unsupervised machine learning method capable of mapping graph structures onto a fixed dimensional display space.

2.1   Self-Organizing Maps for Structured Data

Approaches to enable SOMs to map graph structured information were proposed relatively recently in [3,8]. The approach in [3] extends the classical SOM method by processing individual nodes of a graph, and by incorporating topological information about a node's offsprings in a directed acyclic graph. Nodes in the graph can be labeled so as to encode properties of objects which are represented

---

[1] This section does not contain any new material, but simply pulls together information which were published by us [3,7,8] in a coherent and self consistent manner to explain the basic motivation of using GraphSOM in this paper.

by the node. One of the main advantages is that the SOM-SD is of linear computational complexity when processing graphs with a *fixed* out-degree, and hence, the SOM-SD is capable of performing tasks such as clustering of graphs and subgraphs in linear time. The SOM-SD is an extension of the standard SOM in that the network input is formed through a concatenation of the node label with the mappings of each of the node's offsprings. This implies that the SOM-SD is restricted to the processing of ordered acyclic graphs (ordered trees), and requires that the trees have a fixed (and relatively small) outdegree. The computational complexity of the SOM-SD grows quadratically with the out-degree[2], and hence, the processing of trees with a large outdegree becomes quickly a very time consuming task. Moreover, the processing of nodes in a tree must be performed in an inverse topological order so as to ensure that the mapping of child nodes is available when processing a parent node.

The approach was extended through the introduction of CSOM-SD [8]. The CSOM-SD incorporates topological information on all the node's neighbors, and hence, the method is capable of processing undirected and cyclic graphs. Both approaches include the standard SOM as a special case, and when applied to graphs, are restricted to learning domains for which the upper bound of any node's connectivity (outdegree) is known a priori (e.g. the maximum number of neighbors for any node in a graph is known a priori). It was found that the computational demand for learning problems involving a high level of connectivity can be prohibitively high for both methods.

In the following, we will explain some of the basic mechanisms of SOM, and SOM-SD as a prelude on the modifications introduced in GraphSOM [7] later. Let us explain the underlying procedures as follows: The basic SOM [2] consists of a q-dimensional lattice of *neurons* representing the display space. Every neuron i of the map is associated with an n-dimensional codebook vector $\mathbf{m}_i = (m_{i1}, \ldots, m_{in})^T$, where T transposes the vector. Figure 1 gives an example of a simple SOM. The neurons are shown with a hexagonal neighborhood relationship; the most commonly used arrangement. This hexagonal neighborhood is used in the training of the SOM.

The SOM is trained by updating the elements of $\mathbf{m}_i$ as follows:

Step 1: One sample input vector $\mathbf{u}$ is randomly drawn from the input data set and its similarity to the codebook vectors is computed. When using the Euclidean distance measure, the winning neuron is obtained through:

$$r = \arg\min_i \|\mathbf{u} - \mathbf{m}_i\| \tag{1}$$

---

[2] As is shown in [7], the computational demand of a SOM-SD is $Nk(p + qn)$, where N is the total number of nodes in the data set, k is the number of neurons on the map, p is the dimension of the data label attached to the nodes in a graph, q is the dimension of the map (typically $q = 2$), and n is the maximum number of neighbors of any node in a graph. N, k, and n are often interdependent. An increase in N, p, or n often requires a larger mapping space k. In many data mining applications, $n \gg k$ which in turn can require a large k. Thus, the computational complexity for large scale learning problems is close to a quadratic one.

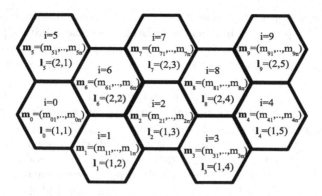

Fig. 1. A simple 2-dimensional map of size 2×5. Each hexagon marks a neuron. Number, associated codebook vector, and coordinate values for each neuron is shown.

Step 2: $\mathbf{m}_r$ itself as well as its topological neighbours are moved closer to the input vector in the input space. The magnitude of the attraction is governed by the learning rate α and by a neighborhood function $f(\Delta_{ir})$, where Δ_{ir} is the topological distance between $\mathbf{m}_r$ and $\mathbf{m}_i$. Here topological distance is used to described the distance between the neurons topologically. In our case we simply used the Euclidean distance to measure the distance topologically. This is the most commonly used method. The updating algorithm is given by:

$$\Delta \mathbf{m}_i = \alpha(t) f(\Delta_{ir})(\mathbf{m}_i - \mathbf{u}), \tag{2}$$

where α is the learning rate decreasing to 0 with time t, $f(.)$ is a neighborhood function which controls the amount by which the codebooks are updated. Most commonly used neighborhood function is the Gaussian function:

$$f(\Delta_{ir}) = \exp\left(-\frac{\|\mathbf{l}_i - \mathbf{l}_r\|^2}{2\sigma(t)^2}\right), \tag{3}$$

where the spread σ is called neighborhood radius which decreases with time t, $\mathbf{l}_r$ and $\mathbf{l}_i$ are the coordinates of the winning neuron and the i-th neuron in the lattice respectively. It is worth noting that $\sigma(t)$ must always be larger than 1 as otherwise the SOM reduces to Vector Quantization and no longer has topology preserving properties [2].

The steps 1 and 2 together constitute a single training step and they are repeated a given number of times. The number of iterations must be fixed prior to the commencement of the training process so that the rate of convergence in the neighborhood function, and the learning rate, can be calculated accordingly.

After training a SOM on a set of training data it becomes then possible to produce a mapping for input data from the same problem domain but which may not necessarily be contained in the training dataset. The level of ability of a trained SOM to properly map unseen data (data which are not part of the training set) is commonly referred to as the *generalization* performance. The generalization performance is one of the most important performance measures. However, in this paper, rather than computing the generalization performance of the SOM, we will evaluate the performance on the basis of micro purity and macro purity. This is done in order to comply with guidelines set out by the INEX-XML mining competition.

To allow for the processing of structured data this training algorithm is extended in [3] by incorporating the information about a node's neighbors. If $\mathbf{u}_i$ is used to denote the label of the i-th node in a graph, then an input vector for the SOM is formed by concatenating the label with the coordinates of the winning neuron of all the present node's neighbors. These coordinates are referred to as the *states* of neighbors. A hybrid input vector, defined as a vector $\mathbf{x}_i = (\mathbf{u}_i, \mathbf{y}_{ch[i]})$ is formed, where $\mathbf{y}_{ch[i]}$ is the concatenated list of states (coordinates of the winning neuron) of all the children of a node i. These states summarise the information which is contained in the child nodes. Note that here we assume a Markov assumption, in that the information on previous child nodes, the child nodes of those child nodes, etc are contained in the states. Since the size of vector $\mathbf{y}_{ch[i]}$ depends on the number of offsprings, and since the SOM training algorithm requires constant sized input vectors, padding with a default value is used for the missing offsprings or for nodes which have less than the maximum outdegree on a graph. Thus, the dimension of $\mathbf{x}_i$ is $p + qw$, where p is the dimension of the data label $\mathbf{u}$, q the dimension of the lattice, and w the maximum outdegree value. The training algorithm of a SOM is altered to account for the fact that an input vector now contains hybrid information (the data label, and the state information of offsprings). Equation 1 and Equation 2 are respectively replaced by the following:

$$r = \arg \min_i \| (\mathbf{x}_j - \mathbf{m}_i)^T \mathbf{\Lambda} \| \tag{4}$$

$$\Delta \mathbf{m}_i = \alpha(t) f(\Delta_{ir})(\mathbf{m}_i - \mathbf{u}) \tag{5}$$

where $\mathbf{\Lambda}$ is a $n \times n$ dimensional diagonal matrix; its diagonal elements $\lambda_{11} \cdots \lambda_{pp}$ are set to μ_1, all remaining diagonal elements are set to μ_2. The constant μ_1 influences the contribution of the data label component to the Euclidean distance, while μ_2 controls the influence of the states on the same distance measure. Thus, if μ_1 is large relative to μ_2 then the contribution of data labels is more important in the Euclidean distance measure relative to that exerted by the states (past information contained in the child nodes), and vice versa. In reality, it is the ratio $\frac{\mu_1}{\mu_2}$ that is important rather than their relative values. For simplicity, $\mu_2 = 1 - \mu_1, \quad 0 \le \mu_1 \le 1$ is normally used. Then, the SOM-SD adds a new step to the training algorithm [3]:

Step 3: The coordinates of the winning neuron are passed onto the parent node which in turn updates its vector $\mathbf{y}$ accordingly.

The SOM-SD [3] which represents a first attempt in incorporating graph information in the SOM approach requires the processing of data in a strict causal order from the leaf nodes towards the root. Thus strictly speaking it is only applicable to processing tree structures, rather than the more general graph structures which may contain loops (where the Markov assumption that the information in the sub-tree processed so far is contained in the states of the child nodes breaks down). We process nodes in a strict leaf nodes to root node order so that in Step 3 all states of all neighbors are available. The SOM-SD approach has a computational complexity which for large graphs could be quadratic [7]. Hence the SOM-SD approach can only be applied to process tree structured data with small number of nodes.

However, there are many situations in which loops occur in graphs. Hence an extension to circumvent this problem of having to process the information in a strict leaf nodes to root order is necessary. A first attempt was proposed in the CSOM-SD [8]. The CSOM-SD builds on the SOM-SD by adding new steps to the training procedure to address the fact that both ancestors and descendants of the node may need to be taken into account in some applications. There are various ways to do this, but basically these make use of the information stored from previous processing steps to be used as a proxy of the lack of information imposed by the non-causal manner to process the graph as dictated by the graph structure (which may contain loops). Fundamentally the approach is similar to SOM-SD as it introduces yet another constant to take the additional requirement of representing the ancestors (the nodes which are ahead of the current node in the graph structure, and the information of which are represented by the stored states obtained from previous processing steps) into account, and the CSOM-SD can be reduced to the SOM-SD accordingly. Thus, the CSOM-SD suffers the same computational complexity issue which arises in the SOM-SD approach, Nevertheless, for small graphs, it can be applied to situations when there are loops, or un-directed links in the graph structure.

2.2   The GraphSOM

The GraphSOM [7], a very recent development addresses some of the short-comings of SOM-SD. This is made possible by making a key observation in the SOM-SD and CSOM-SD processing of graphs: much of the information presented to the network may be redundant because it concatenates state information of every neighbor to the network input. Redundancies occur when several neighbors are mapped onto the same location on the map. This likelihood increases with increasing n, and becomes unavoidable when $n > k$. The GraphSOM processes the graph structured data by concatenating the data label with the activation of the map when mapping all of a node's neighbors. Note that here it is the activation of the map rather than the states of the map that is being presented to the GraphSOM inputs. Since the dimension of the map remains static, independent of the size of a training set, and independent of the outdegree of graphs, this implies that the GraphSOM's computational complexity is reduced to a linear one with respect to the outdegree of graphs. In other words, a GraphSOM can process graphs with a large outdegree much more efficiently than a SOM-SD. It is found that the GraphSOM includes the SOM-SD as a special case, and hence, one can expect that the clustering performances of the GraphSOM can be at least as good as that for the SOM-SD.

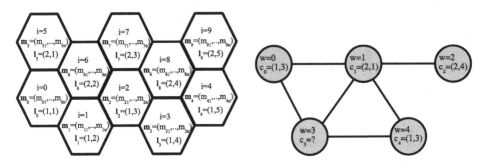

Fig. 2. A 2-dimensional map of size 2×5 (left), and an undirected graph (right). Each hexagon is a neuron. ID, codebook, and coordinate value for each neuron is shown. For each node, the node number, and coordinate of best matching codebook is shown.

Thus, the GraphSOM approach is to simply concatenate the activation (state) of the map with respect to all of the node's neighbors [7]. Formally, when processing a node i the input to the SOM is $\mathbf{x}_i = (\mathbf{u}_i, \mathbf{M}_{ne[i]})$, where $\mathbf{u}_i$ is defined as before, and $\mathbf{M}_{ne[i]}$ is a k dimensional vector containing the activation of the map $\mathbf{M}$ when presented with the neighbors of node i. An element M_j of the map is zero if none of the neighbors are mapped at the j-th neuron location, otherwise it is the number of neighbors that were mapped at that particular location.

It is much easier to express the underlying concepts in the GraphSOM[3] by considering the following example as shown in Figure 2. This figure shows a SOM. For simplicity, we assume further that no data label is associated with any node in the graph. Then when processing node $w = 3$ the network input is the k dimensional vector $\mathbf{x}_3 = (0, 0, 2, 0, 0, 1, 0, 0, 0, 0)$. This is because two of the neighbors of node 3 are mapped at the coordinate $(1, 3)$ which refers to the $2-$nd neuron, and the third neighbour of node 3 is mapped at $(2, 1)$ which refers to the $5-$th neuron. No other neuron is activated by a neighbour of node 3. These activations are listed in sequential order so as to form the vector $\mathbf{x}_3$. It can thus be observed that $\mathbf{x}_3$ summarizes the mappings of the neighbors of node 3 by listing the two activations: one at coordinate $(1, 3)$, the third neuron, and one at coordinate $(2, 1)$, the sixth neuron. Then the mapping of node 3 can be computed, and the information be passed onto its neighbors in the same way as is done for the SOM-SD. Note that the input remains unchanged regardless of the order of the neighbors; and that the dimension of $\mathbf{x}$ remains unchanged regardless of the maximum outdegree of any node in the dataset.

In comparison, the dimension of input vectors for the same graph when processed using SOM-SD is 6 (because $q = 2$ and $n = 3$). Thus, the number of computations with the GraphSOM is similar to that of SOM-SD when processing small graphs. But, the GraphSOM becomes more and more efficient as the connectivity (and consequently the size) of a graph increases. When dealing with large maps then the GraphSOM approach [7] can be approximated quite well

---

[3] This is the same example used in [7] to illustrate the developing of the thought process behind the proposed GraphSOM approach.

by consolidating neighboring neurons into groups. This approximation is valid since it exploits a general property with SOMs where only those nodes, which are mapped in close proximity to one another, share the greatest similarities, and hence, little information is lost by grouping them together. This leaves the granularity of the mappings unchanged but allows for a more compact representation of $\mathbf{M}_{ne[i]}$. For example, when using a 2×5 grouping for the map in Figure 1 then the network input for node 3 becomes $\mathbf{x}_3 = (1, 2, 0, 0, 0)$. A more coarse grouping can be chosen to achieve greater speed gains at the cost of some inaccuracy in the representation of the state of the map.

A summary of differences between the GraphSOM when compared to SOM-SD (and CSOM-SD) can be given as follows: (1) GraphSOM has a lower computational complexity when dealing with large graphs than SOM-SD; (2) GraphSOM's computational complexity can be scaled arbitrarily through a grouping of neurons at the cost of some inaccuracy in the representation of the graph while for SOM-SD it grows quadratically; (3) the input dimension of GraphSOM is independent of the level of connectivity in a graph while in SOM-SD it is a variable; and (4) the GraphSOM allows the processing of cyclic, positional, and non-positional graphs while SOM-SD can only be applied to trees, and CSOM-SD can be applied to cyclic graphs, graphs with un-directed links with low dimensions.

In fact, the GraphSOM is capable of processing the same classes of graphs as the GNN [1] with the exception that the labeling of links is currently not supported. It is obvious to note that the GraphSOM includes the SOM-SD as a special case.

We note in passing that a related idea has been proposed by some researchers in the context of sequence processing. In particular, the Merge SOM model [12] stores compressed information of the winner in the previous time step, whereby the winner neuron is represented via its content rather than its location within the map. The content of a neuron i consists of the weight w_i and the context c_i. These two characteristics of neuron i are stored in a merged form, i.e. as a linear combination of the two vectors. It should be stressed that this approach is different from the GraphSOM approach [7].

3   Experiments

This paper applies the SOM-SD and the GraphSOM to a large dataset consisting of structured documents from the web as contained in the INEX 2007 dataset (for the XML mining competition task). More specifically, the methods are applied to cluster a subset of documents from Wikipedia. The task is to utilize any of the features (content or structure) of the documents in order to produce a given grouping (clustering). We decided to investigate the importance of structural information to the clustering of these documents. Such an approach had been very successful at previous INEX clustering events. The documents are formatted in XML, and, hence, are naturally represented as tree structures. The dataset is made available in two parts, a training set and a test set. Each data in these sets is labeled by one of 21 unique numeric labels. The label indicates the desired grouping of the document. Hence, the task is to cluster pages into 21 groups. The machine learning method used for the experiments is trained *unsupervised*. This means that the data label is not used when training the SOM. The labels are used to evaluate whether the trained SOM groups the training data as desired. The SOMs' ability to perform the desired grouping of the training data is maximised by tweaking the various training parameters through trial and error. After determining the best set of training parameters, the SOM is used to map the unseen data in the test set. The performances reported in this paper are computed based on the mappings of the test set.

Since the data is represented in a tree form, and hence, the application of the SOM-SD is possible. However, it will be found that the outdegree of the dataset is prohibitively large. The GraphSOM has a reduced computational complexity relative to the SOM-SD. In applying the GraphSOM, the computational time required, compared with the ones using SOM-SD, is reduced significantly. This is especially important, as the INEX 2007 contains a relatively larger dataset.

The importance of this application is manifold:

- XML is an increasingly popular language for representing many types of electronic documents.
- An application to data mining tasks can help to demonstrate the capabilities of the GraphSOM, or the lack of it, over previous machine learning methods, e.g. SOM-SD approach which are capable of clustering graphs.
- The datasets considered (viz. the INEX Wikipedia dataset) is a benchmark problem used INEX 2007.

This paper gives some preliminary results[4]. Results presented here were obtained from training the SOMs for eight runs each. The best result is presented in this paper. Note that under normal circumstances, a SOM would have to be run under possibly hundreds of training conditions in order to determine its peak performance. This is due to the fact that a number of training parameters need to be determined through trial and error (for any SOM training algorithm). Amongst these parameters are the dimensionality of the map, the geometry of the map, the type of neighborhood relationship between the codebook entries of the map, a learning rate, the number of training iterations, weighting measures for the data label and structural component of the inputs, and several others. A suitable choice of training parameters is essential in obtaining a well performing GraphSOM (similar to all SOM approaches of which GraphSOM is one).

A first set of experiments utilizes the XML structure of the documents. A node in the graph represents an XML tag, the links represented the encapsulation of the tags. For example, the XML sequence `<a><b></b><c></c></a>` produced a graph with a root node representing the tag `<a>` and its two offsprings `<b>` and `<c>`. We assigned a unique numerical ID number to each unique XML tag, then added the ID number as a label to the corresponding node in the graph. For example, if we assign the ID number 101 to tag `<a>`, then the node representing this tag will have been assigned the numeric label 101. In other words,

---

[4] The GraphSOM has been developed very recently[7]. Time constraints and implementation issues prevented us from conducting experiments more thoroughly.

Table 1. Results when clustering the test dataset into 21 clusters

SOM-SD	
Micro average purity	0.262457
Macro average purity	0.26159

GraphSOM	
Micro average purity	0.26885
Macro average purity	0.26635

the SOMs are trained to cluster XML formatted documents solely on topological information, and on the type of XML tag embedded in the document. No further information is provided to the training algorithm.

The results when clustering the dataset into 21 clusters are summarized in Table 1. In comparison, the performances of the clustering task obtained by other participants of the competition is shown in Table 2. It is observed that (1) despite successes of the SOM-SD method in earlier INEX competitions [4,6] the performances obtained here are well below those obtained by others, and (2) both the GraphSOM and the SOM-SD perform about equally well. A difference between the latter is the training times needed. Given that the training dataset contained 48, 306 tree structures, one for each document, with a maximum out-degree of 1,945, this size of outdegree would require an estimated 40 years of training time for the SOM-SD on a top end desktop computer with 2.8 GHz clock rate! To avoid this, and to enable the use of the SOM-SD for comparison purposes, we pruned the graphs to have a maximum outdegree of 32 by truncating nodes with a larger outdegree. This reduced the training time for the SOM-SD to a more reasonable 36 hours[5]. In comparison, the GraphSOM is capable of processing the graphs without pruning in about 48 hours by using a 8×8 grouping of nodes. The results shown were obtained when using for both networks a size of 160×128, the number of training iterations equal to 200, $\alpha(0) = 0.8$, the weights μ_1, μ_2 were $(0.05, 0.95)$, and $\sigma(0) = 20$.

Table 2. Results when clustering into 21 clusters obtained by competing groups

Name	Micro avg. purity	Macro avg. purity
Guangming Xing	0.62724	0.571855
Jin YAO & Nadia ZERIDA	0.51530897	0.61035

Pruning can have a negative impact on the clustering performance since some relevant information may be removed. The GraphSOM allows the processing of large graphs without requiring pruning, and hence, can be expected to produce performances which are at least as good as those obtained by a SOM-SD if not better. While this has been confirmed by these experiments, it is also found that

---

[5] This is one of the major disadvantages of using SOM-SD in processing larger datasets. In order to ensure a reasonable turn-around time it is necessary to prune the data back so that the training time is kept to reasonable duration. In doing the pruning, some information would be lost. In contrast, the GraphSOM does not require as extensive pruning or no pruning at all, and hence it is far more efficient than the SOM-SD.

the pruning did not reduce the SOM-SD performance. These observations (generally relative poor performance, and pruning without effect on results) caused us to suspect that the XML structure of the documents may not be a feature which leads to a desired clustering result as was defined by INEX-2007.

Thus, one of the reasons for the relatively poor performance may be that the desired clustering of the documents may not rely on the XML structure as much as the learning tasks in previous years did. A difference with the datasets used in previous years is that this year's dataset contained embedded hyperlinks which pointed from one document to another document. It may be that a successful clustering requires the encoding of the link structure of the documents rather than the XML structure. To verify this hypothesis, we created a new graph representation of the dataset where a node represents a document as a whole, and the links represent the hyperlinks pointing from one document to another. This produced one directed graph for each; the training set and the test set containing $48,306$ and $48,305$ nodes respectively. The graphs contained cycles. The maximum outdegree for each of the two datasets were 410 and 544 respectively. Note that some documents have several (redundant) links to another document. We consolidated these by listing only its first occurrence. This was performed since we assumed that such redundancy may not significantly influence the results, and to help reduce the turn around time for the experiments. Had we included redundant links, the maximum outdegree would have increased to 471 and 579 respectively. Note also, that only those links were used which pointed to documents within the same dataset (i.e. for the training set we used only those links which pointed to other documents in the training set). Links to documents outside the same dataset were discarded. Nevertheless, we added the total number of (unique) hyperlinks as a data label to each node in the graph. It is interesting to note that the maximum outdegree of any node in the datasets is $2,118$ for the training set and $2,088$ for the test set if all links (internal, external, and redundant links) are considered. This implies (a) that most hyperlinks are to pages not within the same dataset, and (b) that Wikipedia document hyperlink structure is relatively rich.

For the purpose of training the SOM-SD, we again had to resort to pruning. Here, pruning was done selectively by identifying links which are involved in cycles and pruning these first. This allowed the generation of an acyclic graph with sufficiently small outdegree which allowed the training of a SOM-SD in a timely fashion. No such pruning was necessary for the training of the GraphSOM. Maps were trained with the same parameters as before on these datasets. The results are summarized in Table 3.

It can be observed that this produced an improvement in the performance by 3 to 9 percentage points. It is also observed that the GraphSOM performs slightly better than the SOM-SD which may be attributed to that the GraphSOM is trained on a non-positional graph which was not pruned. The finding indicate that information which is useful for obtaining the desired partitioning of the test data set is embedded in the hyperlink structure of the documents, and that the task does not appear to rely on positional information about the links (i.e. the order of the hyperlinks does not appear to be important). Nevertheless, the performances obtained still fall short of those obtained by other participants of the competition. It shows that structural information may not be a key ingredient to obtain the desired clustering. It is thus likely that textual information within the documents is of importance for this given task. This can be addressed through a suitable labeling of the nodes in the graph so as to encode textual information. To surmise one of the reasons why the addition of content to link structure may not result in improved accuracy is that there are many ways in which information on content can be extracted. For example, in an XML document, content could mean the extraction of information on each paragraph or on each document. Since such information is contained in words, there will need to be a good way to include the information and meaning conveyed by words, or combination of words. This is not a particularly easy task to perform as it is well known that some combination of words may lead to quite different meaning. However, without an accurate representation of the contents, the SOM may not receive sufficient information for it to discriminate web pages. This points to possible ways in which our approach can be improved: through a better representation of the content information of the documents. This is left for a future task and will not be addressed further in this paper.

4 Conclusions

The paper applied the newly developed GraphSOM [7] for the first time to a relatively large clustering application represented by the INEX 2007 competition dataset. Comparisons with its predecessor (the SOM-SD) revealed that the added capabilities of the GraphSOM can help to produce improved results. But most importantly, it was demonstrated that the lower computational complexity of the GraphSOM and its ready scalability to larger datasets allow the processing of graphs without requiring pre-processing measures such as pruning in order to reduce time requirements for the training procedures. This represents a major advance on the SOM-SD approach in terms of its practicality, in that GraphSOM is shown to be able to handle relatively large datasets without any pruning, thus preserving the information content in the dataset. In contrast, the SOM-SD approach is required to prune the training dataset so as to keep the training time to reasonable duration, and thus may lead to a loss of information.

This paper presented some initial experimental results which do not measure up to results obtained by others on this clustering task. It can be assumed that by fine-tuning training parameters, and by providing a richer set of information to the GraphSOM learning procedure will help to improve the performances. For example, the data label attached to nodes may contain a description of the content found in the document. Such tuning may not improve the performance significantly if the decisive criteria for the clustering task is based on structural components. One could, moreover, attempt to train the GraphSOM on a hybrid structure which combines both the underlying XML structure of individual documents as well as the hyperlink structure between documents. This may improve the performance of the method as it takes into account the nature of a dataset which contains structured documents, and a structural dependency between documents. There are various

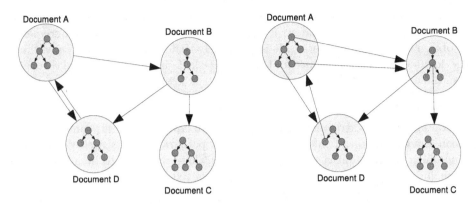

Fig. 3. Hybrid structures. When considering hyperlinks to point from one document to another (left), or when considering hyperlinks to point from within an embedded XML-tag to another document (right).

ways in which such combination can be effected. One way in which this can be effected is that if we consider hyperlinks to be a binary relation between two documents as depicted in Figure 3 (left). Each document is represented by its XML (tree-)structure. A SOM could be trained on these tree-structures as was carried out in this paper. Then, once the SOM has converged to a final mapping, a second SOM is trained on the underling hyperlink structure of the dataset where, to each node, a data label which summarizes the activation of the first SOM (trained on XML information) is attached. Since the XML representation of documents is done through a tree structure, and since processing can be done in an inverse topological order, and hence, the root of the XML tree structure can be a representation of the tree structure as a whole. Thus, it would suffice to solely utilize the mapping (state) of the root as a label for the corresponding node in the hyperlink structure.

Alternatively, one may wish to take account of the fact that hyperlinks are embedded within a section defined by an XML tag (Figure 3 on the right), and hence, the source of a hyperlink can be any node in the XML tree. Thus, it may be more appropriate to use the mapping of the XML-tag which is the source of a hyperlink for the label for the associated node.

One may observe that for both of these approaches, the dimensionality of the node labels depends on the outdegree of all nodes in the XML tree. For instance, in Figure 3 (right), Document A has two nodes with outgoing hyperlinks to Document B. To account for the various activations of the nodes involved, concatenation could be used to obtain the associated data labels. However, to avoid problems which would arise with large outdegrees, one could use the Graph-SOM approach for forming the node labels (i.e. to use the activation of the map rather than concatenating individual activations). The second map, once trained on the underlying hyperlink structure, would produce a mapping of documents according to the overall hybrid structure.

These instances show that one may not ignore the underlying structure of a document, whether the inherent structure, or the extended structure as provided

Table 3. Results when clustering into 21 clusters using the hyperlink structure

SOM-SD	
Micro average purity	0.253804
Macro average purity	0.337051

GraphSOM	
Micro average purity	0.27423
Macro average purity	0.353241

through the hyperlink structure if one wishes to cluster the documents. The validation of these hypotheses and intuition is left as a future research task.

Acknowledgment. The first and second authors wish to acknowledge financial support by the Australian Research Council through its Discovery Project grant DP0452862 (2004 - 2006) and DP077148 (2007 - 2009) to support the research reported in this paper.

References

1. Scarselli, F., Yong, S., Gori, M., Hagenbuchner, M., Tsoi, A., Maggini, M.: Graph neural networks for ranking web pages. In: Web Intelligence Conference (2005)
2. Kohonen, T.: Self-Organizing Maps. Springer Series in Information Sciences, vol. 30. Springer, Berlin (1995)
3. Hagenbuchner, M., Sperduti, A., Tsoi, A.: A self-organizing map for adaptive processing of structured data. IEEE Transactions on Neural Networks 14(3), 491–505 (2003)
4. Hagenbuchner, M., Sperduti, A., Tsoi, A., Trentini, F., Scarselli, F., Gori, M.: Clustering xml documents using self-organizing maps for structures. In: Fuhr, N., Lalmas, M., Malik, S., Kazai, G. (eds.) INEX 2005. LNCS, vol. 3977, pp. 481–496. Springer, Heidelberg (2006)
5. Hagenbuchner, M., Sperduti, A., Tsoi, A.: Contextual self-organizing maps for structured domains. In: Workshop on Relational Machine Learning (2005)
6. KC, M., Hagenbuchner, M., Tsoi, A., Scarselli, F., Gori, M., Sperduti, S.: Xml document mining using contextual self-organizing maps for structures. LNCS. Springer, Berlin (2007)
7. Hagenbuchner, M., Sperduti, A., Tsoi, A.: Self-organizing maps for cyclic and unbound graphs. In: European symposium on Artificial Neural Networks, April 23-25 (2008) (to appear)
8. Hagenbuchner, M., Sperduti, A., Tsoi, A.: Contextual processing of graphs using self-organizing maps. In: European symposium on Artificial Neural Networks. Poster track, Bruges, Belgium, April 27-29 (2005)
9. Sperduti, A., Starita, A.: A memory model based on LRAAM for associative access of structures. In: Proceedings of IEEE International Conference on Neural Networks, June 2-6, 1996, vol. 1, pp. 543–548 (1996)
10. Neuhaus, M., Bunke, H.: Self-organizing maps for learning the edit costs in graph matching. IEEE Transactions on Systems, Man, and Cybernetics, Part B 35(3), 503–514 (2005)
11. Rahman, M.K.M., Yang, W.P., Chow, T.W.S., Wu, S.: A flexible multi-layer self-organizing map for generic processing of tree-structured data. Pattern Recogn. 40(5), 1406–1424 (2007)
12. Strickert, M., Hammer, B.: Neural gas for sequences. In: WSOM 2003, pp. 53–57 (2003)

Document Clustering Using Incremental and Pairwise Approaches

Tien Tran, Richi Nayak, and Peter Bruza

Information Technology, Queensland University of Technology,
Brisbane, Australia
{t4.tran,r.nayak,p.bruza}@qut.edu.au

Abstract. This paper presents the experiments and results of a clustering approach for clustering of the large Wikipedia dataset in the INEX 2007 Document Mining Challenge. The clustering approach employed makes use of an incremental clustering method and a pairwise clustering method. The approach enables us to perform the clustering task on a large dataset by first reducing the dimension of the dataset to an undefined number of clusters using the incremental method. The lower-dimension dataset is then clustered to a required number of clusters using the pairwise method. In this way, clustering of the large number of documents is performed successfully and the accuracy of the clustering solution is achieved.

Keywords: Clustering, structure, content, XML, INEX 2007.

1   Introduction

Majority of the electronic data on the Web is presented in the format of semi-structured data. The popular semi-structured data representation languages are XML (Extensible Markup Language), HTML (Hypertext Markup Language) and XHTML (Extensible HyperText Markup Language). XML has become a standard for information representation within the enterprise as well as in its exchange. XML is used because of its flexibility and self-describing nature to use the structure in storing content. XML documents have a great acceptance in many industries such as government, entertainment, education, e-business etc. In recent years XML have also been used by researchers as data input such as in Web services, Bioinformatics, etc.

Data represented in XML format do not follow a strict structure. This results in a heterogeneous collection of the XML data in content as well as in structure. In other words, a collection of XML documents may vary in terms of contents within the documents as well as in terms of the structures used to store the contents of the documents. The proliferation of XML documents has raised many issues concerning with the management of XML data. Researchers have proposed tasks such as classification and clustering of XML documents according to their structure and content. XML document clustering (or grouping) is important in many applications such as information retrieval, database indexing, data integration and document engineering to identify similar documents.

N. Fuhr et al. (Eds.): INEX 2007, LNCS 4862, pp. 222–233, 2008.

Clustering approaches can be classified into two types: pairwise and incremental. A pairwise clustering approach, in this case, groups the data according to the pairwise distance matrix [1,2,3]. The pairwise distance matrix represents the distance (or similarity) between each pair of data (or documents) in the dataset. Whereas, incremental clustering approach groups the data by measuring the distance between the input data and existing cluster representatives. Approaches such as graph clustering method [4] uses a pairwise distance matrix for clustering of documents. The computation time to generate a pairwise distance matrix can be very expensive when dealing with a large dataset. On the other hand, incremental clustering approaches [5,6] can deal with a large dataset more efficiently. However, the incremental clustering approaches can suffer poor accuracy and sensitivity to the input order due to not calculating the distance between each pair of documents. The trade-off between the pairwise clustering and incremental clustering is the accuracy of the clustering solution generated and the scalability of the clustering process.

In this paper, we use a clustering approach that first performs an incremental clustering method on the dataset to reduce the dimension of the dataset. A pairwise distance matrix is computed between the intermediate clusters to merge the clusters together to obtain a required number of clusters. The incremental method progressively groups the large dataset into a number of clusters by comparing each coming document with the existing cluster representatives. The cluster representative is represented simply by the document that first form the cluster. After the grouping of documents performed by the incremental method, a pairwise distance matrix is computed to measure the similarity between the cluster representatives. A graph clustering method is then applied to the matrix to merge the clusters together so that the required number of clusters according to the user-defined criterion are generated.

The sequential combination of the incremental clustering with the clustering based on pairwise distance matrix enables the clustering approach to perform clustering of large datasets with good accuracy. The structural similarity between documents is computed by measuring the common elements occurring in the nested paths of XML documents. The semantic similarity between document contents is computed by using a Latent Semantic Kernel (LSK) with a subset of the input dataset. This subset is selected from the clusters of XML documents that are formed based on structural similarity. This is done with a hypothesis that the documents describing similar information will have similar structure. We present the experiments and results conducted with the Wikipedia dataset of the INEX 2007 Document Mining Challenge [7] in this paper. The Wikipedia dataset contains 48035 XML documents.

This paper is structured as follows. The following section gives an overview of the clustering approach. Section 3 explains the pre-processing stage by discussing how the structure and the content of the dataset are extracted and represented. Section 4 explains the clustering process. The results of the clustering process are presented and analyzed in Section 5.

2   Overview of the Clustering Approach

Figure 1 illustrates the clustering approach employed in this paper. The first stage of the clustering approach begins with the pre-processing of the input dataset. The pre-processing of the documents in the dataset is necessary to remove irrelevant data which may degrade or contribute little to the clustering process. The output of the pre-processing stage is the documents' features and their representations. In the case of XML document, two features are extracted: the structure and the content. These two features become input to the incremental clustering method. The incremental clustering performs the first run of the clustering process. After the incremental clustering stage, documents are grouped into an undefined number of clusters (generated at run-time) depending on the clustering threshold set by the user. A pairwise distance matrix is then computed which contains the similarity between intermediate cluster representatives. The graph clustering method [4] uses this matrix to achieve the final clustering solution. In this approach, the number of clusters generated by the incremental method is always higher than the number of clusters generated by the graph clustering method.

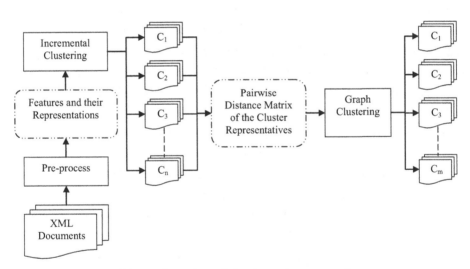

Fig. 1. Overview of the clustering approach

3   Pre-processing

3.1   Structure Mining

In this study, the structure of the XML document is represented as a collection of paths. Each path contains the element names from the root to a leaf node (node contains the textual content). For example in fig. 2, the following paths are extracted from the document: a/c/d, a/c/d, a/c/e, a/b/f, a/b/f/e. Each path is considered as an individual item in the XML document structure. Therefore the order of the paths is ignored. Duplicated paths are eliminated. For instance, from the example document in fig. 2, there are 5 paths extracted from the document, two paths are duplicated i.e. a/c/d, therefore one path is removed and only 4 paths are used to represent the document structure. The final output representation of the document structure is thus a summary representation.

```
⟨a⟩
 ⟨c⟩
   ⟨d⟩ t₅, t₄, t₇ ⟨/d⟩
   ⟨d⟩ t₅, t₂, t₃ ⟨/d⟩
     ⟨e⟩ t₅, t₂, t₁ ⟨/e⟩
 ⟨/c⟩
 ⟨b⟩
   ⟨f⟩ t₅, t₂, t₁ ⟨/f⟩
   ⟨f⟩
     ⟨e⟩ t₂, t₁, t₁, t₄ ⟨/e⟩
   ⟨/f⟩
 ⟨/b⟩
⟨/a⟩
```

Fig. 2. Example of an XML document

The similarity between two XML document structures is determined using a function called $CPSim$ which is defined by Nayak and Tran [5]. $CPSim$ is defined as:

$$CPSim(d_x, d_y) = \frac{\sum_{i=1}^{|P_x|} \max(\int_{j=1}^{|P_y|} Psim(p_i, p_j))}{\max(|P_x|, |P_y|)} \quad (1)$$

$CPSim$ is the common path similarity coefficient between two XML documents d_x and d_y. It ranges from 0 to 1 (1 is the highest). It computes the sum of the best path similar coefficient ($Psim$) of all the paths in d_x with d_y with respect to the maximum number of paths in the two documents. $Psim$ measures the degree of similarity between two paths. It is defined as:

$$Psim(p_i, p_j) = \frac{\max(CNC(p_i, p_j), CNC(p_j, p_i))}{\max(|p_i|, |p_j|)} \quad (2)$$

$Psim$ of paths p_i and p_j is the maximum similarity of the two CNC functions with respect to the maximum number of elements in both paths. CNC, the common node coefficient, is sum of the elements occurring between two paths p_i and p_j in hierarchical order. Consider two paths p_1 and p_2 that contain the following elements a/b/c/d and a/b/d respectively. The CNC function to calculate the common number of nodes from p_1 to p_2 starts at the leaf element. If the leaf elements of both paths match then both paths will move up one level. The element matching is again started at these levels. Otherwise the leaf element of

p_1 will match with the element in the next level of p_2. The CNC function will continue until it reaches the root of p_2 or an element in p_1 can not find a match with p_2. The CNC function is not transitive. In the above example, the CNC measure of p_1 to p_2 is 1 and the CNC of p_2 to p_1 is 3.

3.2   Content Mining Using Latent Semantic Kernel

Latent Semantic Kernel. We utilise the idea of Latent Semantic Kernel(LSK) [8] to learn and measure the document contents. The latent semantic kernel is constructed based on latent semantic analysis (LSA) and singular value decomposition (SVD) [9]. Given a term-document matrix $TD_{|T| \times |D|}$, $|T|$ is the number of unique terms extracted from a corpus and $|D|$ is the number of documents in the corpus. Each cell in the matrix represents the frequency of a term in a document. Using this matrix, the SVD is applied to decompose the matrix into 3 matrices, U, S and V. V and U are orthonormal columns of left and right singular vectors respectively. S is a diagonal matrix of singular values in which the diagonal values are ordered in decreasing magnitude. The SVD model can optimally approximate matrix TD with a smaller sample of matrices by selecting k largest singular values and setting the rest of the values to zero, such that, matrix U can be redefined to matrix U_k of $|T| \times k$, V to matrix V_k of $|D| \times k$, and S to matrix S_k of $k \times k$. This transforms the matrix TD from the higher-order to a lower $k-$dimensional document space. Readers can refer to Landauer et al. [9] for more technical details on SVD and LSK. U_k is now used as a kernel to find semantic similarity between document contents in $k-$dimensional document space.

Given two document content vectors, d_x and d_y, the semantic similarity between the document contents is measured using cosine measure which is defined as follows:

$$contentScore(d_x, d_y) = \frac{d_x^T P P^T d_y}{|P^T d_x||P^T d_y|}. \tag{3}$$

Matrix U_k is represented by P. It is used as a mapping function to calculate the content similarity of documents d_x and d_y.

Construction of the LSK and Content Representation. Previous researchers have built latent semantic kernels either on a large data set [10] or a small domain dependant data set [11]. The construction of the kernel on a large corpus can be expensive in terms of memory space and processing time, particularly the use of SVD algorithm. To deal with the scalability problem, we propose building the LSK by selecting a small subset of documents from the input documents. In order to select the diverse samples of input documents, we propose to first cluster the documents based on the structural similarity using the incremental clustering approach as outlined in figure 1. We hypothesize that the documents describing similar information will have a similar structure. Considering that each cluster generated by the incremental clustering approach contains similar documents, a small subset of documents from each cluster is randomly selected to construct the LSK. In the cases where a cluster does not contain sufficient number of documents to be used for the development of the kernel, additional documents are selected from other large size clusters.

Given a collection of XML documents $\{d_1, d_2, ..., d_n\}$ denoted by D, the content of a document, d_i, is modelled as a vector $\{td_{i1}, td_{i2}, ..., td_{im}\}$ where m is number of distinct terms (after the stop-word removal and stemming [12]) extracted from the selected subset of documents (the same subset used to build the latent semantic kernel in this study). Each value of the vector represents by a binary number, 1 or 0, 1 indicating that the term exists in the document.

4   Clustering Approach

As mentioned earlier, the clustering approach involves two clustering methods: incremental and pairwise. The incremental clustering process used in this approach adapts the hierarchical clustering method. It starts with no cluster. The first document of the input data set forms the first cluster. The document becomes the cluster representative. When the next document comes in, it is compared with the existing cluster representative(s). If the similarity between the new document and an existing cluster representative has the largest similarity value and it exceeds the user-defined clustering threshold, the document is assigned to that cluster. Otherwise the document forms a new cluster and the document becomes the cluster representative. The similarity between the document and the clustering representative is measured using either the $CPSim$ measure (equation 1) for the structure-only clustering or the $contentScore$ measure (equation 3) for the content-only clustering.

After the initial grouping of the documents into an undefined number of clusters, an iteration process is executed. The iteration process runs the incremental clustering process again by comparing each document with all the existing cluster representatives. The purpose is to reassign each document to a cluster with the maximum similarity value without using the clustering threshold. Depending on the number of clusters generated by the incremental clustering process, a merging process may take place. The merging process is used to reduce the dimension of the cluster solution which is generated by the incremental method. This is particularly true for the cases where the number of clusters generated by the incremental process is too large for the pairwise clustering to process. The merging stage is used to merge all the clusters that contain only 1 document with the existing clusters that contain more than 1 document. It proceeds by merging the cluster, having only 1 documents, to an existing cluster, containing more than 1 document, having the highest similarity value between their representatives. A pairwise distance matrix is computed by measuring the similarity between each pair of cluster representatives. The similarity between the cluster representatives is measured in the same way as the measuring between a document and a cluster representative in the incremental method.

The pairwise distance matrix is then fed into the graph clustering method [4]. The graph clustering method merges the clusters generated by the incremental clustering method into the final number of clusters defined in this INEX document mining task.

5   Experiments and Discussion

This section presents the clustering experiments and results with the 48035 documents of Wikipedia collection used in the INEX 2007 document mining challenge [7].

The experiments are set up to measure two features of XML documents for the clustering task: structure and content. The first set of clustering results is evaluated to check the clustering performance using the structure-only measure for the document similarity. The structure-only clustering uses the structure information without the content information to cluster the documents. The document structure is a set of distinct paths extracted from the document which is used for the structure-only clustering using the $CPSim$ measure as described in section 3.1. The paths contain the element (tag) names from the root node to a leaf node are used to calculate the CNC function.

The second set of clustering results is evaluated to check the clustering performance using the content-only measure for the document similarity. The content-only clustering uses the content information without the structure information to cluster the documents. To perform the clustering on the content-only information, a LSK is built with 1024 documents selected from the input documents. These documents are chosen from the 586 clusters that were generated by the incremental clustering process (without the merging stage) based on the structural similarity using the clustering threshold of 0.3. A total of 41935 terms were extracted from these 1024 documents after stop-word removal (words such as is, the, or etc.) and stemming [12] to construct a term-document matrix for the construction of the kernel. The 41935 terms are also used to represent the content for the documents in the dataset. The content of a document, d_i, is represented by a vector $\{td_{i1}, td_{i2}, ..., td_{im}\}$ where m equals to the 41935 terms. Each cell of the vector contains 1 or 0 (1 indicates that a term exists in a document) to indicate the existence of a term in a document. The cosine measure (equation 3) is used to measure the document content similarity.

The experiments are conducted to generate the two different number of clusters. These are 10 clusters and 21 clusters which are required by the INEX 2007 challenge. Since the computation for the structure-only measure employed in this approach is expensive, a low clustering threshold value of 0.3 is chosen for the structure-only clustering. Whereas, the content-only clustering uses the clustering threshold of 0.9 because from the extensive empirical analysis, it has shown that using a higher clustering threshold produces a better clustering solution for the content-only clustering.

5.1   Structure-Only Comparisons

Tables 1 and 2 present the results and compare them with other participants in the INEX 2007 document mining challenge for the clustering task based on the structure-only. The structure clustering solution results are compared with Kutty et al. [13] and Hagenbuchner et al. [14]. Kutty et al. approach is based on generation of closed frequent subtrees from the Wikipedia dataset. Documents

Table 1. Comparing the clustering results for structure-only on Wikipedia dataset with 10 clusters

Approaches	Micro F1	Macro F1
Hagenbuchner et al.	0.251	0.257
Kutty et al.	0.251	0.263
Our Approach	0.251	0.252

Table 2. Comparing the clustering results for structure-only on Wikipedia dataset with 21 clusters

Approaches	Micro F1	Macro F1
Hagenbuchner et al.	0.269	0.266
Kutty et al.	0.253	0.269
Our Approach	0.251	0.253

are represented with the closed frequent subtrees that they contain. The XML documents are progressively grouped into a defined number of clusters using the degree of similarity that share according to the closed frequent subtrees. Hagenbuchner et al. approach is based on self-organizing maps (SOM). The SOM clustering is performed using graph representations of input documents. It is a type of neural network model for data analysis based on unsupervised learning where the result is a topology preserving mapping from a high-dimensional input space to usually 2-dimensional map space.

Results in tables 1 and 2 show that the performance of clusters considering the structural similarity only do not vary much among participants. Amongst the clustering solution results for all participants in the clustering task, the structure-only results are worse in comparison to the clustering solution results by considering the content of documents. The reason is as follows. XML documents in the Wikipedia collection conform to the same structure definition. All documents are very similar in structure (including the tag names) that is used to describe different content topic. For example, datasets such as IEEE where the same schema is used to represent various conference (or topic) papers. It is hard to infer any unique structure representation for each individual category.

Based on these results (tables 1 and 2) for the clustering with structure-only similarity, it can be ascertained that the structure of the testing corpus in INEX 2007 challenge does not play a significance role in the clustering task. In another word, no matter what approaches are used, the clustering solution based on the structure-only will possibly stay somewhere in the range of 0.2 and 0.3.

5.2 Content-Only Comparisons

Figures 3 and 4 display the clustering solution results based on the content-only. Our content-only results are compared with Yao et al. [15]. For the content-only clustering, Yao et al. has extracted 130969 terms (or words) from the documents in the Wikipedia collection after the stop-word removal, stemming and the

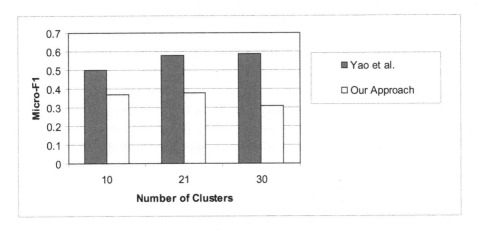

Fig. 3. Clustering micro-f1 results in INEX 2007 document mining challenge based on content-only similarity

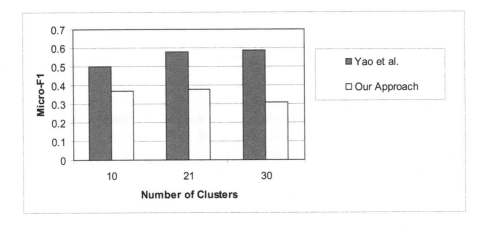

Fig. 4. Clustering macro-f1 results in INEX 2007 document mining challenge based on content-only similarity

removal of words which length are lesser than 3. The content of the input documents is represented as a vector space model [16] in which each cell in the vector contains the tf*idf (term frequency * inverse document frequency) appearing in a document. Using the vector space model, the approach then applied a agglomerative clustering method proposed by Zhao et al. [17,18] which combines the features from both partitional and agglomerative approaches for the clustering of the document contents. On the other hand, we use the latent semantic kernel to measure the content similarity. Results in figures 3 and 4 show that the clustering solutions performed by Yao et al. achieve higher accuracy than ours. One prominent reason is that the document content used in our method is based only on 41935 terms extracted from 1024 documents (which is also used to build the kernel) which is around 3 times lesser than Yao et al. The results do not illustrate that the content learning based on the latent semantic kernel is worse than the vector space model, however, it shows that even though the kernel is built using the terms from only a small subset of the input documents it can still infer semantic relationships between the input document content to a degree.

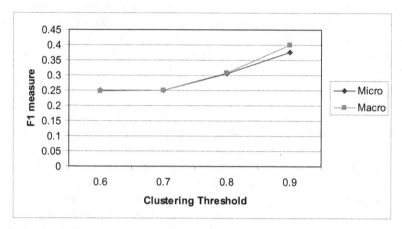

Fig. 5. Clustering solutions with different clustering thresholds based on content-only similarity with 21 clusters

Table 3. The number of clusters generated by the incremental clustering process using different clustering thresholds based on the content-only similarity

Clustering thresholds	#Clusters after the first run	#Clusters after the merging stage
0.6	1033	973
0.7	4782	4162
0.8	24128	11370
0.9	45419	2097

The experiments and result analysis show that the clustering threshold used for the incremental clustering process has a great impact on the clustering solutions. Figure 5 shows the micro-f1 and macro-f1 values of the clustering solutions with different clustering thresholds from 0.6 to 0.9. The clustering solutions are for 21 clusters using the content-only similarity. It demonstrates that the use of a higher clustering threshold for the incremental clustering is better than a lower threshold. This is due to the fact that a higher clustering threshold allows the incremental clustering process to generate a large amount of clusters at the first run. Based on this large amount of clusters generated, the iteration stage can relocate the input documents to the clusters that share the maximum similarity value with the document. This helps the incremental clustering process to reduce the order dependency to the input documents and to correctly group the input documents with the existing clusters. Table 3 shows the number of clusters generated by the incremental clustering process using different clustering thresholds. It shows that using the clustering threshold of 0.9 has generated around 45419 clusters in the first run of the incremental clustering process which is very close to the number of the input documents which is 48305. This number is significantly reduced after applying the merging phase as many clusters had only 1 document in it. The merging phase reduces the number of clusters by merging them with the most similar cluster. The clustering threshold of 0.9 has produced a better clustering solution result (as shown in figure 5) than the other clustering thresholds. This emphasises that the order of the input documents is an important aspect in an incremental clustering method since the more number of distinct clusters generated from the first run tends to produce a better clustering solution result.

6   Conclusions and Future Work

This paper presents the performance of the clustering approach that we employed to participate in the INEX 2007 Document Mining Challenge. Our clustering approach is based on a sequential combination of the incremental clustering and the pairwise-distance based clustering. From the results based on the Wikipedia dataset, it can be ascertained that the consideration of structural similarity between the documents does not improve the clustering performance significantly. Our clustering results along with other participants show that the structure information of the Wikipedia collection plays a small role in determining the true categories of the Wikipedia dataset. Clustering based on the structure-only information is good for heterogeneous structured domain documents. In a XML corpus, the structure of the XML documents should also play a significant role in clustering along with the content, for example, tag names should carry significant semantic information to use them them into clustering.

The experiments and results also reveal that the LSK built only on a subset of terms from the corpus can infer semantic relationships to a degree which can be used to cluster the document content. In future work, we would explore in more depth on how LSK can be used in the incremental approach more effectively and how it can be used in parallel with the structure information.

Acknowledgments. Computational resources and services used in this work were provided by the HPC and Research Support Unit, Queensland University of Technology, Brisbane, Australia.

References

1. Do, H.H., Rahm, E.: Coma - a system for flexible combination of schema matching approaches. In: 28th VLDB, Hong Kong, China, August) propose a hybrid matching algorithm using the modulation of veraious approaches. They support user feedback and reuse previous matchings (one to one matching) (2002)

2. Lee, L.M., Yang, L.H., Hsu, W., Yang, X.: Xclust: Clustering xml schemas for effective integration. In: 11th ACM International Conference on Information and Knowledge Management (CIKM 2002), propose a clustering method that computes a similarity between XMl schemas (one to one matching), Virginia (November 2002)
3. Lian, W., Cheung, D.W., Maoulis, N., Yiu, S.M.: An efficient and scalable algorithm for clustering xml documents by structure. IEEE TKDE 16(1), 82–96 (2004)
4. Karypis, G.: Cluto - software for clustering high-dimensional datasets karypis lab
5. Nayak, R., Tran, T.: A progressive clustering algorithm to group the xml data by structural and semantic similarity. IJPRAI 21(3), 1–21 (2007)
6. Nayak, R., Xu, S.: Xcls: A fast and effective clustering algorithm for heterogenous xml documents. In: Ng, W.-K., Kitsuregawa, M., Li, J., Chang, K. (eds.) PAKDD 2006. LNCS (LNAI), vol. 3918. Springer, Heidelberg (2006)
7. Fuhr, N., Lalmas, M., Trotman, A., Kamps, J.: Focused access to xml documents. In: 6th International Workshop of the Initiative for the Evaluation of XML Retrieval, INEX 2007, Revised and Selected Papers, Dagstuhl Castle, Germany. Springer, Heidelberg (2007) (to appear 2008)
8. Cristianini, N., Shawe-Taylor, J., Lodhi, H.: Latent semantic kernels. Journal of Intelligent Information Systems (JJIS) 18(2) (2002)
9. Landauer, T.K., Foltz, P.W., Laham, D.: An introduction to latent semantic analysis. Discourse Processes (25), 259–284 (1998)
10. Kim, Y.S., Cho, W.J., Lee, J.Y.: An intelligent grading system using heterogeneous linguistic resources. In: Gallagher, M., Hogan, J.P., Maire, F. (eds.) IDEAL 2005. LNCS, vol. 3578, pp. 102–108. Springer, Heidelberg (2005)
11. Yang, J., Cheung, W., Chen, X.: Learning the kernel matrix for xml document clustering. In: e-Technology, e-Commerce and e-Service (2005)
12. Porter, M.: An algorithm for suffix stripping. Program 14(3), 130–137 (1980)
13. Kutty, S., Tran, T., Nayak, R., Li, Y.: Clustering xml documents using closed frequency subtrees - a structure-only based approach. In: 6th International Workshop of the Initiative for the Evaluation of XML Retrieval, INEX 2007, Dagstuhl Castle, Germany, December 17-19 (2007)
14. Hagenbuchner, M., Tsoi, A., Sperduti, A., Kc, M.: Efficient clustering of structured documents using graph self-organizing maps. In: 6th International Workshop of the Initiative for the Evaluation of XML Retrieval, INEX 2007, Dagstuhl Castle, Germany, Decemeber 17-19 (2007)
15. Yao, J., Zerida, N.: Rare patterns to improve path-based clustering. In: 6th International Workshop of the Initiative for the Evaluation of XML Retrieval, INEX 2007, Dagstuhl Castle, Germany, December 17-19 (2007)
16. Salton, G., McGill, M.J.: Introduction to modern information retrieval. McGraw-Hill, New York (1983)
17. Zhao, Y., Karypis, G.: Empirical and theoretical comparisons of selected criterion functions for document clustering. In: Machine Learning, pp. 311–331 (2004)
18. Zhao, Y., Karypis, G.: Hierarchical clustering alogrithms for document datasets. Data Mining and Knowledge Discovery 10(2), 141–168 (2005)

XML Document Classification Using Extended VSM

Jianwu Yang and Fudong Zhang

Institute of Computer Sci. & Tech., Peking University, Beijing 100871, China
{yangjianwu,zhangfudong}@icst.pku.edu.cn

Abstract. Structured link vector model (SLVM) is a representation recently proposed for modeling XML documents, which was extended from the conventional vector space model (VSM) by incorporating document structures. In this paper, we describe the classification approach for XML documents based on SLVM and Support Vector Machine (SVM) in INEX 2007 Document Mining Challenge. The experimental results on the challenge's data set show that it outperforms any other approach on XML document classification task at the challenge.

Keywords: XML Document, Classification, Support Vector Machine (SVM), Vector Space Model (VSM), Structured Link Vector Model (SLVM).

1 Introduction

XML is the W3C recommended markup language for semi-structured data. Its structural flexibility makes it an attractive choice for representing data in application domains[1], including news items (NewsML), mathematical formulae (MathML), vector graphics (SVG), as well as some proprietary designs used by specific enterprises and institutions. Among the different possible XML-based documents, the focus of this paper is on those with elements containing textual descriptions.

The recent proliferation of XML adoption in large digital archives [1,2] calls for new document analysis techniques to support effective semi-structured document management, sometimes down to the level of the composing elements. Even though the tasks of interest are still clustering, classification and retrieval, conventional document analysis tools developed for unstructured documents [3] fail to take the full advantage of the structural properties of XML documents.

To contrast with ordinary unstructured documents, XML documents represent their syntactic structure via (1) the use of XML elements, each marked by a user-specified tag, and (2) the associated schema specified in either DTD or XML Schema format. In addition, XML documents can be cross-linked by adding IDREF attributes to their elements to indicate the linkage. Thus, techniques designed for XML document analysis normally take into account the information embedded in both the element tags as well as their associated contents for better performance. For example, the structural similarity between a pair of IDREF-free XML documents can be defined as

---

[1] Hundreds of different XML applications can be found at http://xml.coverpages.org/xmlApplications.html

N. Fuhr et al. (Eds.): INEX 2007, LNCS 4862, pp. 234–244, 2008.
© Springer-Verlag Berlin Heidelberg 2008

some edit distance between unordered labeled trees[2], i.e., to compute the minimum number of operators needed to edit the tree from one from to another. In the literature, different tree edit distances have been proposed for measuring XML document dissimilarity [4, 5], which are equivalent in principle except for the edit operators allowed and whether repetitive and optional XML elements were considered. However, computing the edit distance between a pair of unordered labeled trees is NP-complete [6] in general and yet the distance is not optimal in any sense. This is undesirable for large-scale applications. An alternative is to measure the depth difference with reference to the root element for defining structural dissimilarity between a pair of XML elements [7, 8]. The depth differences can then be aggregated for estimating the overall document dissimilarity. While the associated computational cost is low, the accuracy is limited. Other than trees, XML documents have also been represented as time series [9], with each occurrence of a tag corresponding to an impulse. Document similarity was then computed by comparing the corresponding Fourier coefficients of the documents. This approach does not take into account the order in which the elements appear and is adequate only when the XML documents are drastically different from each other, i.e., they have very few tags in common. In [10], WordNet -- an ontology of general concepts [11] has been used to measure the semantic similarity of the elements' names and their values. However, in many applications, domain-specific knowledge is needed instead, which is sometimes not easy to be captured.

Structured Link Vector Model (SLVM), which forms the basis of this paper, was originally proposed in [12] for representing XML documents. It was extended from the conventional vector space model (VSM) by incorporating document structures (represented as term-by-element matrices), referencing links (extracted based on IDREF attributes), as well as element similarity (represented as an element similarity matrix).

Table 1 shows a more complete list of related works and their comparison in terms of representation and the nature of similarity considered.

Table 1. A comparison of related works in the literature with XML similarity considered

References	Structural similarity	Semantic similarity	Remarks
[12]	Yes	No	Extending VSM
[7,8]	Yes	No	Tree edit distance
[9]	Yes	No	Fourier coefficients
[10]	No	Yes	Ontology–based
[13]	Yes	No	Tree-based generative language model
[14, 15]	Yes	No	Extending VSM
[16]	No	Yes	Extending query relaxation
[17]	Yes	No	Bayesian network model
[18]	Yes	No	A mixture Language model
[19]	Yes	Yes	Queries in natural language

---

[2] A labeled unordered tree is a tree structure with all its nodes labeled but the order of the children of any parent node not maintained. The use of unordered trees for representing XML documents is justified by the fact that two documents with identical contents but different orderings of their sibling elements should be considered as semantically equivalent.

2 Structured Link Vector Model (SLVM)

2.1 Basic Representation

Vector Space Model (VSM) [20] has long been used to represent unstructured documents as document feature vectors which contain term occurrence statistics. This bag of terms approach assumes that the term occurrences are *independent* of each other.

Definition 2.1. *Assume that there are n distinct terms in a given set of documents D. Let doc_x denote the xth document and d_x denote the* **document feature vector** *such that*

$$d_x = [d_{x(1)}, d_{x(2)} \cdots\cdots, d_{x(n)}]^T$$
$$d_{x(i)} = TF(w_i, doc_x) \, IDF(w_i)$$

where $TF(w_i, doc_x)$ is the frequency of the term w_i in doc_x, $IDF(w_i) = log(|D|/DF(w_i))$ is the inverse document frequency of w_i for discounting the importance of the frequently appearing terms, $|D|$ is the total number of the documents, and $DF(w_i)$ is the number of documents containing the term w_i.

Applying VSM directly to represent XML documents is not desirable as the document syntactic structure tagged by their XML elements will be ignored. For example, VSM considers two documents with an identical term appearing in, say, their "title" fields to be equivalent to the case with the term appearing in the "title" field of one document and in the "author" field of another. As the "author" field is semantically unrelated to the "title" field, the latter case should be considered as a piece of less supportive evidence for the documents to be similar when compared with the former case. Using merely VSM, these two cases cannot be differentiated.

Structured Link Vector Model (SLVM), proposed in [12], can be considered as an extended version of vector space model for representing XML documents. Intuitively speaking, SLVM represents an XML document as an array of VSMs, each being specific to an XML element (specified by the `<element>` tag in a DTD).[3]

Definition 2.2. *SLVM represents an XML document doc_x using a* **document feature matrix** *$\Delta_x \in R^{n \times m}$, given as*

$$\Delta_x = [\Delta_{x(1)}, \Delta_{x(2)}, \cdots\cdots, \Delta_{x(m)}]$$

where m is the number of distinct XML elements, $\Delta_{x(i)} \in R^n$ is the TFIDF feature vector representing the ith XML element (e_i), given as $\Delta_{x(i,j)} = TF(w_j, doc_x.e_i) \cdot IDF(w_j)$ for all j=1 to n, and $TF(w_j, doc_x.e_i)$ is the frequency of the term w_j in the element e_i of doc_x.

---

[3] In the current version of SLVM, only the elements corresponding to the leaf nodes of the XML DOM tree are modeled.

Definition 2.3. *The* **normalized document feature matrix** *is defined as*

$$\tilde{\Delta}_{x(i,j)} = \Delta_{x(i,j)} / \sum_k \Delta_{x(i,k)}$$

where the factor caused by the varying size of the element content is discounted via normalization.

Example 2.1. Figure 1 shows a simple XML document. Its corresponding document feature vector d_x, document feature matrix Δ_x, and normalized document feature matrix $\tilde{\Delta}_x$ are shown in Figure 2-4 respectively. Here, we assume all the terms share the same *IDF* value equal to one.

The form of SLVM studied in this paper is only a simplified one where only the leaf-node elements in the DTD are incorporated without considering their positions in the document DOM tree and their consecutive occurrence patterns. In addition, the inter-connectivity between the documents based on IDREF is also not considered. One obvious advantage is that this simplification can make the subsequent similarity learning much more tractable. Also, this kind of unigram-like approach makes it applicable to most of the unseen XML documents as long as there are no newly encountered terms. If the consecutive occurrence patterns of the elements are to be taken into consideration, the most extreme case is to have each possible path of the DOM tree corresponds to one column in Figure 3. This however will increase the computational complexity exponentially. Also, the generalization capability will be poor (e.g,. a book with three authors cannot be modeled if a maximum of two authors are assumed in the SLVM's document feature matrix).

```
<article>
    <title>Ontology Enabled Web Search</name>
    <author>John</author>
    <conference>Web Intelligence</conference>
</article>
```

Fig. 1. An XML document

$$d_x = \begin{bmatrix} 1 \\ 1 \\ 2 \\ 1 \\ 1 \\ 1 \end{bmatrix} \begin{matrix} \text{Ontology} \\ \text{Enabled} \\ \text{Web} \\ \text{Search} \\ \text{John} \\ \text{Intelligence} \end{matrix}$$

thisDocument

Fig. 2. The document feature vector for the example shown in Figure 1

$$\Delta_x = \begin{array}{c} \begin{matrix} \text{title} & \text{author} & \text{Coference} \end{matrix} \\ \begin{bmatrix} 1 & 0 & 0 \\ 1 & 0 & 0 \\ 1 & 0 & 1 \\ 1 & 0 & 0 \\ 0 & 1 & 0 \\ 0 & 0 & 1 \end{bmatrix} \end{array} \begin{matrix} \text{Ontology} \\ \text{Enabled} \\ \text{Web} \\ \text{Search} \\ \text{John} \\ \text{Intelligence} \end{matrix}$$

Fig. 3. The document feature matrix for the example in Figure 1

$$\tilde{\Delta}_x = \begin{array}{c} \begin{matrix} \text{title} & \text{author} & \text{Conference} \end{matrix} \\ \begin{bmatrix} 0.5 & 0 & 0 \\ 0.5 & 0 & 0 \\ 0.5 & 0 & \sqrt{2}/2 \\ 0.5 & 0 & 0 \\ 0 & 1 & 0 \\ 0 & 0 & \sqrt{2}/2 \end{bmatrix} \end{array} \begin{matrix} \text{Ontology} \\ \text{Enabled} \\ \text{Web} \\ \text{Search} \\ \text{John} \\ \text{Intelligence} \end{matrix}$$

Fig. 4. The normalized document feature matrix for the example in Figure 1

2.2 Similarity Measures

Using VSM, similarity between two documents doc_x and doc_y is typically computed as the cosine value between their corresponding document feature vectors, given as

$$sim(doc_x, doc_y) = \frac{d_x d_y}{\| d_x \| \| d_y \|} = \tilde{d}_x \tilde{d}_y^{\,T} = \sum_{i=1}^{k} \tilde{d}_{x(i)} \tilde{d}_{y(i)} \tag{1}$$

where n is the total number of terms and $\tilde{d}_x = d_x / \| d_x \|$ *denotes normalized* d_x. *So, the similarity measure can also be interpreted as the inner product of the normalized document feature vectors.*

For SLVM, with the objective to model semantic relationships between XML elements, the corresponding document similarity can be defined with an element similarity matrix introduced.

Definition 2.4. The ***SLVM-based document similarity*** *between two XML documents* doc_x *and* doc_y *is defined as*

$$sim(doc_x, doc_y) = \sum_{j=1}^{m} \sum_{i=1}^{m} M_{e(i,j)} \cdot (\tilde{\Delta}_{x(i)}^{\,T} \bullet \tilde{\Delta}_{y(j)}) \tag{2}$$

where M_e *is a matrix of dimension m×m and named as the* ***element similarity matrix***.

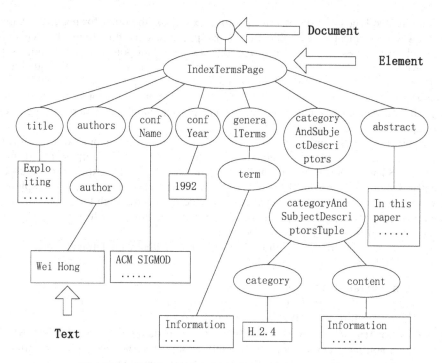

Fig. 5. The DOM tree of the ACMSIGMOD Record dataset

The matrix M_e in Eq. (2) captures both the similarity between a pair of XML elements as well as the contribution of the pair to the overall document similarity (*i.e.*, the diagonal elements of M_e are not necessarily equal to one). An entry in M_e being small means that the two corresponding XML elements should be unrelated and same words appearing in the two elements of two different documents will not contribute much to the overall similarity of them. If M_e is diagonal, this implies that all the XML elements are not correlated at all with each other, which obviously is not the optimal choice.

The structural similarity between a pair of XML documents can thus be computed based on different tree edit distances [4,5] which differ from each others in terms of the set of allowed edit operators and their support for repetitive and optional XML elements. It has been proved in [6] that computing the edit distance between a pair of unordered labeled trees is NP-complete in general and yet the distance is not optimal in any sense related to the elements' semantics. In [12], the element similarity was pre-set to be related to the path difference between two elements as well as their depth difference with reference to the root derived from the document schema. To obtain an optimal M_e for a specific type of XML data, we proposed in [21] to learn the matrix using pair-wise similar training data in an iterative manner.

3 SVM for XML Documents Classification

SVM was introduced by Vapnik in 1995 for solving two-class pattern recognition problems using the Structural Risk Minimization principle [22]. Given a training set containing two kinds of data (one for positive examples, the other for negative examples), which is linearly separable in vector space, this method finds the decision hyper-plane that best separated positive and negative data points in the training set. The problem searching the best decision hyper-plane can be solved using quadratic programming techniques [23]. SVM can also extend its applicability to linearly non-separable data sets by either adopting soft margin hyper-planes, or by mapping the original data vectors into a higher dimensional space in which the data points are linearly separable [22, 23, 24].

Joachims [25] first applied SVM to text categorization, and compared its performance with other classification methods using the Reuters-21578 corpus. His results show that SVM outperformed all the other methods tested in his experiments. Subsequently, Dumais [26], Yang [27], Cooley [28], and Bekkerman [29] also explored how to solve text categorization with SVM respectively. Although based on different document collections, their experiments confirmed Joachim's conclusion that SVM is the best method for classifying text documents.

SVM success in practice is drawn by its solid mathematical foundations which convey the following two salient properties:

- **Margin maximization:** The classification boundary functions of SVM maximize the margin, which in machine learning theory, corresponds to maximizing the *generalization* performance given a set of training data.
- **Nonlinear transformation of the feature space using the kernel trick:** SVM handle a nonlinear classification efficiently using the kernel trick which implicitly transforms the input space into another high dimensional feature space.

In SVM, the problem of computing a margin maximized boundary function is specified by the following quadratic programming (QP) problem:

minimize: $W(\alpha) = -\sum_{i=1}^{n} \alpha_i + \frac{1}{2}\sum_{i=1}^{n}\sum_{j=1}^{n} \alpha_i \alpha_j y_i y_j k(x_i, x_j)$

subject to: $\sum_{i=1}^{n} y_i \alpha_i = 0; \ \forall i : 0 \le \alpha_i \le C$

where n The number of training data, α is a vector of n variables, where each component α_i corresponds to a training data (x_i, y_i). C is the soft margin parameter controlling the influence of the outliers (or noise) in training data.

The classification function is:

$$f(x) = \text{sgn}\{\sum_{i,=1}^{n} \alpha_i y_i k(x_i, x) + b\}$$

where b is a threshold for categorization.

The kernel $k(x_i, x_j)$ could be regarded as the similarity function between two data points. For linear boundary, the kernel function is $x_i \cdot x_j$, a scalar product of two data points. The nonlinear transformation of the feature space is performed by replacing $k(x_i, x_j)$ with an advanced kernel, such as polynomial kernel $(x^T x_i + 1)^p$ or RBF kernel $exp(-\frac{1}{2\delta^2} \| x - x_i \|^2)$.

In SLVM, the similarity between two XML documents is defined as definition 2.4, so we consider the kernel $k(x_i, x_j)$ for XML documents classification based on SLVM as:

$$k(x_i, x_j) = sim(doc_x, doc_y) = \sum_{j=1}^{m} \sum_{i=1}^{m} M_{e(i,j)} \cdot (\tilde{\Delta}_{x(i)}^{T} \bullet \tilde{\Delta}_{y(j)}).$$

4 Experiments

In the experiments, all the algorithms were implemented by us in C++, except the SVM algorithm in SVMTorch [30]. All experiments were run on a PC with a 3.0 GHz Intel CPU and 512M RAM.

4.1 Initial Experiments

Test data were not available until shortly before the conclusion of the XML classification competition. As a consequence, the initial approaches addressed in this section evaluate performances on the training data. Performance evaluations on test data will be given in Section 4.2. Thus, initially we resorted to splitting the available data (the original training data set) into two sub-sets:

- **Training Set:** Part of the original training data set is selected to be used for training purposes.
- **Test Set:** The remaining is used as test data.

The number of different elements is one of key factors in efficiency, but the most of elements' occurrence times in the data set are less than 10. Thus we eliminate those elements whose occurrence times in the data set are less than 10 in the experiment, and the remaining elements are about 15% of all elements. The experimental result show that the time cost is reduced evidently and the effect is not nearly influenced by the elimination.

As a basic format, the matrix M_e in Eq. (2) is set as diagonal, this means that all the XML elements are not correlated at all with each other, which obviously is not the optimal choice. In [12], the element similarity (the entry of the matrix M_e) was pre-set to be related to the path difference between two elements as well as their depth difference with reference to the root derived from the document schema, but the experimental results show it is useless in the data set. Fallaciously, the experimental results also show it is useless that the element similarity estimated using the edit distance [4,5] in the data set. So, the matrix M_e is set as diagonal in the advance experiments.

4.2 Advanced Experiments

According to the experimental results, those elements whose occurrence times are more than 10 in the train data set are considered as available, and the matrix M_e is set as diagonal in the advance experiment.

Table 2. The Results for SVM Classification Based on SLVM

Class	1	2	3	4	5	6	7	8	9	10	11	12	13	14	15	16	17	18	19	20	21	Size	Recall
1	560	3	1	3	9	3	1	5	13	1	14	0	0	40	0	4	1	0	0	2	0	660	0.8485
2	0	945	27	10	34	1	0	14	1	0	5	0	0	67	1	14	11	0	0	0	0	1130	0.8363
3	2	6	546	0	1	0	0	1	0	0	5	1	0	42	0	3	2	0	0	7	1	617	0.8849
4	10	21	2	1087	47	2	7	17	3	0	24	8	0	253	0	55	9	2	0	38	3	1588	0.6845
5	5	22	3	25	7655	0	1	41	45	0	51	3	16	229	11	243	41	2	4	21	0	8418	0.9094
6	0	0	0	1	7	142	0	0	0	0	0	0	0	39	0	10	1	1	0	0	0	201	0.7065
7	0	0	0	0	2	0	421	0	0	0	0	0	0	60	0	1	0	0	0	0	0	484	0.8698
8	1	9	1	5	40	0	0	1984	52	2	10	0	0	76	2	34	8	2	0	8	0	2234	0.8881
9	6	2	2	1	59	0	0	42	1099	0	7	7	2	53	2	16	10	2	0	18	1	1329	0.8269
10	0	0	0	0	0	0	0	0	0	577	2	0	0	12	0	0	0	0	0	0	0	591	0.9763
11	1	4	4	13	46	0	0	6	6	0	6962	2	0	179	6	24	8	1	0	5	0	7267	0.958
12	0	3	0	6	16	0	1	1	2	0	4	2134	1	75	0	3	3	0	0	65	0	2314	0.9222
13	0	1	0	0	26	0	0	0	0	0	2	0	242	27	0	5	1	0	0	0	0	304	0.7961
14	12	46	30	89	290	3	35	69	46	11	191	54	18	10767	26	248	50	3	0	113	4	12105	0.8895
15	0	0	0	1	19	0	0	4	0	0	4	7	0	62	172	2	0	0	1	3	0	275	0.6255
16	4	9	3	32	320	1	4	25	20	0	72	3	4	319	0	3010	24	3	2	28	1	3884	0.775
17	5	15	0	3	115	2	0	21	11	0	47	10	8	126	0	55	780	1	3	14	3	1219	0.6399
18	0	0	0	0	14	0	0	1	2	0	0	0	0	5	0	3	0	204	0	1	0	230	0.887
19	0	0	0	0	3	0	0	0	0	0	0	0	0	4	0	1	2	0	231	0	0	241	0.9585
20	2	3	9	24	103	0	1	14	53	1	20	89	5	173	5	49	7	0	0	2088	13	2659	0.7853
21	0	1	1	2	4	0	0	2	0	0	2	0	0	7	0	2	0	0	0	6	528	555	0.9514

Micro average recall = 0.8722; Macro average recall = 0.8390

The classification result based on SLVM is presented in Table 2, which is based on utilizing 100% of the original training set for training purposes, and the test dataset is provided by INEX 2007. Each row of the table is the real class and each column is the found class by the algorithm.

5   Conclusion and Future Works

In this paper, we studied in detail a proposed extension of VSM called SLVM for representing XML documents so that term semantics, element similarity, as well as elements' relative importance for a given set of documents can all be taken in account. And we applied SLVM and SVM to XML documents classification. The proposed method was demonstrated to outperform any other approach on XML document classification task at INEX 2007 Document Mining Challenge.

For future work, we are interested to study how the similarity matrix obtained via the machine learning approach and support multiple word sense identification which serves as an important component for automatic ontology generation.

Acknowledgment

The work reported in this paper was supported by the National Natural Science Foundation of China Grant 60642001.

References

1. Early Americas Digital Archive,
   http://www.mith2.umd.edu:8080/eada/intro.jsp
2. Contemporary Culture Virtual Archives in XML, http://www.covax.org/
3. Berry, M.: Survey of Text Mining: Clustering, Classification, and Retrieval. Springer, Heidelberg (2003)
4. Zhang, Z.P., Li, R., Cao, S.L., Zhu, Y.Y.: Similarity Metric for XML Documents. In: Proceedings of the 2003 Workshop on Knowledge and Experience Management (FGWM 2003), Karlsruhe (2003)
5. Nierman, A., Jagadish, H.V.: Evaluating Structural Similarity in XML Documents. In: Proceedings of the Int. Workshop on the Web and Databases (WebDB), Madison, WI (2002)
6. Zhang, K., Statman, R., Shasha, D.: On the editing distance between unordered labeled trees. Information Processing Letters 42(3), 133–139 (1992)
7. Abolhassani, M., Fuhr, N., Malik, S.: HyREX at INEX. In: Proceedings of the 2003 INEX Workshop, Schloss Dagstuhl (2003)
8. Azevedo, M.I.M., Amorim, L.P., Ziviani, N.: A Universal Model for XML Information Retrieval. In: Fuhr, N., Lalmas, M., Malik, S., Szlávik, Z. (eds.) INEX 2004. LNCS, vol. 3493, pp. 311–321. Springer, Heidelberg (2005)
9. Flesca, S., Manco, G., Masciari, E., Pontieri, L., Pugliese, A.: Detecting structural similarities between xml documents. In: Proceedings of the International Workshop on the Web and Databases (WebDB), Madison, WI (2002)
10. Schenkel, R., Theobald, A., Weikum, G.: XXL @ INEX 2003. In: Proceedings of the 2003 INEX Workshop, Schloss Dagstuhl (2003)
11. Fellbaum, C.: WordNet: An Electronic Lexical Database. MIT Press, Cambridge (1998)
12. Yang, J., Chen, X.: A semi-structured document model for text mining. Journal of Computer Science and Technology 17(5), 603–610 (2002)
13. Ogilvie, P., Callan, J.: Language Models and Structured Document Retrieval. In: Proceedings of the 2002 INEX Workshop, Schloss Dagstuhl (2002)
14. Mass, Y., Mandelbrod, M., Amitay, E., Carmel, D., Maarek, Y., Soffer, A.: JuruXML – an XML Retrieval System at INEX 2002. In: Proceedings of the 2002 INEX Workshop, Schloss Dagstuhl (2002)
15. Crouch, C., Mahajan, A., Bellamkonda, A.: Flexible XML Retrieval Based on the Extended Vector Model. In: Proceedings of the 2004 INEX Workshop, Schloss Dagstuhl (2004)
16. Liu, S., Chu, W.: Cooperative XML (CoXML) Query Answering at INEX 2003. In: Proceedings of the 2003 INEX Workshop, Schloss Dagstuhl (2003)

17. Vittaut, J., Piwowarski, B., Gallinari, P.: An algebra for Structured Queries in Bayesian Networks. In: Proceedings of the 2004 INEX Workshop, Schloss Dagstuhl (2004)
18. Sigurbjornsson, B., Kamps, J., Rijke, M.: The University of Amsterdam at INEX 2004. In: Proceedings of the 2004 INEX Workshop, Schloss Dagstuhl (2004)
19. Woodley, A., Geva, S.: NLPX at INEX 2004. In: Proceedings of the 2004 INEX Workshop, Schloss Dagstuhl (2004)
20. Salton, G., McGill, M.J.: Introduction to Modern information Retrieval. McGraw-Hill, New York (1983)
21. Yang, J.W., Cheung, W.K., Chen, X.O.: Integrating Element Kernel and Term Semantics for Similarity-Based XML Document Clustering. In: Proceedings of 2005 IEEE/WIC/ACM International Conference on Web Intelligence (WI 2005), Compiegne, France (2005)
22. Vapnic, V.: The Nature of Statistical Learning Theory. Springer, New York (1995)
23. Cortes, C., Vapnik, V.: Support Vector networks. Machine Learning 20, 273–297 (1995)
24. Osuna, R.F., Girosi, F.: Support vector machines: Training and applications. In: A.I. Memo. MIT A.I. Lab (1996)
25. Joachims, T.: Text Categorization with Support Vector Machines: Learning with Many Relevant Features. In: Nédellec, C., Rouveirol, C. (eds.) ECML 1998. LNCS, vol. 1398, pp. 137–142. Springer, Heidelberg (1998)
26. Dumais, S., Platt, J., Heckerman, D., Sahami, M.: Inductive learning algorithms and representations for text categorization. In: Proceedings of the 1998 ACM CIKM International Conference on Information and Knowledge Management, pp. 148–155 (1998)
27. Yang, Y., Liu, X.: A re-examination of text categorization methods. In: 22nd Annual International ACM SIGIR Conference on Research and Development in Information Retrieval (SIGIR 1999), pp. 42–49 (1999)
28. Cooley, R.: Classification of News Stories Using Support Vector Machines. In: Proceedings of the 16th International Joint Conference on Artificial Intelligence Text Mining Workshop (1999)
29. Bekkerman, R., Ran, E.Y., Tishby, N., Winter, Y.: On feature distributional clustering for text categorization. In: Proceedings of the 24th ACM SIGIR International Conference on Research and Development in Information Retrieval, pp. 146–153 (2001)
30. Collobert, R., Bengio, S.: SVMTorch: support vector machines for large-scale regression problems. Journal of Machine Learning Research 1, 143–160 (2001)

Overview of the INEX 2007 Entity
Ranking Track

Arjen P. de Vries[1,2,*], Anne-Marie Vercoustre[3], James A. Thom[4],
Nick Craswell[5], and Mounia Lalmas[6,**]

[1] CWI, Amsterdam, The Netherlands
[2] Technical University Delft, Delft, The Netherlands
[3] INRIA-Rocquencourt, Le Chesnay Cedex, France
[4] RMIT University, Melbourne, Australia
[5] Microsoft Research Cambridge, Cambridge, UK
[6] Queen Mary, University of London, London, UK

Abstract. Many realistic user tasks involve the retrieval of specific entities instead of just any type of documents. Examples of information needs include 'Countries where one can pay with the euro' or 'Impressionist art museums in The Netherlands'. The Initiative for Evaluation of XML Retrieval (INEX) started the XML Entity Ranking track (INEX-XER) to create a test collection for entity retrieval in Wikipedia. Entities are assumed to correspond to Wikipedia entries. The goal of the track is to evaluate how well systems can rank entities in response to a query; the set of entities to be ranked is assumed to be loosely defined either by a generic category (entity ranking) or by some example entities (list completion). This track overview introduces the track setup, and discusses the implications of the new relevance notion for entity ranking in comparison to ad hoc retrieval.

1   Introduction

Information retrieval evaluation assesses how well systems identify information objects relevant to the user's information need. TREC has used the following working definition of relevance: 'If you were writing a report on the subject of the topic and would use the information contained in the document in the report, then the document is relevant.' Here, a document is judged relevant if any piece of it is relevant (regardless of how small that piece is in relation to the rest of the document).

Many realistic user tasks seem however better characterised by a different notion of relevance. Often, users search for specific entities instead of just any type of documents. Examples of information needs include 'Countries where one can pay with the euro' or 'Impressionist art museums in The Netherlands', where the

* This work is partially supported by the Dutch ICES/KIS III. BSIK project Multi-mediaN.
** This work is partially supported by DELOS, the European Network of Excellence in Digital Libraries.

N. Fuhr et al. (Eds.): INEX 2007, LNCS 4862, pp. 245–251, 2008.
© Springer-Verlag Berlin Heidelberg 2008

entities to be retrieved are countries and museums; articles discussing the euro currency itself are not relevant, nor are articles discussing Dutch impressionist art.

To evaluate retrieval systems handling these *typed* information needs, the Initiative for Evaluation of XML Retrieval (INEX) started the XML Entity Ranking track (INEX-XER), with the aim to create a test collection for entity retrieval in Wikipedia. Section 2 provides details about the collection and assumptions underlying the track. Section 3 summarizes the results of the participants. Section 4 presents some findings related to the modified working definition of relevance, comparing entity ranking to ad hoc retrieval.

2   INEX-XER Setup

The main objective in the INEX-XER track is to return *entities* instead of 'just' web pages. The track therefore concerns triples of type <category, query, entity>. The category (that is *entity type*), specifies the type of 'things' to be retrieved. The query is a free text description that attempts to capture the information need. Entity specifies a (possibly empty) list of example instances of the entity type.

The usual information retrieval tasks of document and element retrieval can be viewed as special instances of this more general retrieval problem, where the category membership relates to a syntactic (layout) notion of 'text document', or, in the case of INEX ad hoc retrieval, 'XML element' or 'passage'. Expert finding uses the semantic notion of 'people' as its category, where the query would specify 'expertise on $\mathcal{T}$' for expert finding topic $\mathcal{T}$.

2.1   Data

The general case of retrieving entities (such as countries, people and dates) requires the estimation of relevance of items (i.e., instances of entities) that are not necessarily represented by text content other than their descriptive label [2]. INEX-XER 2007 approached a slightly easier sub-problem, where we restricted candidate items to those entities that have their own Wikipedia article. This decision simplifies not only the problem of implementing an entity ranking system (ignoring the natural language processing requirement of the general case), but, importantly, it also simplifies evaluation – as every retrieved result will have a proper description (its Wikipedia entry) to base the relevance judgement on.

The Wikipedia category metadata about entries has been exploited to loosely define entity sets. This category metadata is contained in the following files:

- categories_name.csv which maps category ids to category names
- categories_hcategories.csv which defines the category graph (which is not a strict hierarchy!)
- categories_categories.csv which maps article ids (that is pages that correspond to entities) to category ids

The entities in such a set are assumed to loosely correspond to those Wikipedia pages that are labeled with this category (or perhaps a sub-category of the given category). For example, considering the category 'art museums and galleries' (10855), an article about a particular museum such as the 'Van Gogh Museum' (155508) may be mapped to a sub-category like 'art museums and galleries in the Netherlands' (36697). Obviously, the correspondence between category metadata and the entity sets is far from perfect, as Wikipedia articles are often assigned to categories inconsistently. Since the human assessor of retrieval results is not constrained by the category assignments made in the corpus when making his or her relevance assessments, track participants have to handle the situation that the category assignments to Wikipedia pages are not always consistent, and also far from complete. correct answers may belong to other categories *close to* the provided one in the Wikipedia category graph, or may not have been categorized at all by the Wikipedia contributors. The challenge is to exploit a rich combination of information from text, structure and links for this purpose.

2.2 Tasks

In 2007, the track has distinguished two tasks, Entity Ranking and List Completion.

The motivation for the Entity Ranking task is to return entities that satisfy a topic described in natural language text. In other words, in the entity ranking task, the information need includes which category (entity type) is desired as answers. An Entity Ranking topic specifies the category identifier and the free-text query specification.[1] Results consist of a list of Wikipedia pages (our assumption is that all entities have a corresponding page in Wikipedia). For example, with 'Art museums and galleries' as the input category and a topic text 'Impressionist art in the Netherlands', we expect answers like the 'Van Gogh museum' and the 'Kröller-Müller museum'.

In the List Completion task, instead of knowing the desired category (entity type), the topic specifies between one and three correct entities (instances) together with a free-text context description. Results consist again of a list of entities (Wikipedia pages). As an example, when ranking 'Countries' with topic text 'European countries where I can pay with Euros', and entity examples such as 'France', 'Germany', 'Spain', then the 'Netherlands' would be a correct completion, but the 'United Kingdom' would not. Because the problem is to complete the partial list of answers, the given examples are considered non-relevant results in the evaluation of this task.

2.3 Topics

Figure 1 shows an example topic, developed by a sailing enthusiast. The INEX-XER topics can be used for both entity ranking and list completion tasks. When evaluating methods for entity ranking, the example entities given in the topic are of course not to be known by the entity ranking system. Likewise, in the list completion task, the category information would not be provided.

---

[1] Multiple categories are allowed per topic.

```
<inex_topic topic_id="60" query_type="XER">
<title>olympic classes dinghy sailing</title>
<entities>
  <entity id="816578">470 (dinghy)</entity>
  <entity id="1006535">49er (dinghy)</entity>
  <entity id="855087">Europe (dinghy)</entity>
</entities>
<categories>
  <category id="30308">dinghies</category>
</categories>
<description>
The user wants the dinghy classes that are or have been olympic classes,
such as Europe and 470.
</description>
<narrative>
The expected answers are the olympic dinghy classes, both historic and
current. Examples include Europe and 470.
</narrative>
</inex_topic>
```

Fig. 1. Example topic

As mentioned before, Wikipedia categories define the entity type only loosely. Relevant entity answers may not belong to the specified category (in the corpus). Looking into the relevance assessments of the 2007 XER topics, we find that only 221 Wikipedia entries out of the total 996 relevant topic-entity pairs have at least one of the categories as given in the topic assigned in their metadata. For example, when ranking explorers in response to the information need 'Pacific navigators Australia explorers' (topic 65), some of the relevant Wikipedia entries have been labelled with categories 'explorers of australia' or 'explorers of the pacific' instead of topic category 'explorers'. Other relevant entities may have no category information at all. The category given in the topic should therefore be considered no more than an indication of what is expected, not a strict constraint (like in the CAS title for the ad hoc track).

2.4   The 2007 Test Collection

The created INEX-XER test collection provides training topics and testing topics.

A training set of 28 topics, based on a selection of 2006 ad hoc adapted to the entity task, has been kindly made available by INRIA (who developed this data) for participants to develop and train their systems. The relevance assessments have been derived from the articles judged relevant in 2006, limiting the set to the corresponding 'entities'. Of course, this procedure gives no guarantee that all the relevant entities have been assessed; this depends on completeness of the ad hoc pool. Also, notice that the original title, description and narrative fields have not been updated to reflect the new entity ranking interpretation of the training topics.

Table 1. Entity ranking results, the run from each of the groups with the best MAP, sorted by MAP

Team	Run	MAP
utwente	qokrwlin	0.306
inria	ER_comb-Q-TC-n5-a1-b8	0.293
uopen	er01	0.258
ukobe	qlm50_wwswitchlda800_fixed	0.227
utampere	er_2v2	0.210
uceg	ceger	0.191
unitoronto	single_nofilter	0.130
uhannover	qcs	0.123

The testing data consists of two parts. Topics 30–59 have been derived from the ad hoc 2007 assessments, similar to the way that the training data have been produced. For these topics, description and narrative may not be perfect, but they should be similar to the training topics. These topics have been assessed by track organizers (i.e., not by the original topic authors), with pools consisting of the articles that contained relevant information in the INEX ad hoc 2007 assessments. Of the originally proposed set of ad hoc derived topics, seven topics have been dropped because the ad hoc pools on which to base the XER assessments did not exist, and two topics have been dropped because their answer sets contained more than 50 relevant entities (and therefore we do not trust the original pools to be sufficiently complete). The final set consists of *21 ad hoc derived entity ranking test topics with assessments*.

Topics 60–100 are the genuine XER topics, created by participants specifically for the track. Almost all topics have been assessed by the original topic authors. From the originally proposed topics, we have dropped topics 93 (because it was too similar to topic 94) and topic 68 (because the underlying information need was identical to that of topic 61). Nine more topics were dropped because their answer sets contained more relevant entities than (or just about as many as) the pool depth (of 50), and two topics have been dropped because their assessments were never finished. The final set consists of *25 genuine entity ranking test topics with assessments* (that could eventually be expanded to 35 should we decide to perform more assessments).

3   Results

The eight participating groups submitted in total eighteen runs for the entity ranking task, and six participants submitted another ten list completion runs. Pools were constructed from the top 50 results for the two highest priority entity ranking and the two highest priority list completion runs. The pools contained on average about 500 entities per topic.

Table 1 presents the best results per group on the entity ranking task, in reverse order of mean average precision. Participants reported that runs exploiting the rich structure of the collection (including category information, associations

Table 2. List completion results, the run from each of the groups with the best MAP, sorted by MAP. The additional columns detail the number of examples that have been removed from the submitted run, and the MAP of the original submission.

Team	Run	MAP	#ex	MAP (uncorrected)
inria	LC_comb-Q-a2-b6	0.309	0	0.309
utwente	qolckrwlinfeedb	0.281	115	0.246
utampere	lc_1	0.247	1	0.246
unitoronto	single_EntityCats_d0_u0	0.221	102	0.198
uceg	ceglc	0.217	1	0.217
uopen	lc01	0.207	101	0.168

between entities, and query-dependent link structure) have performed better than the baseline of plain article retrieval (see e.g. [1], as well as participant papers in this volume).

Table 2 summarizes the list completion results. The given topic examples are regarded non-relevant in the evaluation of the list completion task. Because several teams had by mistake included these given topic examples in their submissions, the Table lists the mean average precision of runs after removing the given examples from the submitted ranked lists. The number of topic examples removed and the original scores are given in the two remaining columns. Notice a minor anomaly: because we discovered in the judging phase (after submission) that one of the examples provided for topic 54 had been incorrect (i.e., non-relevant), the runs of some teams included the replacement entity (WP2892991, now a topic example but not at the time of submission).

4   Relevance in Entity Retrieval

Many ad hoc topics can serve as entity ranking topics, but the articles containing relevant passages must be re-assessed. These relevance assessments for the ad hoc derived training and testing topics provide therefore a basis to compare the notion of relevance for entity ranking to that used in ad hoc retrieval. On the eighteen ad hoc 2007 topics that were re-assessed as XER topics, only about 35% of the originally relevant documents have been assessed relevant. Depending on the topic, often surprisingly many articles that are *on topic* for the ad hoc track are not relevant entities. Also, articles that contain hubs (e.g., the Wikipedia 'list of ...' pages) are not entities, and not considered relevant.

Looking at two specific example topics to illustrate, only 6 out of the 129 'French president in the 5th republic' relevant ad hoc results (XER topic 35 and ad hoc topic 448) are actually presidents, and only 32 out of the 267 'Bob Dylan songs' relevant ad hoc results (XER topic 54 and ad hoc topic 509) are actually songs. The latter result set includes many articles related to Bob Dylan himself, The Band, albums, a documentary, a speech given at the March on Washington, singer-songwriters that play covers or tributes, cities where Bob Dylan lived, etc. Even though the ad hoc results do often contain a passage that mentions a song title, the XER model of the information need seems (arguably) closer to the 'real' user need; and, even in the ad hoc retrieval scenario, one would expect the entity results to be of higher relevance value (for this particular information need) than the remaining relevant pages.

While the track has not investigated the pool quality in-depth, the participants (and topic-authors) missed only few entities that they knew about (this could be a bigger issue with the ad hoc derived topics). In some specific cases, we validated the pool completeness using manually identified hubs in the collection, but no missing entities were found. We have found that runs for the list completion task contribute different relevant entities to the pools than the runs for entity ranking; an additional benefit from defining the two tasks from one entity retrieval problem.

5 Conclusions

The INEX 2007 XML Entity Ranking track has build the first test collection for the evaluation of information retrieval systems that support users that search for specific entities rather than just any type of documents. We developed a set of 28 training and 21 testing topics derived from ad hoc 2006 and 2007 topics, as well as a set of 25 genuine XER topics. The differences between the ad hoc relevance assessments and those of the entity ranking interpretation of the same topics demonstrate that the XER 2007 test collection captures (as expected) a different user need than ad hoc search, and a distinct interpretation of relevance. We would like to investigate in more detail whether system evaluation using genuine XER topics differs significantly from using the ad hoc derived topics, but this will require first the acquisition of larger topic sets; the main goal for the 2008 edition of the track. Aside from acquiring a larger number of topics for the current tasks on the Wikipedia collection, we will investigate how to evaluate searching for relations between entities. Another useful extension of the current track setup would be to allow ranking arbitrary passages as result entities, instead of limiting the possible answers to Wikipedia entries only.

References

1. Pehcevski, J., Vercoustre, A.-M., Thom, J.: Exploiting locality of wikipedia links in entity ranking. In: Macdonald, C., Ounis, I., Plachouras, V., Ruthven, I., White, R.W. (eds.) ECIR 2008. LNCS, vol. 4956, pp. 258–269. Springer, Heidelberg (2008)
2. Zaragoza, H., Rode, H., Mika, P., Atserias, J., Ciaramita, M., Attardi, G.: Ranking very many typed entities on wikipedia. In: Proceedings of the 16th ACM CIKM International Conference on Information and Knowledge Management, Lisbon, Portugal, pp. 1015–1018 (2007)

L3S at INEX 2007: Query Expansion for Entity Ranking Using a Highly Accurate Ontology

Gianluca Demartini, Claudiu S. Firan, and Tereza Iofciu

L3S Research Center
Leibniz Universität Hannover
Appelstrasse 9a D-30167 Hannover, Germany
{demartini,firan,iofciu}@L3S.de

Abstract. Entity ranking on Web scale datasets is still an open challenge. Several resources, as for example Wikipedia-based ontologies, can be used to improve the quality of the entity ranking produced by a system. In this paper we focus on the Wikipedia corpus and propose algorithms for finding entities based on query relaxation using category information. The main contribution is a methodology for expanding the user query by exploiting the semantic structure of the dataset. Our approach focuses on constructing queries using not only keywords from the topic, but also information about relevant categories. This is done leveraging on a highly accurate ontology which is matched to the character strings of the topic. The evaluation is performed using the INEX 2007 Wikipedia collection and entity ranking topics. The results show that our approach performs effectively, especially for early precision metrics.

1 Introduction

Entity search is becoming an important step over the classical document search as it is done today on the Web. The goal is to find entities relevant to a query more than just finding documents (or parts of documents) which contain relevant information. Ranking entities according to their relevance with respect to a given query is crucial in scenarios where the amount of information is too large to be managed by the final user. That is, a correct ranking scheme can help the system in presenting the user only entities of interest, and avoiding the user having to analyse the entire set of retrieved entities.

As a first step in this direction, we present, in this paper, our approach to ranking entities in Wikipedia, we evaluate it on the Wikipedia XML corpus provided within the INEX 2007 initiative, and we investigate how extended category information influences the results. The main contribution of this paper is a methodology for refining the user query using semantic information such as an ontology based on Wikipedia and Wordnet[1]. This refinement is done matching the category information in the topic with both the Wikipedia hierarchy

---

[1] http://wordnet.princeton.edu/

N. Fuhr et al. (Eds.): INEX 2007, LNCS 4862, pp. 252–263, 2008.

structure and the YAGO ontology. The query is then expanded using the additional category information. The experimental results show that using additional information about the category structure improves early precision by 8%.

The rest of the paper is organized as follows. In section 2 we describe the general architecture of the developed retrieval system. In section 3 we present an ontology, based on Wikipedia and WordNet, that we use to improve the effectiveness of our entity ranking algorithms. In section 4 we present the generated inverted index for the XML Wikipedia collection. In section 5 we formalise the ranking algorithms we propose. In section 6 we present the experimental evaluation results and a comparison among the proposed algorithms. In section 7 we present and compare the previous approaches in entity search and ranking. Finally, in section 8 we describe future improvements and conclude the paper.

2   Architecture Overview

In this section we describe the architecture of the Entity Ranking System we developed for creating the runs submitted to INEX 2007. The architecture design is presented in figure 1. The first step is the creation of the inverted index out of the XML Wikipedia document collection. Starting from the raw structured XML documents, we created a Lucene[2] index[3] with one Lucene document (i.e., a vector in the Vector Space) for each Wikipedia document (see more details in section 4). We first parse the document collection using standard Java libraries[4]. After this, we create an index with different fields (acting as separate inverted indexes, which can be combined for retrieval) for the title, text, and category of Wikipedia entities.

After the creation of the inverted index, the system is able to process the INEX Entity Ranking 2007 topics. Two different approaches are adopted (see details in section 5):

1. The INEX topic is first processed in order to create a disjunctive Lucene query using the title and description information (i.e., a disjunction of all the terms appearing in the title and description parts of the topic). A document is thus retrieved if any query term is found in any of the index fields.
2. A possible extension is done using the category field of the topic together with information from the YAGO[7] ontology (see section 3 and figure 6) in addition to the Lucene query obtained after this first step.

After the generation of the query, the fields of the index can be queried and a ranked list of entities is retrieved merging the ranked lists coming from the different fields. The ranking of the retrieved entities is done according to cosine similarity with the query using the standard TFxIDF scoring function.

---

[2] http://lucene.apache.org/

[3] The IR model used by Lucene is the Vector Space Model with standard cosine similarity.

[4] We used the Java 6 `javax.xml.stream.*` classes.

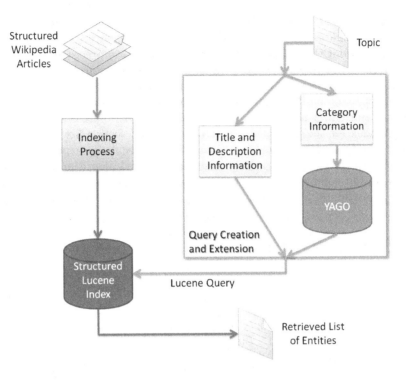

Fig. 1. Architecture of the Entity Ranking System

3   Using YAGO

YAGO[5] [7] is a large and extensible ontology that builds on entities and relations from Wikipedia. Facts in YAGO have been automatically extracted from Wikipedia and unified with WordNet, using rule-based and heuristic methods. It contains more than 1 million entities and 5 million facts and achieves an accuracy of about 95%. All objects (e.g., cities, people, URLs) are represented as entities in the YAGO model. The ontology is constructed in such a way as to be able to express entities, facts (the triple [entity, relation, entity] is called a fact), and even relations between facts and properties of relations.

The creation of YAGO focuses on integrating entities from Wikipedia with semantics from WordNet. Each Wikipedia page title is a candidate to become an entity in YAGO, and the Wikipedia categories of that page become its containing classes. Wikipedia categories are organized in a directed acyclic graph, which yields a hierarchy of categories. This hierarchy, however, reflects merely the thematic structure of the Wikipedia pages. Thus, WordNet is used to establish the hierarchy of classes, as WordNet offers an ontologically well-defined taxonomy of synsets (i.e., sets of synonyms). Each synset of WordNet becomes

---

[5] Available for download at http://www.mpii.mpg.de/~suchanek/yago/

a class of YAGO and the *subClassOf* based hierarchy of classes is taken from the hyponymy relation of WordNet. This gives for each Wikipedia page a set of conceptual categories arranged in a taxonomic hierarchy. More data is gathered exploiting WordNet synsets as synonyms and exploiting Wikipedia redirects as alternative names for the entities. For the purpose of this article we used the MySQL export of YAGO and combined it with the INEX Wikipedia dataset. This allows us to make use of the *subClassOf* relation in YAGO, providing us with semantic concepts describing Wikipedia entities.

Figure 2 shows how the type of a Wikipedia page can be found through Wikipedia and YAGO relations. Starting from the the TV comedy series (an entity in our case) "`Married... with Children`", we extract from the Wikipedia taxonomy the containing Wiki category - "Sitcoms". Then, leveraging YAGO's WordNet knowledge (i.e., the *subClassOf* relationship) we find that "Sitcoms", and thus all entities in this Wikipedia category, is of the type "Situation Comedy". When looking at a certain Wiki category, we find that not all of the subcategories in Wikipedia are of the same type as the parent category, and we can thus filter some out. For example, the Wikipedia category "Sitcoms" which is of the YAGO type "Situation Comedy" contains the subcategory "Sitcom Characters", of the YAGO type "Fictional Character", which we can avoid considering as proper subcategory.

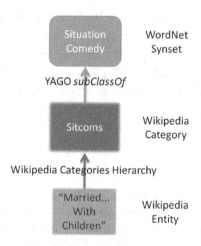

Fig. 2. Retrieving the type of an entity using YAGO

4 Index Structure

Given the XML document collection, we created an entity-driven inverted index in order to enable the search and ranking of entities. We have chosen to use a Lucene index because of the possibility of generating a structured inverted index with fields which are searchable in parallel. The Lucene index fields act as separate inverted indexes. The final result of a query is obtained merging (i.e., doing the union of) the ranked lists of results from each field. The Retrieval Status Values are normalized and they can be compared because of the homogeneous use of cosine similarity as function for ranking vectors in all the fields.

Following the structure of the XML Wikipedia documents, in the inverted index, we store content divided into the following fields:

- *id*, the unique identifier of the article;
- *title*, the title of the article;
- *text*, the entire textual content of the article;
- *categories*, the categories listed at the bottom of the article.

5   Algorithms

We have implemented two approaches for entity ranking. Both approaches extend the traditional IR vector space model, enriching it with semantic information. Additionally to textual information from Wikipedia articles we also keep context information (i.e., category information) either extracted from Wikipedia or inferred using YAGO. The examples in the following sections are based on the following topic:

Table 1. INEX Entity Ranking topic example

Topic	#78
Title	European fruit trees
Description	I want a list of European fruit tree sorts.
Narrative	Each answer should be an article about the the specific fruit tree.
Category	trees

5.1   Naïve Approach

As a baseline approach for constructing the query, we consider only the information given in the title and description parts of the topic[6], as presented in figure 3. For search we use the Vector Space Model and ranking is done using standard cosine similarity and TFxIDF weighting scheme[7]. We construct a disjunctive query containing both textual and contextual information (i.e., keywords and category information). For the textual part of the query we consider the keywords from the title and the description of the topic which we run against the *title* and *text* fields in the index. In the contextual part of the query we consider the category information from the topic which we run against the *categories*

---

[6] The narrative part of the topic contains too many non-specific keywords that might over-relax the query and, therefore, is not included.

[7] All search and ranking settings were left as default in Lucene.

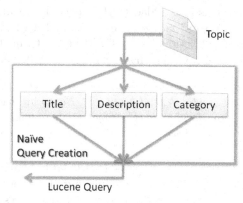

Fig. 3. Query creation using only topic information

field. For example, for the topic described in table 1, the query resulting after stopword removal is the following:

> *title:(European fruit trees) text:(European fruit trees) title:(I want list European fruit tree sorts) text:(I want list European fruit tree sorts) category:(trees)*

Due to the fact that a disjunctive query is built, if a Wikipedia entity does not belong to the category in the topic, but is relevant to the topic title and description, it is still retrieved with a lower rank.

5.2 Categories Based Search

While the category information which is present in the topic should contain most of or all the retrievable entities, this is for many topics not the case. Wikipedia is constructed manually by different contributors, so that the category assignments are not always consistent. Many categories are very similar and in some of these cases the difference is very subtle so that similar entities are sometimes placed in different categories by different contributors (e.g., hybrid powered automobiles are either in the "hybrid vehicles" or the "hybrid cars" category, inconsistently, and very seldom they are in both).

In the previous approach the category given in the topic was used to make the query more likely to retrieve entities from within that category. The method described here constructs an additional list of categories closely linked to the ones given in the topic description. This extended list of categories is then used instead of the topic categories in query construction. The simplest starting point would be using merely Wikipedia subcategories looking at the Wikipedia categories hierarchy. Apart from this, we use two different types of category expansion, *Children* and *Siblings*.

Subcategories. Wikipedia itself has a hierarchical structure of categories. For each category we are presented with a list of *Subcategories*. This list of

Subcategories is taken as-is and added to the query. For example, some of the subcategories for the "Actors" category are: "Animal actors", "Child actors", "Actors with dwarfism", "Fictional actors". More in detail, for this approach and the selected topic (see table 1), the query has the following form:

title:(European fruit trees) text:(European fruit trees) title:(I want list European fruit tree sorts) text:(I want list European fruit tree sorts) category:("trees" "conifers" "wood" "fagales" "sapindales" "individual trees" "palms" "trees of africa" "trees of new zealand" "trees of hawaii")

Children. The *Children* list of categories is created by starting from the *Subcategories* list and filtering inappropriate ones out. It is more effective not to include all the Wikipedia subcategories in our *Children* list as some of them are not real subcategories, that is, they are not of the same type. As subcategories for a country, it is possible to have categories about presidents, movie stars, or other important persons for that country. This means that although we have as a starting category a country we end up having people as subcategories, which is not what we want in the entity retrieval context. The solution to this is selecting only those subcategories having the same class as the initial category. As described in section 3, YAGO contains also class information about categories. We make use of this *subClassOf* information to identify suitable categories of the same type. Thus, a Wikipedia subcategory is included in the *Children* list only if the intersection between its ancestor classes and the ancestor classes in YAGO (excluding top classes like *entity*) of the initial category is not empty. The final list of *Children* will therefore contain only subcategories of the same type as the category given in the topic. Figure 4 presents an example of the *Children* list of the category "Sitcoms".

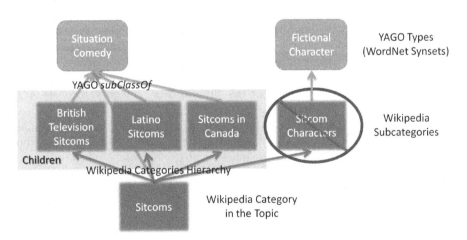

Fig. 4. Example of *Children* identification starting from the "Sitcoms" category

For the selected topic (see table 1), due to the fact that all the *Children* categories have the same type as the topic category, none of them are filtered and the query looks the same as for the *Subcategories* approach.

Siblings. Using YAGO we can also retrieve categories of the same type as the starting category, not restricting just to the Wikipedia subcategories. We first determine the type of the starting category using the *subClassOf* relation in YAGO. Knowing this type we construct a list of all the categories of the same type and add them to the *Siblings* set. *Siblings* are, thus, all the categories of the exact same type as the initial category. Figure 5 shows how, starting from the category "Sitcoms", a list of *Siblings* is created.

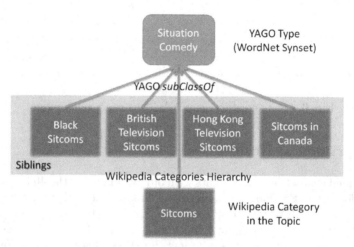

Fig. 5. Example of *Siblings* identification starting from the "Sitcoms" category

Figure 6 depicts the inclusion of *Children* and *Siblings* in the query creation process. Constructing the query is done similarly to the naïve approach setting. The difference relies in the category matching part. In the naïve approach we had only the categories given within the topic while in this case we have the additional two lists of *Children* and *Siblings*. For the selected topic (see table 1) the query after the *Siblings* extension is:

> title:(European fruit trees) text:(European fruit trees) title:(I want list European fruit tree sorts) text:(I want list European fruit tree sorts) category:("trees" "individual trees" "trees of africa" "trees of hawaii" "trees of new zealand")

The resulting expanded list of categories is then matched against the *categories* field of the index. These extensions allows to find relevant entities with category information (e.g., "conifers" using the *Subcategory* or *Children* approach) different from the one which is present in the topic (e.g., "trees").

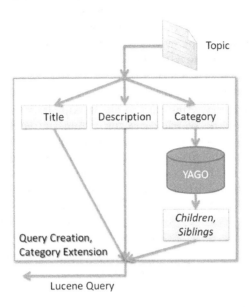

Fig. 6. Query creation using YAGO category information

6   Experimental Results

We performed evaluation experiments with our system using the 46 testing topics provided for the INEX 2007 Entity Ranking Track. Out of these, 21 topics were derived from the INEX 2006 ad hoc collection and the remaining 25 topics were developed by the track participants. We note that the ad hoc part of the dataset and, in particular, the relevance assessments for these topics, were not done under the assumption of a user searching for entities, thus influencing the effectiveness metrics values. In table 2 we can see that the effectiveness of our system is higher on the ad hoc topics. Apart from the ad hoc task being easier than the entity ranking task, the training of our system was done using only ad hoc topics, as no entity ranking topics were available at that time.

Table 2. Mean Average Precision values of different algorithms for different parts of the topic set (i.e., Ad Hoc topics and/or Entity Ranking topics (xer))

Algorithms	ad hoc	xer	all
Naïve	0,1839	**0,0708**	**0,1225**
Wiki subcategories	0,1765	0,0617	0,1141
Children	**0,1980**	0,0574	**0,1215**
Siblings	0,0767	0,0538	0,0643

The results (presented in table 3) show that according to early precision metrics, the use of Wikipedia subcategories outperforms the naïve approach (P@10) and that filtering these subcategories (i.e., the *Children* approach) at query time yields best performance (P@5 and P@10). Extending the query with additional category information improves the results by 8% for P@10 and by 3% for P@5 over the baseline. In terms of MAP, the official metric of the INEX 2007 Entity Ranking Track, we can see that the naïve approach still shows the best results, although very close to the *Children* approach. We can see that the use of *Children* categories improves over the *Subcategories* approach, showing that type filtering on Wikipedia subcategories is necessary. The use of *Siblings* in the categories extension yields bad results overall. This can be explained by the fact that the expansion contains a large number of categories (sometime even up to 1,000 categories), many of them being just noise; therefore, the disjunctive query becomes too vague and the effectiveness decreases.

Table 3. Results of effectiveness evaluation according to different metrics

Algorithms	MAP	P@5	P@10	bpref
Naïve	**0.1225**	0.1435	0.1283	0.1820
Wiki subcategories	0.1141	0.1348	0.1326	**0.1954**
Children	**0.1215**	**0.1478**	**0.1391**	0.1865
Siblings	0.0643	0.1043	0.0913	0.1026

7 Related Work

There are only a few systems that deal with entity search and ranking described in the literature. ESTER, presented in [1], combines full-text and ontology search supporting prefix search and joins. It has been tested on the English Wikipedia corpus using the *isA* and *subClassOf* relations of the YAGO ontology [7]. ESTER focuses on efficiency, the recall is high while the precision is reasonable and it becomes higher when using only Wikipedia data, without additional information from the ontology.

Another framework focusing on effectiveness and efficiency, presented in [3], adopts methods for finding different types of entities (e.g., phone number, email address, etc.) on the Web. While their main accent is on scaling on Web-size datasets, our approach can better manage an heterogeneous set of entity types.

A related field is the one of Expert Search (ES) where the aim is to find people (i.e., a specific type of entity) who are experts on the given topic. The topic of ES is a relatively new one but already several systems have been proposed. The systems proposed in the past use several information and features like Social Network information [2]; co-occurrences of terms and changes in the competencies of the people [9]; rule-based models and FOAF[8] data [5]; and using posts on Web Forums [8]. One of the first approaches is the Enterprise PeopleFinder [6] also known as P@noptic Expert [4]. This system first builds a candidate profile attaching all documents related to the user, giving different weights to the documents based on their type (e.g., a homepage is more important than other web

---

[8] http://www.foaf-project.org/

pages), in one big document which represents the candidate. In contrast with ES systems, our methodology can deal with multiple entity types.

8   Conclusions and Future Work

In this paper we proposed algorithms to rank entities in Wikipedia. Our approaches use a structured inverted index to represent the entities which are present in Wikipedia and use the YAGO ontology in order to rewrite the user's query for improving the effectiveness of the results. We described how the category part of the topic is extended by matching the user input with the Wikipedia categories hierarchy and filtering the results of this matching using a highly accurate ontology. The evaluation experiments show that the approach based on extending the query with additional category information improves the results over the baseline in terms of early precision metrics.

There are a couple more approaches that we will investigate in the future. A first approach deals with the disambiguation of the query. This can be done, for example, by extracting adjectives and nouns from the topic's title, description and narrative. A more complex approach would be, for a given topic, to extract the existing Wikipedia entities that are specified in the topic and see if these entities (or entities linking to them) belong to the topic's categories. As in the Wikipedia pages one can often find lists of other entities, another approach to automatically enrich the category information would be to assume that if the majority of entities in a "list page" belongs to a category, then the rest of entities in the list should belong to the same category. Moreover, we have shown that filtering out Wikipedia subcategories of unrelated types improves the results.

Acknowledgments. This work was partially supported by the Okkam project funded by the European Commission under the 7th Framework Programme (IST Contract No. 215032).

References

1. Bast, H., Chitea, A., Suchanek, F., Weber, I.: Ester: efficient search on text, entities, and relations. In: SIGIR 2007: Proceedings of the 30th annual international ACM SIGIR conference on Research and development in information retrieval, pp. 671–678. ACM, New York (2007)
2. Campbell, C.S., Maglio, P.P., Cozzi, A., Dom, B.: Expertise identification using email communications. In: Proceedings of the 12th ACM Conference on Information and Knowledge Management (CIKM 2003), pp. 528–531 (2003)
3. Cheng, T., Yan, X., Chang, K.C.-C.: Entityrank: Searching entities directly and holistically. In: VLDB, pp. 387–398 (2007)
4. Craswell, N., Hawking, D., Vercoustre, A., Wilkins, P.: P@noptic Expert: Searching for Experts not just for Documents, Ausweb (2001)
5. Li, J., Boley, H., Bhavsar, V.C., Mei, J.: Expert finding for eCollaboration using FOAF with RuleML rules. In: Montreal Conference on eTechnologies (MCTECH) (2006)

6. McLean, A., Vercoustre, A.M., Wu, M.: Enterprise PeopleFinder: Combining Evidence from Web Pages and Corporate Data. In: Proceedings of Australian Document Computing Symposium (2003)
7. Suchanek, F.M., Kasneci, G., Weikum, G.: Yago: a core of semantic knowledge. In: Proceedings of the 16th international conference on World Wide Web, pp. 697–706 (2007)
8. Zhang, J., Ackerman, M.S., Adamic, L.: Expertise networks in online communities: structure and algorithms. In: Proceedings of the 16th international conference on World Wide Web, pp. 221–230 (2007)
9. Zhu, J., Gonçalves, A.L., Uren, V.S., Motta, E., Pacheco, R.: Mining Web Data for Competency Management. In: Web Intelligence 2005, pp. 94–100 (2005)

Entity Ranking Based on Category Expansion

Janne Jämsen[1], Turkka Näppilä[1], and Paavo Arvola[2]

[1] Department of Computer Sciences, Kanslerinrinne 1,
FI-33014 University of Tampere, Finland
{janne.jamsen;turkka.nappila}@cs.uta.fi
[2] Department of Information Studies, Kanslerinrinne 1,
FI-33014 University of Tampere, Finland
paavo.arvola@uta.fi

Abstract. This paper introduces category and link expansion strategies for the XML Entity Ranking track at INEX 2007. Category expansion is a coefficient propagation method for the Wikipedia category hierarchy based on given categories or categories derived from sample entities. Link expansion utilizes links between Wikipedia articles. The strategies are evaluated within the entity ranking and list completion tasks.

1 Introduction

Entity ranking is an emerging field of information retrieval (IR) which aims to retrieve and rank entities that match a given query. The entities in question may be concrete objects, such as cities, persons, organizations, or purely abstract notions, such as mathematical formulas. This differs from traditional IR in which the focus is on finding topically relevant documents or their parts [3,5].[1],[2]

The new XML Entity Ranking (XER) track of INEX uses the Wikipedia XML Collection. The corpus consists of approximately 659,000 articles (XML documents) which are classified using approximately 113,000 categories. Approximately 13,900,000 (wiki-)links exist between the articles.[3] In addition, the collection features a category hierarchy (a directed graph) which records the subcategory–parent category relationships. For the purposes of entity ranking, each entity is assumed to correspond to a Wikipedia article.

The topics in the XER track which are used to evaluate the performance of the participants' ranking schemes are represented in XML. Fig. 1 contains a sample topic. Each topic consists of a title, description, narrative, and lists of

---

[1] The relatively late awakening to the entity ranking problem may seem surprising taking into account that the related problem of (named) entity recognition has already been studied for years [12].

[2] In the Entity Ranking Track of INEX, an entity is assumed to correspond to a Wikipedia article. Nevertheless, there may be several Wikipedia articles that refer to a relevant entity (i.e., a Wikipedia article) possibly making them topically relevant but not relevant entities as such.

[3] This figure does not contain multiple links (pointing to the same direction) between each document pair.

N. Fuhr et al. (Eds.): INEX 2007, LNCS 4862, pp. 264–278, 2008.
© Springer-Verlag Berlin Heidelberg 2008

```
<inex_topic>
  <title> European countries where I can pay with Euros </title>
  <description> I want a list of European countries where I can pay with Euros. </description>
  <narrative> Each answer should be the article about a specific ... </narrative>
  <entities>
    <entity id="10581">France</entity>
    <entity id="11867">Germany</entity>
    <entity id="26667">Spain</entity>
  </entities>
  <categories>
    <category id="185">european countries</category>
  </categories>
</inex_topic>
```

Fig. 1. An example of the INEX 2007 topic

sample entities and categories related to the topic. However, of these only the title and the entity and category lists were utilized in the XER track.

The track is comprised of two tasks: entity ranking (ER) and list completion (LC). In both tasks, the aim is to return a ranked list of entities based on a keyword query (the title of a topic). In addition, in the ER task, the participants may use the list of sample categories and, in the LC task, the list of sample entities.

We discovered three components of the Wikipedia XML Collection to be useful in entity ranking:

1. the textual content of the Wikipedia articles,
2. the category hierarchy, and
3. the link structures between Wikipedia articles.

We start our query evaluation with a standard result list generated by a partial match-based information retrieval system. For that purpose, we use TRIX (Tampere Retrieval and Indexing for XML) with document order based (DoOrBa) scoring method to process the query titles against the documents.

The essential contribution of our paper is on utilizing the category hierarchy. In order to get more precise answers for the queries, we use a category expansion method which propagates descending coefficients especially to the nearby categories (i.e., parents, children, and siblings) of the given categories in the hierarchy. In addition, in the ER task, we experimented with a related link expansion method which propagates document scores (yielded by TRIX) along links.

The rest of the paper is organized as follows. In the next section, we review some related work. Section 3 introduces TRIX (which participated also in the INEX 2007 Ad Hoc track [1]) and the category and link expansion methods. In Section 4, the runs and preliminary results are presented. Finally, the results are discussed and the conclusions are drawn in Sections 5 and 6 respectively.

2   Related Work

Expert search is one of the earliest forms of entity ranking in the focus of academia.[4] In an early 1998 paper [14], Mattox introduced a software application

---

[4] Some early approaches to expert search utilized hand-crafted expertise databases but we do not discuss them here.

called Expert Finder that returns lists of experts for keyword queries utilizing a
company's internal documents. The later system P@noptic Expert by Craswell
et al. [5] builds on a similar idea. (Neither of the papers reports any evaluation
results.) More recently, TREC (Text REtrieval Conference) has started to drain
IR researchers attention to the problem by its Enterprise Track which, since 2005,
has incorporated an expert search task. The task features a collection of corporate
documents, and the participants are left with the task of finding the occurrences
of employees within those documents and using this information to rank the employees by their expertise on specified topics (described by queries) [6,17].

A common approach in expert search (as in TREC) is first to match documents with a query using some document-centric IR method and then to infer
the rankings of individual employees from the resulting document scores (or
rankings). Some of the most popular methods for inferring the employee rankings consist basically of calculating an aggregate (e.g., sum or maximum) of the
scores for related documents per each employee (see, the review in [17]). Some
of the more complex methods exploit the inherent graph structure spanned by
documents, experts (in practice, their names and email addresses) and the con-
tainment relationships between them. To this end, it is possible to apply methods
analogous to Web search algorithms that exploit the hyperlink structure of the
Web [17,16].

As a restricted form of entity ranking, expert search makes it relatively easy
to craft heuristic rules that take into account the special properties of human
experts and their usual roles in the documents [8]. The more general problem
studied in the XER track, which is not restricted to any particular entity type,
has received considerably less attention. In 2006, Chakrabarti et al. [3] discussed
efficient algorithms for ranking a fixed number of entities (of any type) occurring
in a document collection that best match to a given keyword query. Their work
was based on coupling database technologies and document retrieval systems.
However, they did not evaluate the ranking results.

In a broader sense, entity ranking has already found its way into the search engine markets. Notable commercial endeavours include Google's Product Search
[11] (formerly known as Froogle) and Google's Maps search engine [10]. Product
Search allows anyone to submit sellable products by providing the required information on its attributes including its price, condition, brand, description and
product type. The actual search engine accepts keyword queries and includes options for sorting the list of matching products by relevance, price or rating. The
Maps search engine ranks geospatial entities (also shown as "placemarks" on the
map interface) based on a keyword query. The entities may be user-defined and
annotated locations or entities mined from heterogeneous sources such as digital
yellow and white pages, building plans and targeted databases [4].

3   Approach

As a baseline for our experiments, we matched topic titles against the textual
contents of Wikipedia articles using TRIX. In addition, we implemented and

```
<p>
It was rumoured ... left the band (and formed a band called
<collectionlink xlink:href="221501.xml"> Soulfly </collectionlink>
), and the others announced that they would continue under...
</p>
```

Fig. 2. A branch element containing a text element

tested two complementary methods: category expansion (used in both tasks) and link expansion (used in the ER task only). Short descriptions of these methods can be found in the following sections (3.1 - 3.3).

3.1 TRIX

The motivation for the TRIX DoOrBa approach is to emphasize the importance of the first descendant elements in document order (shortly, ido). The first descendant elements mean, e.g., titles for the sections and headings, abstracts and keywords for the whole documents. As a result, the weight of these elements should affect more on the retrieval status value of the ancestor element and consequently of the whole document. The DoOrBa scoring method propagates recursively the element scores for the ancestors. This is done by giving decreasing values for the descendant elements based on their position (ido).

Indexing. The DoOrBa scoring is based on an inverted file, where the locations of keys are denoted by structural indices presented in [15], also known as Dewey numbers. The idea of the indices in the context of XML is that the topmost (root) element is indexed by $\langle 1 \rangle$ and its children by $\langle 1, 1 \rangle$, $\langle 1, 2 \rangle$, $\langle 1, 3 \rangle$, etc. Further, the children of the element with the index $\langle 1, 2 \rangle$ are labelled by $\langle 1, 2, 1 \rangle$, $\langle 1, 2, 2 \rangle$, and so on. This kind of indexing enables analyzing of the relationships among elements in a straightforward way. For example, the ancestors of the element labelled by $\langle 1, 2, 2, 1 \rangle$ are associated with the indices $\langle 1, 2, 2 \rangle$, $\langle 1, 2 \rangle$, and $\langle 1 \rangle$. In turn, any descendant related to the index $\langle 1, 2 \rangle$ is labelled by $\langle 1, 2, \xi \rangle$ where ξ is a non-empty part of the index. Moreover, because the labelling for the siblings is executed in the document order the indexing works well in figuring out the preceding–following relationships between known indices as well. As an illustration of this, we can say that the element $\langle \xi, i \rangle$ is the ith child of the element ξ, and thus preceding an element $\langle \xi, i + 1 \rangle$, if it exists. As a remark of the space efficiency of using such a method, the size of the content-only inverted file for the Wikipedia collection (4.6 GB) is 739 MB, calculated after stemming and stopword removal.

Scoring. Similarly to, e.g., GPX [9], in the DoOrBa the scores are calculated separately for leaf elements and branch elements. This is done so that the leaf scores have been delivered upwards to the branch elements. A leaf element is considered here to be an element which contains directly a text element. It is worth noting that an element is considered to have no more than one text element directly. In other words, the text element means all the direct text content of an element. A branch element is an element having child elements (other than text elements). Due to these definitions an element can be a leaf element, a branch element, or even both. For example, the paragraph in Fig. 2 contains both text elements and is also a branch element (it has a child: *collectionlink*).

In Fig. 2, the text presented in italics form the content of the element p. The score of an element is a sum of leaf element and branch element scores. Since the text elements tend to be short, although of varying length and importance, the score of the text element is basically the sum of the *idf* (inverse document frequency) values of query terms in the element. The leaf score (text score) is calculated with the following equation, in which m is the number of (unique) terms in the query expression:

$$textScore(q, \xi) = \sum_{t=1}^{m} idf_t \tag{1}$$

The score for the branch element is calculated recursively based on the scores of its child elements. This has been done so that the scores of the child elements are considered in relation to their positions (ido). The primary goal is to emphasize scores of child elements appearing early (ido) in the child list. This is done by applying a specific *child score vector* (as a variable, CS) for the element weighting.

The child score vector is filled with constant values, which are used to express the contribution each child has in branch element weighting. The position of the value in the vector corresponds to the child number (ido), and the smaller the value, the more important is the corresponding child. We use $CS[i]$ to denote the ith component of the vector. For example, applying $CS = \langle a, b, c \rangle$ for the element ξ, means that a is for $\langle \xi, 1 \rangle$, b for $\langle \xi, 2 \rangle$, c for $\langle \xi, 3 \rangle$. On the basis of this, we get a following general matching formula, which combines element's branch score (if it has any descendants) with element's text score (if it contains any text):

$$score(q, \xi) = \sum_{i=1}^{min(n, len(CS))} \left(\frac{score(q, \langle \xi, i \rangle)}{v \cdot (a + CS[i])} \right) + textScore(q, \xi) \tag{2}$$

where

- $score(q, \xi)$ is the score of the element ξ in relation to the query q,
- n is the number of child elements,
- $len(CS)$ is the length of CS,
- i is the child element position in the element's child list,
- v and a are constants for tuning.

Decreasing the value of a and v emphasizes the effect of the vector CS. Consequently, the equation $v \cdot (a + CS[i])$ is actually used as a substitute of a length normalization component. For example, if we have a vector $CS = \langle 1, 2, 3, 4, 5 \rangle$, $a = 0$, and $v = 1$, the weight of the first child is taken into account as a whole, the score of the second child increases the element's score by the $1/2$ of the child's score, the third by $1/3$, and so on.

For the INEX 2007 Ad Hoc track as well as for the XER track quite basic settings have been used. For every run, we used the CS as an infinite vector $CS = \langle 1, 2, 3, \ldots \rangle$. However, the early precision results for the Focussed task of the Ad Hoc track were satisfactory. TRIX DoOrBa reached 15th, 17th and 19th positions in the precision at 5 % recall, with runs from 8 institutes ahead.

3.2   Category Expansion

Category expansion, as understood in this paper, stands for the act of deriving from a set of initial categories (specific to a topic) an expanded set of categories that covers the relevant entities more or less accurately. Each category in the expanded set (or in the hierarchy as a whole if also zero scores are used) can be assigned a numeric coefficient, a *matching score*, that describes its conformance to the initial categories (the greater the score, the more closely the category matches to the initial categories).[5] The scores can be allocated to individual articles, e.g., by taking the scores of their best-ranking direct categories.

In a classical, well-defined is–a hierarchy (e.g., found in thesauri and in many programming and modelling languages), members of a subcategory (i.e., a specialization) are implicitly members of the corresponding supercategory, too. For example, each art museum is necessarily a museum. As a result, given that we want only museums to be included in an answer, we can prioritize the entities that have been assigned to the category *museums* and/or one of its (direct or indirect) subcategories. Provided the categories have a full coverage (i.e., there are no museums besides those under the category *museums* and its subcategories), we can restrict ourselves to these entities.

Unfortunately, the semantics of the category hierarchy of the Wikipedia XML Collection follows neither of these principles in detail.[6] This makes it practically impossible to make a binary distinction between matching and non-matching categories. For this, we use the relative positions and proximity within a category hierarchy as the determining factor in approximating the extent of match between two categories. The extent may lessen both in moving upwards or downwards in the hierarchy but possibly at different rates.

Our model for category expansion consists of the following components:

- The set of all categories in the Wikipedia XML Collection is denoted by C.
- The matching score of a category c_j relative to a category c_i $(c_i, c_j \in C)$ will be denoted by $M_i(c_j)$.
- The category hierarchy is conceptualized as a directed (ideally, but not necessarily, acyclic) graph. To this end, we adopt the conventional parent–child terminology to denote the hierarchical relationships among categories. Formally,

---

[5] Note that the notion of graph-oriented expansion in the context of IR is not novel to this paper. For example, Järvelin, Kekäläinen, and Niemi [13] introduce a tool for ontology-based query expansion. Also noteworthy are the various spreading activation-based techniques for keyword search and related IR tasks (see, e.g., [2,7]) as well as many hyperlink-based IR methods.

[6] For example, the article for the Finnish author Tove Jansson is assigned to a subcategory of the category *Countries*. Obviously, this does not imply an is–a relationship.

the hierarchy is represented a set H (a subset of $C \times C$) consisting of tuples of the form $\langle p, c \rangle$ where p is the parent category and c is the child category.

- The starting point of category expansion is a set of initial categories denoted by $I(I \subseteq C)$. For each topic in the ER task, this is the set of given categories that specify the desired type of entities in an answer. For each topic in the LC task, a set of initial categories, which may or may not be relevant, is obtained indirectly from the provided sample entities by taking each category that has at least one (explicitly assigned) member among them.[7]

The user-provided parameters (shared by all topics) include:

- d: decay down, a coefficient in the range [0,1] that determines the rate the matching scores diminish during the downward expansion (i.e., from parent to child);
- u: decay up, a coefficient in the range [0,1] that determines the rate the matching scores diminish during the upward expansion (i.e. from child to parent);
- t: threshold, a constant in the range, [0,1] that constraints the expansion.

Once the set of initial categories I is established, one category expansion is executed for each included category. Fig. 3 demonstrates this in the case of two initial categories. The two expansions are depicted in Fig. 3(a) and 3(b). The category hierarchy is interpreted from the top down, ancestors shown above descendants. The decay down coefficient is 0.9 and the decay up coefficient 0.5.

Formally, for each category c_i ($c_i \in I$) the function *category_expansion* below is called (the function call has the form *category_expansion*(c_i, C, H, u, d, t)). The function returns a set of tuples of the form $\langle c_j, s_j \rangle$ where c_j is a category ($c_j \in C$) and $s_j = M_i(c_j)$ (the matching score of c_i is 1).

For functional definitions we use the standard mathematical (and set-theoretical) notation. The signature of the function is denoted by $f : \alpha \rightarrow \beta$ where α is the domain and β is the range. In what follows, R denotes the set of (non-negative) real numbers.

Definition 1. Given an initial category c, all categories C, a category hierarchy H, a decay-up coefficient u, a decay-down coefficient d and a threshold t, the matching scores of categories in C relative to c are given by the function *category_expansion* ($C \times P(C) \times P(C \times C) \times R \times R \times R \rightarrow P(C \times R)$):

$$category\_expansion(c, C, H, u, d, t) = expansion(C - \{c\}, \{\langle c, 1 \rangle\}, H, u, d, t). \tag{3}$$

The actual functionality is included in the function *expansion* below. Here ID denotes the set of indentifiers for categories (i.e., $ID = C$). (In Section 3.3, ID = A.)

---

[7] In order to better emphasize categories that are shared by multiple sample entities, multiple occurrences of the same category might be allowed in the set (not possible for pure sets).

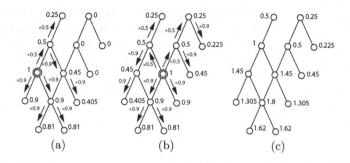

Fig. 3. Category expansion with two initial categories

Definition 2. Given the so-far unvisited categories U (in Section 3.3 Wikipedia articles), the so-far accumulated scores S for visited ones, the category hierarchy G (in Section 3.3 the wikilink network among the Wikipedia articles), a decay-up coefficient u, a decay-down coefficient d and a threshold t, the function *expansion* $(P(ID) \times P(ID \times R) \times P(ID \times ID) \times R \times R \times R \rightarrow P(ID \times R))$ gives the matching scores of categories (in Section 3.3 Wikipedia articles):

$$
expansion(U, S, G, u, d, t) = \begin{cases} expansion(U - N, S \cup S_N, G, u, d, t) \\ \quad \text{,if } N \neq \emptyset \wedge \neg \exists x, s : \langle x, s \rangle \in N \wedge s < t \\ \quad \text{where} \\ \quad\quad N = \{x \mid \exists s : \langle x, s \rangle \in S_N\} \text{ and} \\ \quad\quad S_N = \{\langle x, s_x \rangle \mid x \in U \wedge s_x = \\ \quad\quad max\_score(x, S, G, u, d) \wedge \neg \exists y : \\ \quad\quad (y \in U \wedge max\_score(y, S, G, u, d) > s_x)\} \\ S \cup \{\langle x, 0 \rangle \mid x \in U\} \\ \quad \text{,otherwise.} \end{cases}
$$

$$(4)$$

The function *expansion* calls the function *max_score* defined in the following.

Definition 3. Given a category x (in Section 3.3 a Wikipedia article), the scores accumulated so far S, the category hierarchy D (in Section 3.3 the Wikilink network among the Wikipedia articles), a decay-up coefficient u and a decay-down coefficient d, the function *max_score* $(ID \times P(ID \times R) \times P(ID \times ID) \times R \times R \rightarrow R)$ gives the maximum available score for x taking into account the so-far given scores for its parents and children:

$$
max\_score(x, S, D, u, d) = \begin{cases} max(R), \text{if } R \neq \emptyset \\ \quad \text{where} \\ \quad\quad R = \{s_y \cdot d \mid \exists y : (\langle y, s_y \rangle \in S \wedge \langle y, x \rangle \in D)\} \cup \\ \quad\quad \{s_y \cdot u \mid \exists y : (\langle y, s_y \rangle \in S \wedge \langle x, y \rangle \in D \wedge \\ \quad\quad sp\_constraints(y))\} \\ 0, \text{otherwise.} \end{cases}
$$

$$(5)$$

The function *sp_constraints* can be used to introduce additional constraints that further limit the upward expansion (based on our experiences, the upward expansion results easily in a massive expansion throughout the hierarchy). In our current implementation, we require that only categories that are ancestors of the initial category may initiate an upward expansion (see, Fig. 3).

After the matching scores are calculated for each initial category, the total matching score of an arbitrary category c can be calculated using either of the formulas 6 or 7:

$$M(c) = \max_1^n M_i(c), \tag{6}$$

$$M(c) = \sum_1^n M_i(c). \tag{7}$$

Especially in the LC task, the formula 6 can be assumed to bring better out the categories that are shared by multiple sample entities (and which are therefore more likely to be relevant). In order to balance the summing effect, we also experimented with the logarithm of the formula 6 and a weighted average of the formulas 6 and 7. In our example, the final matching scores (calculated simply as sums) are depicted in Fig. 3(c).

3.3   Link Expansion

In the ER task, we experimented with a modification of the above-like expansion where, instead of the category hierarchy, (wiki-)links among Wikipedia articles (i.e., entities) were utilized. The underlying assumption is that the links in encyclopaedia articles usually point out to other articles that are somewhat closely related to them. This is reminiscent of the modified *tf–idf* schemes in which the content of the neighbouring hyperlinked pages is taken into account [18]. Because encyclopaedia articles are usually designed to avoid extensive overlapping, this sort of strategy could be assumed to work even better for Wikipedia articles than for random web pages.

We have the following components that are analogous to category expansion above:

- The set of all Wikipedia articles A.
- The set L containing the linked article pairs (a subset of $A \times A$). Each of the tuples in the set has the form $\langle v, w \rangle$ where the article v contains one or several links pointing to the article w.
- Top n articles that match a topic query yielded by a text retrieval system (TRIX in our case) together with their associated document scores.

The user-provided parameters d, u and t described in Section 3.2 remain unchanged. (Here the upward expansion refers to expansion along incoming links and the downward expansion to one along outcoming links.)

As in the case of categories, each of the top articles is used separately as a basis for expansion. The expansion mechanism is defined using the function *link_expansion* below.

Definition 4. Given an initial article a and its document score s, all articles A, a link structure L, a decay-up coefficient u, a decay-down coefficient d and a threshold t, the expanded documents scores for related articles are given by the function $link\_expansion$ $(A \times R \times P(A) \times P(A \times A) \times R \times R \times R \rightarrow P(A \times R))$:

$$link\_expansion(a, s, A, L, u, d, t) = expansion(A - \{a\}, \{\langle a, s \rangle\}, L, u, d, t). \quad (8)$$

After the expansions, the accumulated scores are aggregated as in formulae 6 and 7.

For example, evaluating the query "Nordic authors noted for children's literature" using a text-based retrieval system might give a high score to the articles *Nordic countries* and *Children's literature*. An article describing a relevant author, such as Tove Jansson, even if it would not contain the words *Nordic* and *children* might contain a link to the article *Finland* which in turn contains a link to the article *Nordic countries* (or the other way around). The article *Children's literature*, for its part, might contain links that point out directly or intermediately to *Tove Jansson*. Ideally, after the scores gained during the expansions are summed up, articles for Tove Jansson and other relevant authors end up having significant total scores of their owns.

Unfortunately, the graph structure induced by links among documents is remarkably more massive than the category hierarchy. Given the initial high number of final topics and insufficient RAM memory, we were unable to test the expansion to depths greater than 1 in the available time.

4   Runs and Results

In this section, we experiment with the above methods and their combinations, and report the official results. Fig. 4 depicts the interpolated precision–recall curves (recall 0 - 100 %) of some of our experiments with the test data. The similar runs were executed also with the training data with results of similar tendency. The runs include TRIX results (trix), TRIX results accepting only

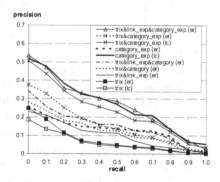

Fig. 4. Recall–precision curves for some of the experimental runs

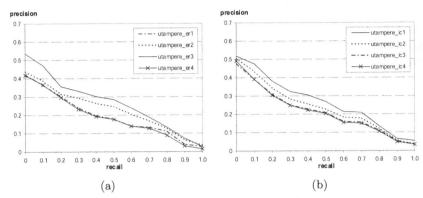

Fig. 5. Recall–precision curves for the final runs

Table 1. Performance scores for the submitted runs in the ER task (for training and test data sets)

	training data				test data			
	P[r]				P[r]			
Run	5	10	R-prec	MAP	5	10	R-prec	MAP
TRIX	0.0357	0.0429	0.0472	0.0302	0.0957	0.0891	0.0926	0.0637
utampere_er_1	0.1714	0.1857	0.1617	0.1198	0.2130	0.2174	0.2119	0.1803
utampere_er_2	0.1714	0.1929	0.1920	0.1729	0.2957	0.2652	0.2444	0.2098
utampere_er_3	0.2000	0.2071	0.2104	0.1783	0.2826	0.2957	0.2758	0.2437
utampere_er_4	0.1643	0.1679	0.1516	0.1138	0.2217	0.2109	0.2081	0.1750

articles belonging (directly) to sample categories (trix&category), category expansion alone (category_exp), TRIX results combined with category expansion (trix&category_exp), TRIX results combined with link expansion (trix&link_exp), TRIX results combined with link expansion accepting only articles belonging (directly) to sample categories(trix&link_exp&category), and finally TRIX results combined first with link expansion and then with category expansion (trix&link_exp&category_exp). As a limitation, experiments that utilized the list of sample categories (available only for the ER task) and those that involved link expansion were executed only for the ER task (er).

The document scores returned by TRIX are considered as the baseline. In the LC task, the sample entities have been removed from the TRIX results. In this setting TRIX alone seems to produce the weakest results, whereas the best results are achieved by combining the different methods and category expansion only (LC). The top-most curves are delivered by TRIX combined with category expansion and TRIX combined with both of category and link expansion (ER) and category expansion only (LC). Due to these findings, we used parameterized runs with combinations of category and link expansion and TRIX for the final runs. In the following, we describe the official results for our official runs submitted to the XER track at INEX 2007. We submitted four runs for the ER

Table 2. Performance scores for the submitted runs in the LC task (for training and test data sets)

Run	training data P[r]				test data P[r]			
	5	10	R-prec	MAP	5	10	R-prec	MAP
TRIX	0.0429	0.0429	0.0472	0.0311	0.0652	0.0717	0.0743	0.0524
utampere_lc_1	0.3643	0.2964	0.2684	0.2221	0.3087	0.2630	0.2739	0.2463
utampere_lc_2	0.3286	0.2607	0.2302	0.2040	0.2870	0.2500	0.2463	0.2188
utampere_lc_3	0.2214	0.2143	0.2058	0.1746	0.2348	0.2130	0.2301	0.1957
utampere_lc_4	0.2143	0.2143	0.1900	0.1654	0.2435	0.2217	0.2255	0.1937

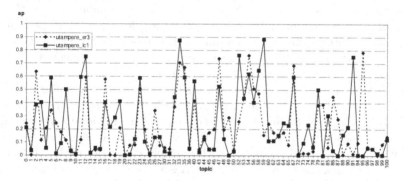

Fig. 6. The average precision scores per topic

task and four for the LC task. Note that, due to a minor error, the best run for the ER task was submitted as the third (utampere_er_3).

4.1 Entity Ranking

The results of our four official runs for the ER task are depicted in the form of precision–recall curves in Fig. 5(a). For the runs utampere_er_2 and utampere_er_3 the initial document scores returned by TRIX were propagated by the link expansion to the depth 1, whereas for the runs utampere_er_1 and utampere_er_4 the initial document scores were used as such. For each run, the resulting scores were combined with the matching scores from the category expansion. In other respects, the runs differ only slightly in parameter values.

As shown in Table 1, the runs utampere_er_2 and utampere_er_3 outperform the runs utampere_er_1 and utampere_er_4 on the training topics as well as on the test topics. However, the difference is statistically significant ($p < 0.05$) between the run utampere_er_3 and the runs utampere_er_1 and utampere_er_4.

4.2 List Completion

Fig. 5(b) and Table 2 report the results of our official runs for the LC task. In the run utampere_lc_1 we used the category expansion method alone, which seems to produce the highest mean average precision rate both on the training and test topics. The runs utampere_lc_3 and utampere_lc_4, which both produced lower early and mean average precision rates than the utampere_lc_1, resemble the run utampere_lc_2 but different parameter values are used in them. The superiority of both utampere_lc_1 and utampere_lc_2 in comparison with both of the utampere_lc_3 and utampere_lc_4 is statistically significant ($p < 0.05$).

5   Discussion and Conclusions

The tested methods and their combinations improve clearly upon the baseline information retrieval system (TRIX), which utilizes both the textual content and XML structure in document matching (see, Fig. 4). This indicates that the utilization of the text content and document structure alone is insufficient. Instead, taking also the category hierarchy and links between documents into account provides a fruitful starting point for entity ranking in the XML Wikipedia Collection.

According to the performance measures, category expansion performs better in the LC task than in the ER task (see, Fig. 5 and Tables 1 and 2). A likely explanation can be derived from the difference in the nature of these tasks. In the ER task only one category is given per topic (there are, however, some topics with two sample categories), whereas in the LC task the provided sample entities (mostly 2 or 3) are usually labelled with multiple categories each. This means that in the LC task the set of initial categories for category expansion is usually more extensive and fine-grained than in the ER task. Thus, it is more straightforward to find related entities (belonging to the expanded set of categories) in the LC task. However, the difference in performance between the best runs in the ER and LC tasks is not statistically significant ($p > 0.05$).

An unexpected finding was that the category expansion alone in the LC task outperformed nearly all the other methods (see, Table 2 and Fig. 5(b)). In other words, taking the topic title into account did not improve the results as it intuitively should. This seems even more surprising as there rarely exists a single category that directly corresponds to the specific information need expressed in the topic title. This demonstrates the high precision achieved by prioritizing categories that are relevant for multiple sample entities (e.g., by using sum as an aggregate function in category expansion).

Link expansion, for its part, improved the results in the ER task. This is evident from Table 1 and Fig. 4. Due to the tight schedule, link expansion was not tested for the LC task. As it seems intuitive that taking account of the link structure in the Wikipedia XML Collection would also improve the results in the LC task, our aim is to utilize it in future experiments.

For some reason in both of the tasks, the test data seems to produce better scores than the training data. The topic-wise analysis (see, Fig. 6) shows that the average precisions of some topics varies a surprisingly lot depending on the task (ER or LC). In principle, it is possible to distinguish the topic type, where the LC task approach might be more helpful than the ER task or vice versa.

In addition to the utilization of the link expansion in the LC task, our focus in future development will be on the improvement of the document matching method in TRIX. The improved results by TRIX will also improve the results in the methods that rely on them.

Acknowledgements

This study was funded in part by the Tampere Graduate School in Information Science and Engineering (TISE) and in part by the Academy of Finland under grant number 115480. The travel and accommodation costs in the conference site were guaranteed by the Nordic Research School in Library and Information Science (NORSLIS).

References

1. Arvola, P.: Document Order Based Scoring for XML Retrieval. In: Pre-Proceedings of INEX 2007, pp. 111–116 (2007)
2. Aswath, D., Ahmed, S.T., D'cunha, J., Davulcu, H.: Boosting Item Keyword Search with Spreading Activation. In: 2005 IEEE/WIC/ACM International Conference on Web Intelligence, pp. 704–707. IEEE Computer Society, Washington (2005)
3. Chakrabarti, K., Ganti, V., Han, J., Xin, D.: Ranking Objects by Exploiting Relationships: Computing Top-K over Aggregation. In: 2006 ACM SIGMOD International Conference on Management of Data, pp. 371–382. ACM Press, New York (2006)
4. Code, K., Jones, M.T., McClendon, B., Charaniya, A.P., Ashbridge, M., Images, V.P., Class, P.: Entity Display Priority in a Distributed Geographic Information System, United States Patent Application No. 20070143345 (2007)
5. Craswell, N., Hawking, D., Vercoustre, A.M., Wilkins, P.: P@noptic Expert: Searching for Experts not just for Documents. In: 7th Australian World Wide Web Conference, pp. 21–25 (2001)
6. Craswell, N., de Vries, A.P., Soboroff, I.: Overview of the TREC-2005 Enterprise Track. In: 14th Text REtrieval Conference, NIST Special Publication 500-266, pp. 199–205 (2005), http://trec.nist.gov/pubs/trec14/t14_proceedings.html
7. Crestani, F.: Application of Spreading Activation Techniques in Information Retrieval. Artificial Intelligence Review 11(6), 453–482 (1997)
8. Fang, H., Zhou, L., Zhai, C.-X.: Language Models for Expert Finding: UIUC TREC 2006 Enterprise Track experiments. In: 15th Text REtrieval Conference, NIST Special Publication 500-272 (2006),
   http://trec.nist.gov/pubs/trec15/t15_proceedings.html
9. Geva, S.: GPX: Gardens Point XML IR at INEX 2006. In: Fuhr, N., Lalmas, M., Trotman, A. (eds.) INEX 2006. LNCS, vol. 4518, pp. 137–150. Springer, Heidelberg (2007)
10. Google Maps, http://www.google.com/maps
11. Google Product Search, http://www.google.com/products
12. Grishman, R., Sundheim, B.: Message Understanding Conference-6: A brief History. In: 16th Conference on Computational Linguistics 1, pp. 466–471. Association for Computational Linguistics, Morristown (1996)

13. Järvelin, K., Kekäläinen, J., Niemi, T.: ExpansionTool: Concept-based Query Expansion and Construction. Information Retrieval 4(3/4), 231–255 (2001)
14. Mattox, D.: Expert Finder. The Edge: The MITRE Advanced Technology Newsletter 2(1) (1998)
15. Niemi, T.: A Seven-tuple Representation for Hierarchical Data Structures. Information Systems 8(3), 151–157 (1983)
16. Rode, H., Serdyukov, P., Hiemstra, D., Zaragoza, H.: Entity Ranking on Graphs: Studies on Expert Finding. Technical Report TR-CTIT-07-81, Centre for Telematics and Information Technology, University of Twente, Enschede (2007)
17. Serdyukov, P., Rode, H., Hiemstra, D.: University of Twente at the TREC 2007 Enterprise Track: Modeling Relevance Propagation for the Expert Search Task. In: 16th Text REtrieval Conference, NIST Special Publication 500-274 (2007), http://trec.nist.gov/pubs/trec16/t16_proceedings.html
18. Sugiyama, K., Hatano, K., Yoshikawa, M., Uemura, S.: Refinement of tf–idf schemes for web pages using their hyperlinked neighboring pages. In: 14th ACM Conference on Hypertext and Hypermedia, pp. 198–207. ACM Press, New York (2003)

Entity Ranking from Annotated Text Collections Using Multitype Topic Models

Hitohiro Shiozaki[1] and Koji Eguchi[2]

[1] Graduate School of Science and Technology, Kobe University,
1-1 Rokkoudai, Nada, Kobe, 657-8501, Japan
hitohiro@cs25.scitec.kobe-u.ac.jp
[2] Graduate School of Engineering, Kobe University,
1-1 Rokkoudai, Nada, Kobe, 657-8501, Japan
eguchi@port.kobe-u.ac.jp

Abstract. Very recently, topic model-based retrieval methods have produced good results using Latent Dirichlet Allocation (LDA) model or its variants in language modeling framework. However, for the task of retrieving annotated documents when using the LDA-based methods, some post-processing is required outside the model in order to make use of multiple word types that are specified by the annotations. In this paper, we explore new retrieval methods using a 'multitype topic model' that can directly handle multiple word types, such as annotated entities, category labels and other words that are typically used in Wikipedia. We investigate how to effectively apply the multitype topic model to retrieve documents from an annotated collection, and show the effectiveness of our methods through experiments on entity ranking using a Wikipedia collection.

1 Introduction

Several topic model-based approaches have been applied to improve the effectiveness of information retrieval [1,2]. For example, retrieval models based on Probabilistic Latent Semantic Indexing (PLSI) [1] or Latent Dirichlet Allocation (LDA) [3] have been studied. Those methods were applied to unstructured documents such as newspaper articles; however, structured documents have different natures, one of which is the richer document representation using multiple types of expressions, such as attributed words, non-attributed words and document metadata that typically appear in Wikipedia. When applying the topic model-based approaches above to such kind of annotated documents, some post-processing is required to distinguish different word types because the models such as PLSI and LDA cannot directly handle more than one word types. Very recently, a multitype topic model was developed [4] to directly handle such multiple word types and to represent topics that captures dependencies between these multiple types of expressions.

In this paper, we propose retrieval models based on the multitype topic model and investigate how to use the multitype topics to improve retrieval performance for annotated documents. We further show the effectiveness of our method for the task of 'entity ranking' using a Wikipedia collection. In the Wikipedia collection, each entity is represented as a document, with an entity ID, that consists of text descriptions, links

N. Fuhr et al. (Eds.): INEX 2007, LNCS 4862, pp. 279–292, 2008.
© Springer-Verlag Berlin Heidelberg 2008

to other entities, and category labels. In our model, the links to other entities are used to specify entity names that appear in link anchor texts. For each document, the three components: the entity names, the other words in the document and document category labels are handled exclusively in three different namespaces.

2 Related Work

Statistical topic models (e.g., [1,3,5,6,7,8]) are based on the idea that a documents is a mixture of topics, where a topic is a probability distribution over words. Hofmann [1] proposed 'Probabilistic Latent Semantic Indexing' (PLSI) model in his pioneering work on the topic modeling. Blei et al. [3] proposed 'Latent Dirichlet Allocation' (LDA) in an extention of the PLSI model by introducing a Dirichlet prior on multinomial distribution over topics for a document. The PLSI model has the overfitting problem and the problem of not generating new documents; however, the LDA model overcomes those problems. A graphical model of the LDA is shown in Figure 1, and the generative process is as follows:

1. For all d documents sample $\theta_d \sim Dirichlet(\alpha)$
2. For all t topics sample $\phi_t \sim Dirichlet(\beta)$
3. For each of the N_d words w_i in document d:
   (a) Sample a topic $z_i \sim Multinomial(\theta_d)$
   (b) Sample a word $w_i \sim Multinomial(\phi_{z_i})$

To estimate the LDA model, they used a Variational Bayesian method. More recently, Teh et al. [9] improved the estimation performance for the LDA model by applying a 'Collapsed' Variational Bayesian method. Instead of using the Variational Bayesian method or its variants, Griffiths et al. [6] applied the Gibbs sampling method to estimate the LDA model. From a viewpoint of the accuracy of the model estimation, the Gibbs sampling method works better than the others above when a sufficient number of iterations are performed. Newman et al. [8] further proposed several variations of the

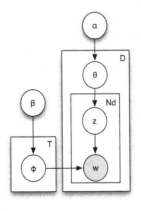

Fig. 1. LDA

LDA, including 'SwitchLDA', that can deal with words and entities. Shiozaki et al. [4] generalized the SwitchLDA model into a multitype topic model, 'GESwitchLDA' [4] that is briefly described in Section 3.1, and applied it to text mining from newspaper articles that were semantically annotated using natural language processing.

One of the important applications of the topic models is ad-hoc retrieval of documents. Hofmann [1] estimated the PLSI model and applied it to the ad-hoc retrieval task. In his experiments, the estimated topic models were used to construct a vector space, where inner product was used to compute similarity between a query and a document. More recently, the LDA model has been applied to the ad-hoc retrieval task. Wei and Croft [10] adopted this method in the framework of language model-based information retrieval, by linearly mixing a conventional language-model representation of a document and a LDA-based document model, as follows:

$$P(w|d) = \lambda \Big(\frac{N_d}{N_d + \mu} P_{ml}(w|d) + \Big(1 - \frac{N_d}{N_d + \mu}\Big) P_{ml}(w|coll)\Big)$$
$$+ (1 - \lambda) P_{lda}(w|d) \qquad (1)$$

where N_d and μ indicate the total number of words in document d and a smoothing parameter, respectively. $P_{ml}(w|d)$ and $P_{ml}(w|coll)$ are obtained by the maximum likelihood estimation of word w in the document d and in the entire collection, respectively. $P_{lda}(w|d)$ is a LDA-based document model that was obtained by marginalizing latent topic variable t over document d, as follows:

$$P_{lda}(w|d) = \sum_t P(w|t)P(t|d) \qquad (2)$$

where $P(t|d)$ and $P(w|t)$ can be estimated such as by the Gibbs sampling, as follows:

$$P(t|d) = \frac{C^{TD}_{td,-i} + \alpha}{\sum_t C^{TD}_{td,-i} + T\alpha} \qquad (3)$$

$$P(w|t) = \frac{C^{WT}_{wt,-i} + \beta}{\sum_w C^{WT}_{wt,-i} + W\beta} \qquad (4)$$

Another line of related work is cluster model, also known as the mixture of unigrams model, which was applied to the ad-hoc retrieval task. In the cluster model, all documents are classified into a set of K clusters that can be deemed as 'topics'. Liu and Croft [2] incorporated the cluster information into a language model of each document at smoothing stage:

$$P(w|d) = \frac{N_d}{N_d + \mu} P_{ml}(w|d) + \Big(1 - \frac{N_d}{N_d + \mu}\Big) P_{ml}(w|cluster) \qquad (5)$$

where $P_{ml}(w|d)$ and $P_{ml}(w|cluster)$ represent a document model and a cluster model, respectively. Main issue of the cluster model is the limitation that each document is generated from a single topic. For long documents and large collections this limitation may hurt the performance.

Wei and Croft [10] reported that the LDA-based retrieval model achieved significant improvements over the cluster model. However, the LDA model can not directly apply to document collections that consist of multiple types of words, such as given by annotations, since the LDA model does not distinguish different word types. This paper focuses on how to apply the multitype topic model previously mentioned to the task of ad-hoc retrieval for annotated documents, typically the Wikipedia collection, and demonstrate its effectiveness over the state-of-the-art LDA-based retrieval model.

3   Retrieval Models for Multitype Documents

3.1   Multitype Topic Models: GESwitchLDA

In an extension of the LDA model, Newman et al. [8] proposed several statistical entity-topic models. Those models, one of which is called 'SwitchLDA', attempted to capture dependencies between entities and topics, where the entities are mentioned in text. They handled a couple of word types: general words and entities; however, Shiozaki et al. [4] developed 'GESwitchLDA' by generalizing the SwitchLDA model to capture dependencies between an arbitrary number of word types, and demonstrated the performance in modeling dependencies between general words, 'who'-entities and 'where'-entities in order to represent factual events. This model is sometimes called a 'multitype topic model'.

A graphical model of the GESwitchLDA is shown in Figure 2, where a multinomial distribution ψ for a topic t with a Dirichlet prior distribution determined by a hyper-parameter γ was introduced. The variable M in Figure 2 denotes the number of word types. The generative process of the GESwitchLDA model is as follows:

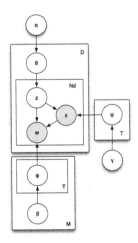

Fig. 2. GESwitchLDA

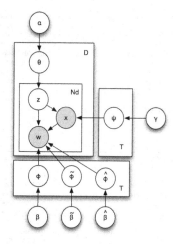

Fig. 3. GESwitchLDA when the number of word types is 3

1. For all d documents sample $\theta_d \sim Dirichlet(\alpha)$
2. For all t topics:
   (a) Sample $\psi_t \sim Dirichlet(\gamma)$
   (b) For each word type $y \in \{0, \cdots, M-1\}$:
       Sample $\phi_t^y \sim Dirichlet(\beta^y)$
3. For each of the N_d words w_i in document d:
   (a) Sample a topic $z_i \sim Multinomial(\theta_d)$
   (b) Sample a flag $x_i \sim Multinomial(\psi_{z_i})$
   (c) For each word type $y \in \{0, \cdots, M-1\}$:
       If $(x_i = y)$ sample a type-y word $w_i \sim Multinomial(\phi_{z_i}^y)$

We can use the Gibbs sampling approach to estimate the GESwitchLDA model [4].

Another graphical model representation of the GESwitchLDA assuming three word types: general word, entity word and category word is shown in Figure 3, and the corresponding generative process is as follows:

1. For all d documents sample $\theta_d \sim Dirichlet(\alpha)$
2. For all t topics sample $\phi_t \sim Dirichlet(\beta), \tilde{\phi}_t \sim Dirichlet(\tilde{\beta}), \hat{\phi}_t \sim Dirichlet(\hat{\beta})$
   and $\psi_t \sim Dirichlet(\gamma)$
3. For each of the N_d words w_i in document d:
   (a) Sample a topic $z_i \sim Multinomial(\theta_d)$
   (b) Sample a flag $x_i \sim Multinomial(\psi_{z_i})$
   (c) If $(x_i = 0)$ sample a word $w_i \sim Multinomial(\phi_{z_i})$
   (d) If $(x_i = 1)$ sample an entity-word $w_i \sim Multinomial(\tilde{\phi}_{z_i})$
   (e) If $(x_i = 2)$ sample a category-word $w_i \sim Multinomial(\hat{\phi}_{z_i})$

Figure 4 is the examples of topics captured by GESwitchLDA from a Wikipedia collection.

software	0.0266	beer	0.0298	game	0.0551
windows	0.0191	wine	0.0278	player	0.0471
system	0.0171	tea	0.0268	card	0.0401
file	0.0161	drink	0.0218	cards	0.0379
version	0.0122	sugar	0.0170	players	0.0288
support	0.0115	coffee	0.0169	play	0.0206
microsoft	0.0114	alcohol	0.0164	games	0.0147
code	0.0106	made	0.0116	played	0.0137
files	0.0095	drinking	0.0107	hand	0.0132
source	0.0093	bottle	0.0106	points	0.0123
linux	0.0093	brewery	0.0095	board	0.0104
program	0.0085	drinks	0.0080	rules	0.0097
operating	0.0080	wines	0.0078	two	0.0086
applications	0.0076	cherry	0.0074	playing	0.0077
user	0.0074	glass	0.0072	deck	0.0074
Microsoft_Windows	0.0254	Wine	0.0303	Poker	0.0335
Linux	0.0213	Beer	0.0219	Board_game	0.0204
Microsoft	0.0154	Grape	0.0170	Card_game	0.0186
Open_source	0.0153	Soft_drink	0.0112	Playing_card	0.0172
Operating_system	0.0143	Coca-Cola	0.0110	Betting_(poker)	0.0122
Unix	0.0132	Coffee	0.0106	The_Price_Is_Right	0.0121
Mac_OS_X	0.0120	Lager	0.0095	Game	0.0116
GNU_General_Public_License	0.0113	Brewery	0.0095	World_Series_of_Poker	0.0107
Computer_software	0.0102	Alcoholic_beverage	0.0089	Contract_bridge	0.0104
Free_software	0.0079	Vodka	0.0088	Gambling	0.0103
Graphical_user_interface	0.0072	Cocktail	0.0086	Go_(board_game)	0.0097
Apple_Computer	0.0071	Tea	0.0085	Texas_hold_'em	0.0095
Java_programming_language	0.0065	Whisky	0.0080	Dice	0.0094
Apple_Macintosh	0.0053	Ale	0.0070	Game_theory	0.0081
IBM	0.0046	Sugar	0.0067	Casino	0.0080
software	0.1259	beverages	0.0857	games	0.1738
computing	0.0483	alcoholic_beverages	0.0560	mental-skill_games	0.1171
free_software	0.0324	food_and_drink	0.0403	tabletop_games	0.0993
application_software	0.0296	beer	0.0381	card_games	0.0515
operating_systems	0.0259	alcohol	0.0366	board_games	0.0414
cultural_movements	0.0255	non-alcoholic_drink	0.0359	playing_cards	0.0389
system_software	0.0228	wine	0.0339	entertainment	0.0368
systems	0.0219	distilled_beverages	0.0254	personal_life	0.0340
engineering	0.0183	beverage_companies	0.0215	poker	0.0334
software_by_operating_system	0.0179	brewers_and_breweries	0.0207	gambling	0.0275
software_engineering	0.0167	beer_by_country	0.0204	poker_players	0.0198
computer_science	0.0167	alcohols	0.0182	culture	0.0191
programming	0.0157	soft_drinks	0.0178	puzzles	0.0169
programming_languages	0.0150	household_chemicals	0.0175	consumer_goods	0.0134
unix	0.0132	fermented_foods	0.0173	wargames	0.0109

Fig. 4. Examples of topics captured by GESwitchLDA from Wikipedia. In each topic we list most likely words and their probability at the top, entities at the middle, and categories at the bottom.

3.2 Multitype Query Likelihood Model

For ad-hoc retrieval task, one of the basic approaches is the query likelihood model [11,12,13]. In this model each document is ranked in order of likelihood of generating a query q by the document model:

$$P(q|d) = \prod_{w \in q} P(w|d)^{c(w,q)} \tag{6}$$

where d denotes a document, q and w denote a query and a term in q, respectively. $c(w, q)$ gives the frequency of w in q. $P(q|d)$ is the likelihood of generating the query words by the document model under the 'bag-of-words' assumption, under which words are independent in a document. Document ranking according to Eq. (6) is equivalent to that in order of the (negative) cross-entropy between a document model and a query model, given by the following:

$$\sum_{w \in q} P(w|q) \log P(w|d) \tag{7}$$

Here we can estimate document model $P(w|d)$ using Dirichlet smoothing [14], as follows:

$$P(w|d) = \frac{N_d}{N_d + \mu} P_{ml}(w|d) + \left(1 - \frac{N_d}{N_d + \mu}\right) P_{ml}(w|coll) \tag{8}$$

where $P_{ml}(w|d)$ is the maximum likelihood estimate of word w in the document d, and $P_{ml}(w|coll)$ is the maximum likelihood estimate of word w in the entire collection. N_d and μ indicate the total number of words in document d and Dirichlet smooting parameter, respectively. In order to apply to the documents that are expressed in multiple word types, we should modify Eq. (7).

Supposing multitype documents, we modify the ranking formula of the query likelihood model indicated in Eq. (7) to the following:

$$\sum_{x \in \mathbf{x}} a_x \sum_{w \in q_x} P(w|x, q) \log P(w|x, d) \quad \text{where } \Sigma a_x = 1 \tag{9}$$

where $\mathbf{x}$ denotes a set of word types and x a specific word type. A weighting parameter a_x allows us to change the balance of the word types in the ranking formula; and its value is selected empirically.

We can estimate document model $P(w|x, d)$ specified for word type x using modified Dirichlet smoothing, as follows:

$$P(w|x, d) = \frac{N_{xd}}{N_{xd} + \mu_x} P_{ml}(w|x, d) + \left(1 - \frac{N_{xd}}{N_{xd} + \mu_x}\right) P_{ml}(w|x, coll) \tag{10}$$

where N_{xd} denotes the total number of words with type x in document d. $P_{ml}(w|x, \cdot)$ is the maximum likelihood estimate of word w with type x in $\cdot$.

3.3 GESwitchLDA-Based Retrieval Model

In this section we suppose to use a Wikipedia collection in order to explain how to apply the multitype topic model to an annotated document collection. $x = 0$, $x = 1$ and $x = 2$ mean the case when the corresponding word w is a general word, the case when the word is an entity, and the case when the word is a category label, respectively. Similarly as when using LDA for ad-hoc retrieval, only using GESwitchLDA may be too coarse for the document representation for ad-hoc information retrieval.

Therefore, we combine the document model of the query likelihood models mentioned above with the GESwitchLDA model in the following two manners:

(a) Calculate $P(w|x, d)$
(b) Calculate $P(w, x|d)$

in order to construct a new GESwitchLDA-based document model. In detail of case (a), we linearly mix the document model in the multitype query likelihood model and the GESwitchLDA model, as follows:

$$P(w|x = 0, d) =$$
$$\lambda\left(\frac{N_{wd}}{N_{wd} + \mu_w}P_{ml}(w|x = 0, d) + \left(1 - \frac{N_{wd}}{N_{wd} + \mu_w}\right)P_{ml}(w|x = 0, coll)\right) +$$
$$(1 - \lambda)P_{tm}(w|x = 0, d)$$
$$P(w|x = 1, d) =$$
$$\lambda\left(\frac{N_{ed}}{N_{ed} + \mu_e}P_{ml}(w|x = 1, d) + \left(1 - \frac{N_{ed}}{N_{ed} + \mu_e}\right)P_{ml}(w|x = 1, coll)\right) +$$
$$(1 - \lambda)P_{tm}(w|x = 1, d)$$
$$P(w|x = 2, d) =$$
$$\lambda\left(\frac{N_{\ell d}}{N_{\ell d} + \mu_\ell}P_{ml}(w|x = 2, d) + \left(1 - \frac{N_{\ell d}}{N_{\ell d} + \mu_\ell}\right)P_{ml}(w|x = 2, coll)\right) +$$
$$(1 - \lambda)P_{tm}(w|x = 2, d) \tag{11}$$

In this paper, we make assumption that $\mu_w = \mu_e = \mu_\ell$, for simplicity. We then rank documents in order of the following values:

$$\sum_{x \in \{0,1,2\}} a_x \sum_{w \in q_x} P(w|x, q) \log P(w|x, d) \quad \text{where } \Sigma a_x = 1 \tag{12}$$

In detail of case (b), we linearly mix the document model in the original query likelihood model and the GESwitchLDA model, as follows:

$$P(w, x = 0|d) =$$
$$\lambda\left(\frac{N_d}{N_d + \mu}P_{ml}(w, x = 0|d) + \left(1 - \frac{N_d}{N_d + \mu}\right)P_{ml}(w, x = 0|coll)\right) +$$
$$(1 - \lambda)P_{tm}(w, x = 0|d)$$
$$P(w, x = 1|d) =$$
$$\lambda\left(\frac{N_d}{N_d + \mu}P_{ml}(w, x = 1|d) + \left(1 - \frac{N_d}{N_d + \mu}\right)P_{ml}(w, x = 1|coll)\right) +$$
$$(1 - \lambda)P_{tm}(w, x = 1|d)$$
$$P(w, x = 2|d) =$$
$$\lambda\left(\frac{N_d}{N_d + \mu}P_{ml}(w, x = 2|d) + \left(1 - \frac{N_d}{N_d + \mu}\right)P_{ml}(w, x = 2|coll)\right) +$$
$$(1 - \lambda)P_{tm}(w, x = 2|d) \tag{13}$$

In this case, we rank documents in order of the following values:

$$\sum_{(w,x) \in q} P(w, x|q) \log P(w, x|d) \tag{14}$$

where q is the set of pairs of word w and type x.

We calculate $P_{tm}(w|x,d)$ and $P_{tm}(w,x|d)$, as follows:

$$P_{tm}(w|x,d) = \sum_t P(w|x,t)P(t|d) \tag{15}$$

$$P_{tm}(w,x|d) = \sum_t P(w,x|t)P(t|d)$$

$$= \sum_t P(w|x,t)P(x|t)P(t|d) \tag{16}$$

We estimate the $P(t|d)$, $P(w|x,t)$ and $P(x|t)$ using the Gibbs sampling, as follows:

$$P(t|d) = \frac{C^{TD}_{td,-i} + \alpha}{\sum_t C^{TD}_{td,-i} + T\alpha}$$

$$P(w|x=0,t) = \frac{C^{WT}_{wt,-i} + \beta}{\sum_w C^{WT}_{wt,-i} + W\beta}$$

$$P(w_e|x=1,t) = \frac{C^{ET}_{et,-i} + \tilde{\beta}}{\sum_e C^{ET}_{et,-i} + E\tilde{\beta}}$$

$$P(w_\ell|x=2,t) = \frac{C^{LT}_{\ell t,-i} + \hat{\beta}}{\sum_\ell C^{LT}_{\ell t,-i} + L\hat{\beta}}$$

$$P(x|t) = \frac{n^x_{t,-i}, +\gamma}{n^{all}_{t,-i} + 3\gamma}$$

where $n^x_t = \sum_{w_x} C^{W_x T}_{w_x t}$ and $n^{all}_t = \sum_x n^x_t$.

4 Experiments

4.1 Task Definition and Evaluation Metrics

In the task of 'entity ranking' in Wikipedia, an entity is represented as a document that explains about it. In other words, each document corresponds to a specific entity, and so the task of entity ranking in Wikipedia is similar to document retrieval with relevance ranking, to some extent. Main difference between the entity ranking and the document retrieval is that a relevant document in the entity ranking needs to define and explain about a specific entity. For example, a document that explains about general informations around an entity without explaining the definition of the entity is deemed as relevant in the document retrieval task, but not relevant in the entity ranking task. We used 28 queries for training and 46 queries for testing. These queries were extracted from the topic titles of the INEX-2007 Entity Ranking Track.[1] We also used the corresponding relevance judgment data for training and for testing, which were constructed in this track.

---

[1] For details of the task, see the overview on the INEX-2007 Entity Ranking Track in this volume of proceedings, or ⟨http://inex.is.informatik.uni-duisburg.de/2007/xmlSearch.html⟩. In [15], you can see details of the Wikipedia collection that was used in this track.

As for evaluation metrics, we used MAP (mean average precision) [16], GMAP (geometric mean average precision) [17] and MRR (mean reciprocal rank) [18]. MAP is a very well accepted evaluation criterion in information retrieval, and is known to be stable and understandable. GMAP is geometric mean of average precision over all queries, instead of using arithmetic mean as in the case of MAP, and so prefers more robust retrieval systems. MRR is the averaged reciprocal value of the best rank of relevant entities for each query, and is often used for evaluating question-answering task. In the training stage, we used the MAP measure to empirically determine the best parameters.

4.2 Experimental Setting

From the Wikipedia collection, we removed the 418 stopwords that were used in 'In-Query' [19], and also removed the general words (other than entities or category labels) that occurred in less than 10 documents. We set the number of topics $T = 400$ and 800. We carried out the Gibbs sampling with a couple of different Markov chains to estimate the GESwitchLDA model, and $P(w, x|t)$ and $P(t|d)$ are averaged, respectively, over the Markov chains for each estimated topic, using a greedy algorithm. We also estimated the LDA model in the same manner above.

We set the Dirichlet smoothing parameter in the original query likelihood model (hereafter 'QL') as $\mu = 250$ that achieved the best results over the training data. We set the Dirichlet smoothing parameter in the multitype query likelihood model (hereafter 'MQL') as $\mu_w = \mu_e = \mu_\ell = 50$ that achieved the best results in the training stage. In case (a) of the GESwitchLDA-based retrieval model (hereafter 'GESI+MQL'), we set $\lambda = 0.6$ and $\lambda = 0.5$ for $T = 400$ and $T = 800$, respectively, each of which achieved the best results over the training data. In case (b) (hereafter 'GESD+QL'), we set $\lambda = 0.6$ and $\lambda = 0.5$ for $T = 400$ and $T = 800$, respectively, each of which worked the best in the training stage. As for the LDA-based retrieval model (hereafter 'LDA+QL'), we set $\lambda = 0.7$ and $\lambda = 0.5$ for $T = 400$ and $T = 800$, respectively, that achieved the best in the training stage. We denote '$a_0{:}a_1{:}a_2$' as ratio of type weights between words, entities and categories.

In the case when $\lambda = 0$, we denote 'LDA' as the LDA-based document model alone, 'GESI' as the GESwitchLDA-based document model alone in case (a), and 'GESD' as the GESwitchLDA-based document model alone in case (b).

4.3 Results

We empirically set the best parameters over the training data using our proposed MQL, GESI+MQL and GESD+QL, as well as QL and LDA+QL for baselines, as mentioned in Section 4.2. Using these best parameters, we obtained the results over the test query data, and computed MAP, GMAP and MRR mentioned in Section 4.1. The testing results are shown in Table 1. From this table, we can observe that our GESI+MQL achieved 25.3% and 38.6% improvements over QL in terms of MAP and GMAP, respectively, over the test data. Comparing with the state-of-the-art baseline: LDA+QL, our GESI+MQL achieved 4.4% and 7.1% improvements in terms of MAP and GMAP, respectively, over the test data. We further performed the Wilcoxon signed-rank test (two-tailed) to the pair of 'GESI+MQL' – 'QL' and the pair of 'GESI+MQL' – 'LDA+QL', the resulting

Table 1. Best Results

	MAP	GMAP	MRR
training			
QL	0.2267	0.0644	0.4892
MQL (1:1:2)	0.2406	0.0645	0.5140
LDA+QL (T=800)	0.2636	0.1004	0.5229
GESD+QL (T=800)	0.2644	0.0946	0.5458
GESI+MQL (T=800, 2:2:3)	0.2866	0.1198	0.5654
testing			
QL	0.2193	0.1056	0.5115
MQL (1:1:2)	0.2298	0.1143	0.5448
LDA+QL (T=800)	0.2633	0.1366	0.5045
GESD+QL (T=800)	0.2623	0.1313	0.5155
GESI+MQL (T=800, 2:2:3)	0.2749	0.1464	0.5580

Table 2. Results (over the training data) when the number of topics was changed (with the fixed type weights)

	MAP	GMAP	MRR
LDA (T=400)	0.0933	0.0154	0.2549
LDA (T=800)	0.1309	0.0256	0.2574
LDA+QL (T=400)	0.2617	0.1025	0.5607
LDA+QL (T=800)	0.2636	0.1004	0.5229
GESD (T=400)	0.0723	0.0124	0.1340
GESD (T=800)	0.1254	0.0213	0.2511
GESI (T=400, 1:1:1)	0.0789	0.0157	0.1724
GESI (T=800, 1:1:1)	0.1281	0.0243	0.2657
GESD+QL (T=400)	0.2497	0.0965	0.5275
GESD+QL (T=800)	0.2644	0.0946	0.5458
GESI+MQL (T=400, 1:1:1)	0.2649	0.1163	0.5305
GESI+MQL (T=800, 1:1:1)	0.2751	0.1146	0.5578

p-values of these pairs were less than 0.05. It means the performance improvements of the GESI+MQL over both QL and LDA+QL were statistically significant. We also observed that the improvement of GESD+QL over QL was statistically significant at 0.05 level, but not over LDA+QL.

Table. 2 shows the results over the training data when the number of topics was changed (with the fixed type weights). Varying the number of topics from 400 to 800, performance of all topic model-based methods was improved in terms of MAP, as well as in terms of MRR except for the case of LDA+QL, at the expense of computational costs.

Table. 3 shows the results over the training data of MQL, GESI and GESI+MQL when the type weights were changed. Table. 4 shows the results over the training data of MQL, GESI and GESI+MQL when the type weight of categories was set to 0. As we mentioned previously, we denote '$a_0:a_1:a_2$' as ratio of type weights between words,

Table 3. Results (over the training data) when the type weights were changed (T=800)

	MAP	GMAP	MRR
MQL (1:1:1)	0.2202	0.0630	0.4889
MQL (1:1:2)	0.2406	0.0645	0.5140
MQL (1:2:1)	0.2007	0.0598	0.4762
MQL (2:1:1)	0.1768	0.0479	0.4566
MQL (1:1:3)	0.2397	0.0601	0.5098
MQL (2:2:3)	0.2374	0.0648	0.4925
GESI (1:1:1)	0.1281	0.0243	0.2657
GESI (1:1:2)	0.1139	0.0188	0.2176
GESI (1:2:1)	0.1273	0.0204	0.2578
GESI (2:1:1)	0.1303	0.0248	0.3045
GESI (1:1:3)	0.0987	0.0152	0.1839
GESI (2:2:3)	0.1207	0.0217	0.2334
GESI+MQL (1:1:1)	0.2751	0.1146	0.5578
GESI+MQL (1:1:2)	0.2864	0.1168	0.5694
GESI+MQL (1:2:1)	0.2615	0.1025	0.5342
GESI+MQL (2:1:1)	0.2316	0.0874	0.4280
GESI+MQL (1:1:3)	0.2830	0.1135	0.5992
GESI+MQL (2:2:3)	0.2866	0.1198	0.5654

Table 4. Results (over the training data) when the type weight of categories was set to 0 (T=800)

	MAP	GMAP	MRR
MQL (1:1:0)	0.1046	0.0247	0.2855
GESI (1:1:0)	0.0933	0.0135	0.2328
GESI+MQL (1:1:0)	0.1530	0.0438	0.3429

entities and categories. From these tables, we observed that MQL was improved when the weight of category labels was increased; on the other hand, GESI was improved when the weight of general words was increased. As for the final GESI+MQL model, it was the best in the case when $a_0 : a_1 : a_2 = 2 : 2 : 3$. When the type weight of categories was 0, the performance turned very low. This mean that the category data play an important role in the entity ranking task in Wikipedia.

Note that, while the results in Table 1 were evaluated over the testing data, all other results above were evaluated over the training data, since we empirically determined the best parameters only using the training data, not using the test data.

5   Conclusions

We proposed a new retrieval model based on statistical topic models for annotated documents. Our model combines the well-accepted query likelihood model and a multitype topic model that can directly handle different word types. We estimated the multitype topic model using the Gibbs sampling. Through experiments on the task of entity ranking in Wikipedia, we compared our model with the query likelihood model and with the state-of-the-art LDA-based retrieval model, and achieved remarkable improvements that were statistically significant.

Incorporating the information from hyperlink structures in Wikipedia remains as a part of our future work.

Acknowledgements

This work was supported in part by the Grant-in-Aid for Scientific Research on Priority Areas "Info-plosion" (#19024055), Young Scientists A (#17680011) and Exploratory Research (#18650057) from the Ministry of Education, Culture, Sports, Science and Technology of Japan.

References

1. Hofmann, T.: Probabilistic latent semantic indexing. In: Proceedings of the 22nd Annual International ACM SIGIR Conference on Research and Development in Information Retrieval, Berkeley, California, USA, pp. 50–57 (1999)
2. Liu, X., Croft, W.B.: Cluster-based retrieval using language models. In: Proceedings of the 27th Annual International ACM SIGIR Conference on Research and Development in Information Retrieval, Sheffield, UK, pp. 186–193 (2004)
3. Blei, D.M., Ng, A.Y., Jordan, M.I.: Latent Dirichlet allocation. Journal of Machine Learning Research 3, 993–1022 (2003)
4. Shiozaki, H., Koji, E., Ohkawa, T.: Entity network prediction using multitype topic models. In: The 12th Pacific-Asia Conference on Knowlede Discovery and Data Mining, Osaka, Japan (to appear, 2008)
5. Ueda, N., Saito, K.: Parametric mixture models for multi-labeled text. Advances in Neural Information Processing Systems 15 (2003)
6. Griffiths, T.L., Steyvers, M.: Finding scientific topics. Proceedings of the National Academy of Sciences of the United States of America 101, 5228–5235 (2004)
7. Steyvers, M., Griffiths, T.: 21: Probabilistic Topic Models. In: Handbook of Latent Semantic Analysis. Lawrence Erbaum Associates (2007)
8. Newman, D., Chemudugunta, C., Smyth, P., Steyvers, M.: Statistical entity-topic models. In: Proceedings of the 12th ACM SIGKDD International Conference on Knowledge Discovery and Data Mining, Philadelphia, Pennsylvania, USA, pp. 680–686 (2006)
9. Teh, Y.W., Newman, D., Welling, M.: A collapsed variational Bayesian inference algorithm for latent Dirichlet allocation. Advances in Neural Information Processing Systems 19 (2007)
10. Wei, X., Croft, W.B.: Lda-based document models for ad-hoc retrieval. In: Proceedings of the 29th Annual International ACM SIGIR Conference on Research and Development in Information Retrieval, Seattle, Washington, USA, pp. 178–185 (2006)
11. Ponte, J.M., Croft, W.B.: A language modeling approach to information retrieval. In: Proceedings of the 21st Annual International ACM SIGIR Conference on Research and Development in Information Retrieval, Melbourne, Australia, pp. 275–281 (1998)
12. Hiemstra, D.: A linguistically motivated probabilistic model of information retrieval. In: Nikolaou, C., Stephanidis, C. (eds.) ECDL 1998. LNCS, vol. 1513, pp. 569–584. Springer, Heidelberg (1998)
13. Song, F., Croft, W.B.: A general language model for information retrieval. In: Proceedings of the 8th International Conference on Information and Knowledge Management, Kansas City, Missouri, USA, pp. 316–321 (1999)

14. Zhai, C., Lafferty, J.: A study of smoothing methods for language models applied to ad hoc information retrieval. In: Proceedings of the 24th Annual International ACM SIGIR Conference on Research and Development in Information Retrieval, New Orleans, Louisiana, USA, pp. 334–342 (2001)
15. Denoyer, L., Gallinari, P.: The Wikipedia XML corpus. ACM SIGIR Forum 40, 64–68 (2006)
16. Baeza-Yates, R., Ribeiro-Neto, B.(eds.): 3: Retrieval Evaluation. In: Modern Information Retrieval, pp. 73–97. Addison-Wesley, Reading (1999)
17. Robertson, S.: On GMAP: and other transformations. In: Proceedings of the 15th ACM International Conference on Information and Knowledge Management, Arlington, Virginia, USA, pp. 78–83 (2006)
18. Voorhees, E.: The TREC-8 Question Answering Track report. In: Proceedings of the 8th Text REtrieval Conference (TREC-8), NIST Special Publication 500-246, pp. 77–82 (1999)
19. Callan, J.P., Croft, W.B., Harding, S.M.: The INQUERY retrieval system. In: Proceedings of the 3rd International Conference on Database and Expert Systems Applications, Valencia, Spain, pp. 78–83 (1992)

An n-Gram and Initial Description Based Approach for Entity Ranking Track

Meenakshi Sundaram Murugeshan and Saswati Mukherjee

Department of Computer Science and Engineering,
College of Engineering, Guindy,
Anna University,
Chennai, India
{msundar_26,msaswati}@yahoo.com

Abstract. The most important work that takes the center stage in the Entity Ranking track of INEX is proper query formation. Both the subtasks, namely Entity Ranking and List Completion, would immensely benefit if the given query can be expanded with more relevant terms, thereby improving the efficiency of the search engine. This paper stresses on the correct identification of "Meaningful n-grams" from the given title and proper selection of the "Prominent n-grams" among them as the utmost important task that improves query formation and hence improves the efficiencies of the overall Entity Ranking tasks. We also exploit the Initial Descriptions (IDES) of the Wikipedia articles for ranking the retrieved answers based on their similarities with the given topic. List completion task is further aided by the related Wikipedia articles that boosted the score of retrieved answers.

Keywords: Entity Ranking, List Completion, n-gram checking.

1 Introduction

INEX is focused on enhancing research on XML retrieval. Entity ranking track, introduced this year, is aimed at retrieving entities (answers) rather than relevant documents. A collection of English Wikipedia documents (659,388 articles) in XML format is used as the corpus. The nature and structure of the corpus used is of special interest and demands special attention. Retrieval using unstructured corpus, consisting of plain text, poses comparatively greater problem over retrieval of semi-structured texts since, while plain text does not give any clue about the contents, both the structure and content of the semi-structured corpus can be exploited for any task in hand. Wikipedia corpus is semi-structured since it has organized the contents in such a way that, a brief summary of what follows in the article is given in the Initial Descriptions (IDES henceforth), and subsequent sections explain the topic in detail. Wikipedia articles also contain several links, which point to other articles that describe the topic. Hence if a system exploits the structure, it would be greatly beneficial.

Efforts have already been made to foster the research on retrieving a list of answers for a given question or topic. The earliest and the most notable effort is the List

N. Fuhr et al. (Eds.): INEX 2007, LNCS 4862, pp. 293–305, 2008.

Question Answering (List QA) introduced in the year 2001 in the Text Retrieval Conference's (TREC) Question Answering track [1]. TREC's Enterprise track, which is the pioneer of the ER track, had a similar kind of task called Expert Search [2], where the goal of the search is to create a ranking of people who are experts in the given topic.

Though List QA and Expert Search are relevant efforts, the tasks in INEX Entity Ranking track stands apart from the earlier efforts and also are more complicated. In List question answering, question clearly specifies the user need, which makes it possible to classify the questions into pre-defined categories and obtain the type of answers expected. In Expert Search of TREC, on the other hand, the expected answers are names of persons alone. This restriction helps to remove irrelevant Named-Entities, if any is retrieved. INEX Entity Ranking track, however, poses more difficulty since only topics are given, which are ambiguous and does not give any clue to the correctness of the retrieved Named-Entities; the answers are not restricted to any particular type as in Expert Search.

Wikipedia articles are grouped into several broad categories and they are further subcategorized. A topic in the Entity Ranking track consists of a title, category, example entities, description and narrative. Here the title is a free text query specification. The "category" is a more generic group to which that title belongs. In some cases, the topic contains two categories where one is the subcategory of the other. The part of the topic that can be used by the two subtasks, *viz.,* Entity Ranking (ER) task and the List Completion (LC) task varies. Given a category and a title, the task in Entity Ranking is to return relevant entities (answers). In List Completion, the task is to complete the partial list of entities, taking the title and a list of example entities as input. This variation also leads to a variation in the approaches to the retrieval of the answers.

The focus of this paper is to address the tasks of Entity Ranking using a good query formation mechanism followed by ranking the retrieved answers. The overall job in the track is to identify the relevant entities (answers). Identifying the relevant entities can be split into two above-mentioned sub-tasks; to form efficient query from the given topic that will help the search engine to retrieve more relevant articles and filtering and ranking such articles based on their similarities with the formed n-grams in the title.

A sample topic from INEX 2007 test-set is given in Figure 1. Here for the Entity Ranking task, the title "Books written by Friedrich Nietzsche" and the categories "books" and "books by friedrich nietzsche" are used to retrieve a list of answers as shown in Figure 2. Whereas for the List Completion task, the title and the given example entities, namely, "Ecce Homo" and "Thus Spoke Zarathustra" are used to complete the rest of the relevant entities as shown in Figure 3. The category information is not used by the List Completion system.

In this paper, the main focus of query formation is to identify and use n-grams. An n-gram denotes a sequence of terms, where "n" specifies the window size. In any natural language, especially in English, it is quite common that, two or more terms combine to give a separate meaning such as are found in Named-Entities and collocations. Most text-processing applications are focused on using meaningful units in text. Since how many n-grams combine to form a Lexical Unit (LU) is not known, this proves to be difficult. Currently, n-grams are used based on window size as contiguous or positional n-grams in many applications such as in automatic evaluation of NLP applications [3, 4].

```
<inex_topic topic_id="33" adhoc_tid="434" query_type="XER" ct_no="40">
<title>Books written by Friedrich Nietzsche</title>
<entities>
<entity id="1795997">Ecce Homo</entity>
<entity id="185614">Thus Spoke Zarathustra</entity>
</entities>
<categories>
<category id="1361">books</category>
<category id="32745">books by friedrich nietzsche</category>
</categories>
<description>The searcher's information needs cover information such as
introduction,
description and reviews of books written by German philosopher Friedrich Nietzsche.
</description>
<narrative>As a student whose major is not philosophy but is interested in Nietzsche's
philosophy, I want to know information about Nietzsche's book.
How many books did Nietzsche write? What is the main content of these books?
And what did other people said about these books?
However, books written by other people about Nietzsche are not among my
information need.
</narrative>
</inex_topic>
```

Fig. 1. A sample topic from INEX 2007's ER track

```
<title>Books written by Friedrich Nietzsche</title>
<categories>
<category id="1361">books</category>
<category id="32745">books by friedrich nietzsche</category>
</categories>
```

Fig. 2. Part of the topic to be used for the Entity Ranking task

```
<title>Books written by Friedrich Nietzsche</title>
<entities>
<entity id="1795997">Ecce Homo</entity>
<entity id="185614">Thus Spoke Zarathustra</entity>
```

Fig. 3. Part of the topic to be used for the List Completion task

Our method is focused on using "Meaningful n-grams" from the given title in the query formation. Some of these n-grams represent the focus of the user need. To capture "Meaningful n-grams", we used the key information available in the corpus, *i.e.*, the names of Wikipedia articles; this combined with an expansion method using WordNet, helps greatly to split the title into "Meaningful n-grams".

Wikipedia articles are organized in such a way that, the name of the article is followed by a short Initial Description (IDES), which provides an overview of the article. Following this description, there are usually several sections, each giving a detailed account on the theme of the article. Each article possibly contains references and links to other related articles.

Since IDES contains important aspects of the topic discussed in the article and act as a summary, our approach uses the IDES as the representation of the document and relied on such descriptions for removing irrelevant answers and ranking retrieved answers.

This paper is organized as follows. In Section 2, we explain the related work. Section 3 gives a detailed description of our approach. Section 4 describes evaluation results and Section 5 concludes the paper.

2 Related Work

Forming queries from the list questions given by TREC, by tagging with part-of-speech (POS) information, and identifying the focus of the question is explored in [5]. In this paper, the question types are determined using predefined search patterns. Various confidence levels are assigned for the focus of the question. Multiple paragraph windows are identified using the occurrence of question keywords. The use of Named-Entities to split the question into unigrams and Named-Entities is discussed in [6]. This paper has classified the web pages into four categories such as, collection page, topic page, relevant page and irrelevant page. The role of collection pages to find the evidence of the retrieved answer is explored in this paper. Lin et al. [4] applied n-gram co-occurrences between the reference summary and the candidate summary for automatically evaluating the summaries. The use of Lucene as an efficient information retrieval engine is demonstrated in several Question Answering systems such as in [7]. This paper uses GATE framework for text processing. Query processing is done using rule based query analyzer. The use of semantic tagger for answer extraction is also explained.

Use of Wikipedia names together with the first sentence for improving the Named-Entity recognition is explored in [8]. In this paper, authors have shown how category labels can be extracted from Wikipedia. The first noun phrase in the definition type sentence is used as the category label. However, this method does not utilize Wikipedia category sections. This is because a particular category may belong to more than one category and selecting appropriate categories from given list of categories needs further processing. Ranking sentences by giving importance to the proper names is discussed in [9]. Assigning different weights for WordNet synonyms and stemmed terms to improve the ranking is also explored.

3 Proposed Approach

Finding relevant answers in the ER tasks heavily depend on the initial set of relevant documents retrieved, which in turn depend on the formation of effective queries from the topic. Our method concentrates on two main areas. First is on query formation, for which we split the title into Lexical Units (LUs). Such units bear immense importance and we call them "Meaningful n-grams". These contribute to the overall efficiency in the relevant document retrieval. The second is the ranking of the retrieved answers.

A very common practice in many applications using NLP is an attempt to expand a given query by using relevant terms. In the literature we find approaches attempting to distinguish between the main (primary) and additional (secondary) query words. Primary keywords are the words that convey the essence of the query. They cannot be ignored. Secondary keywords are the less-relevant words for a particular query. They help to convey the meaning but can be omitted without changing the essence of the meaning. Such terms may be obtained from various sources such as from the Internet or using tools such as WordNet. In this paper, we have used both primary and secondary terms as well as n-grams for an improved retrieval.

Using n-grams in query processing instead of unigrams also helps in improved query formation. While part of the "Meaningful n-grams" in the title are used as primary terms, the rest of the "Meaningful n-grams" that are less important, along with a set of terms retrieved from the contents of the relevant Wikipedia articles are used as secondary terms. The complete set of secondary terms is obtained using part of the "Meaningful n-grams" in the title together with the set of terms retrieved from the names of Wikipedia articles.

The IDES, which act as the summary of the given article in the Wikipedia corpus, are used to filter and rank retrieved answers. Figure 4 shows the overall architecture of the system.

We split the discussion below in two parts: Query Formation and Answer Retrieval and Ranking. For the two subtasks in the Entity Ranking track, part of Query Formation and the whole of Answer Retrieval and Ranking are applied in differing ways. Since the general effort for the two subtasks is the same, we discuss the methods in the general form. Any special emphasis for any subtasks is mentioned during the discussion. If no mention is made, the method is applicable for both the subtasks.

3.1 Query Formation

Query formation from the topic plays a vital role in the effectiveness of the retrieval of the relevant documents by the search engine. The better the queries are able to identify the focus of the given title that represents the user need, the higher is the accuracy of the search engine. Query formation has three subtasks. These are identification of "Meaningful n-grams" in the title, splitting the "Meaningful n-grams" as "Prominent-grams" and "Other n-grams" and retrieval of secondary terms from the corpus, which are discussed in subsequent sub-sections.

3.1.1 Identification of Lexical Units
Our focus in this work is to split the title into "Meaningful n-grams", which we called Lexical Units (LUs)

$$\text{Title} = LU_1 \, LU_2 \ldots LU_n$$

where LU_i is the ith Lexical Unit and each LU is an n-gram of variable size ranging from unigrams to higher order n-grams.

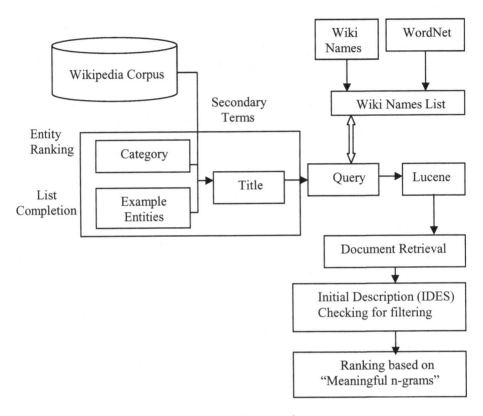

Fig. 4. The overall architecture of our system

The accuracy of the Existing Named-Entity Recognizers (NERs) to retrieve the "Meaningful n-grams" in case of names of persons, places or organizations is quite low. To counter this problem, we exploited the key information available in the large collection of 659,388 articles in Wikipedia corpus. Each of the Wikipedia articles in the corpus has a name that corresponds to its contents. For example, a bi-gram such as "Bob Dylan" has a corresponding Wikipedia article with that name. We formed an initial list of all such names from the corpus. It is possible that a particular article, which is associated with a name, may be referred using other equivalent names. For example, "USA", "United States of America", "United States" all are possible names for the same article, though only one of these is used. To capture all such possible n-grams in the title, we expanded the initial list of Wikipedia article names using WordNet synonyms. We call this expanded list as "Wiki Names List", which forms the base of our n-gram based query processing. Our target is to find all n-grams of maximum length that are present in a given title and in a name in the list. All such n-grams are "Meaningful n-grams". It is also possible to find the full title as one n-gram or not to find any term at all. If none of the terms are present in the "Wiki Names List", all terms in the title are considered as unigrams only. For example, the title of Q91 is "Paul Auster novels". Here the title is checked against the "Wiki Names List". The stemmed form of the common noun "novels" is present in the "Wiki Names List" and is identified as unigram and "Paul Auster", present in the "Wiki Names List" is identified as a "bi-gram". This process, thus, finds two "Meaningful n-grams" in the title. Stopwords are ignored. This method identifies the "Meaningful n-grams".

The next task is to identify the "Prominent n-grams" among these "Meaningful n-grams". The next sub-section describes the method.

3.1.2 The Prominent n-Gram(s) and Other n-Grams

All "Meaningful n-grams" are not equally important. Some of these have the capacity to indicate the focus of the question. To capture this notion, the concept of "Prominent n-grams" is used. "Meaningful n-grams" are split into two sets, *viz*, "Prominent n-grams" and "Other n-grams".

{Meaningful n-grams} = {Prominent n-grams} + {Other n-grams}

An example for "Prominent n-grams" is United States and that for "Other n-gram" is flash memory.

The determination of "Prominent n-grams" relies on the Part-of-Speech (POS) information. We have also used the contents of some of the articles, as is explained next.

There are three possible situations in determination of "Prominent n-grams".

i)   First case occurs, when the title contains proper nouns. There may be more than one proper noun in a given title and also we have to consider the possibility of having proper nouns in the form of n-grams, which may be unigrams or higher order n-grams. All such n-grams are identified as "Prominent n-grams".

ii) The second case occurs when no proper noun has been identified by the POS tagger in the title. Here, we retrieved all such Wikipedia articles that have the name as the n-grams present in the title. Each such Wikipedia article is checked for the presence of the "Other n-grams" in the title. The n-gram, the corresponding article of which contains the maximum number of "Other n-grams" is identified as the "Prominent n-gram".

iii) All other titles, which do not fall in either of the above two cases are considered here. Here all terms are considered to be equally important.

Table 1 shows the general form of all cases with examples. The followings explain the above three cases with examples.

Case (i): The title of Q64 in the test-set is an example for case (i)

Q64 Alan Moore graphic novels adapted to film

Here we have identified "Alan Moore" as a "Prominent n-gram" and is the only primary term used.

The general form of such queries are shown below,

("Prominent n-gram") AND ("Other n-grams" ST_1 ST_2...ST_n)

where "Other n-grams" in this case are "graphic", "novel", "adapted" and "film", and ST_i is a secondary term retrieved from the corpus, the method of which is discussed in Section 3.1.3.

Table 1. General form of all queries with examples

Cases	Example titles	General form	Prominent n-gram/ Primary Terms
Case 1	Alan Moore graphic novels adapted to film	(Prominent n-gram) AND ("Other n-grams" $ST_1\ ST_2...ST_n$)	Alan Moore
Case 2	world volleyball tournaments	(Prominent n-gram) AND ("Other n-grams" $ST_1\ ST_2...ST_n$)	volleyball
Case 3	food additive toxin carcinogen "E number"	$(PT_1\ PT_2....PT_n)$ AND $(ST_1\ ST_2...ST_n)$	"food additive" toxin carcinogen "E number"

Case (ii): The title of Q47 in the test-set is an example for case (ii).

Q47 world volleyball tournaments

In this case, all LUs are unigrams. Three Wikipedia articles are found for "world" and "volleyball" and "tournament". The content of the article with the name as "world" contains none of the other unigrams in the title. Whereas, the contents of the article "volleyball" contains the terms "world" and "tournaments" and the contents of the article "tournament" contains "world" only. Since common nouns such as "tournaments" will not have any article in Wikipedia associated with it, we have used the stemmed word to check for the articles with that name. Hence "volleyball" is identified as a "Prominent n-gram" in this case, since it contains most number of "Other n-grams" in the article corresponding to it.

Case (iii): The title of Q7 in the training-set is an example for case (iii).

Q7 food additive toxin carcinogen "E number"

Here, the Wikipedia article for "food additive" contains the n-gram "E number" and the Wikipedia article for "E number" contains the n-gram "food additive". Since none of the n-grams is identified as "Prominent n-gram", all n-grams in the title are considered as primary terms only.

3.1.3 Secondary Terms Retrieval from Corpus

Entity Ranking task uses category information besides the title as part of the query. To retrieve the secondary terms for this task, we used the given category terms as query and retrieved top n documents, and retrieved terms having high term-frequency (TF), with a minimum document frequency (DF). The threshold values for the term frequency as well as document frequency to retrieve top terms are empirically decided which is explained in the evaluation section. The terms thus retrieved along with the rest of the "Other n-grams" from the title and the unigrams present in the category form the set of secondary terms for Entity Ranking task.

List Completion task uses given partial answer list along with the title as query. We retrieve and analyze the articles that correspond to the answers given in the partial list. If these articles contain some common high frequency terms, those terms are likely to appear in all other relevant answers as well. To capture this intuition, common top "n" terms from the contents of these articles are retrieved. These terms along with "Other n-grams" from the title form the secondary terms.

3.2 Answer Retrieval and Ranking

We used Lucene for indexing the Wikipedia collection and retrieved top "n" documents. These documents are ranked based on the title given and the Initial Descriptions (IDES) of the retrieved articles. If the title contains "Prominent n-grams", IDESs are checked for the presence of the same. The documents that contain the "Prominent n-grams", the IDESS are checked for the presence of "Other n-grams" and ranked accordingly. The documents, which do not contain such "Prominent n-grams", are considered irrelevant and are filtered from further processing. However, if the given title does not contain any "Prominent n-grams", the IDESs are checked for the presence of "Other n-grams" and ranked accordingly.

The higher is the number of "Prominent n-grams" and/or "Other n-grams", the higher is the rank of the document. The final answers are the names of the top ranked Wikipedia articles. If two or more documents contain the same rank, Lucene ranking is used to distinguish such names.

In addition, for the List Completion task, we retrieve the Wikipedia article having the "Prominent n-gram" as its name, if one is found. If this article contains any of the given answers form the provided partial list, we assume that other expected relevant answers should also be present in the same article. Hence, each retrieved answer is checked against this article whose name matches with the "Prominent n-gram" of the title. Any of the retrieved answers being present, the rank of the answer is boosted and hence it automatically moves towards the top.

The following algorithm corresponds to the task of ranking described above.

Input: Title({"Meaningful n-grams"} = {"Prominent n-grams"} + {"Other n-grams"})
    and list of retrieved relevant documents

Output: Ranked documents on the descending value of 'n-gram count'

Algorithm:

```
While (Relevant documents list is not empty) {
//Case 1 and Case 2:
    IF (title contains "Prominent n-grams" )
            {
                Check in the IDES for the presence of "Prominent n-grams"
                IF (IDES contains "Prominent n-grams")
```

```
                                {
                                Check in the IDES for the presence of "Other n-grams"
                                    IF (found)
                                        Increment the identified n-gram count
                                    IF (task is List Completion)
                                            Boost_Score()

                                }
                        ELSE
                                {
                                    Filter the document as irrelevant
                                }
                    }
    //Case 3: title contains no "Prominent n-grams"
            ELSE
                    {
                        Check in the IDES for the presence all "n-grams" in the query.
                        IF (found)
                                Increment the identified n-gram count
                        IF (task is List Completion)
                                Boost_Score()
                    }
    }//while
    Sort the list of retrieved document on the basis of descending n-gram counts
    }//END


Boost_Score()
        {
        IF (name of article == "Prominent n-grams" && content of the article con-
        tains names in the given partial list)
            {
                IF (Contents of the article also contains retrieved answers
                    {
                                Increment the n-gram count of such retrieved an-
                                swer twice.
                    }
                }
        }
```

4 Evaluation

We used Monty tagger[1] that is based on Brill's tagger to find the Part of Speech in-formation (POS).

Primary terms are selected based on the "Prominent n-grams". For obtaining secondary term in Entity Ranking task, we retrieved top 100 documents using the

---

[1] http://web.media.mit.edu/~hugo/montytagger

category as query and retrieved top 5 high TF terms as secondary terms. The threshold values used are experimentally found using different combinations. Similarly for List Completion we retrieved top 5 high TF terms from the articles of the given partial list as secondary terms.

Using Lucene we retrieved top 500 documents, and applied the ranking algorithm to obtain the ranked answer entities.

Mean Average Precision (MAP) is used as the evaluation measure by INEX 2007 Entity Ranking track. Table 2 below shows the overall score obtained by our method for both Entity Ranking and List Completion results using various measures of evaluation.

Table 2. Scores obtained based on different evaluation measures for our approach

	Entity Ranking	List completion
MAP	0.1909	0.2167
R-Precision	0.2337	0.2514
Exact Precision	0.1594	0.1439
Exact Recall	0.3625	0.4419

In the above table, apart from giving the official MAP scores, we have also given the R-precision, exact precision and exact recall scores, where R-Precision measures precision after R documents have been retrieved.

Table 3 below shows the calculated precision values after X documents have been retrieved where X ranges from 5 to 100. Values are averaged over all queries.

Table 3. Comparison of Precision scores at different document levels

Precision	Entity Ranking	List completion
5 docs	0.3304	0.2957
10 docs	0.2522	0.2326
15 docs	0.213	0.1971
20 docs	0.1848	0.1598
30 docs	0.1355	0.1246
100 docs	0.0524	0.0513

The following Figures 5 and 6, show the variation in precision with respect to the number of documents retrieved.

From the figures it is to be concluded that the precision reduces as we consider more number of retrieved documents. With an increase in the number of documents retrieved, there are lesser number of relevant documents and this obviously pushes the precision value lower.

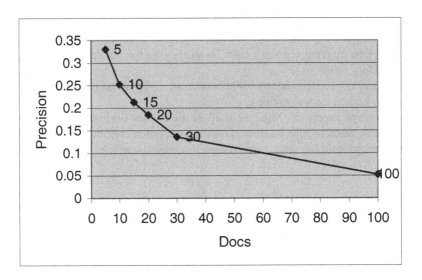

Fig. 5. Precision after X documents for ER task

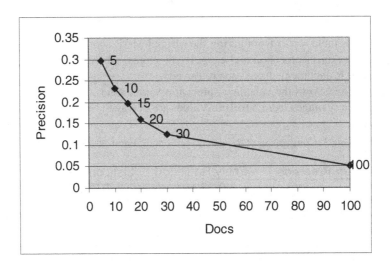

Fig. 6. Precision after X documents for LC task

5 Conclusion

The main focus of this paper is improving the queries in such a way to represent "Meaningful n-grams" and thereby forming effective queries. This has helped to retrieve relevant answers to a great extent. The proposed method relies on the identification of the "Prominent n-grams" and efficiently forms queries that help Lucene to retrieve relevant documents. It takes the help of the Initial Descriptions (IDES) given in a Wikipedia article as the representation of the document. The proposed method further ranks the retrieved answers.

Although, the method proposed proves to be effective in most cases, there are some drawbacks in this method. Since the success of the retrieval mainly depends on the identification of "Prominent n-grams", its effectiveness is dependent on the presence of such "Prominent n-grams". Future work includes exploring possible ways to form a better query. The use of the "collection pages" that contain a list of answers should also be considered in future work.

References

1. Voorhees, E.M.: Overview of the TREC 2001 Question Answering Track. In: Proceedings of the 10th Text REtrieval Conference (2001)
2. Craswell, N., de Vries, A.P., Soboroff, I.: Overview of the TREC 2005 Enterprise Track. In: Proceedings of the 14th Text REtrieval Conference (2005)
3. Soricut, R., Brill, E.: A Unified Framework for Automatic Evaluation using N-gram Co-Occurrence Statistics. In: Proceedings of the Association for Computational Linguistics (ACL) (2004)
4. Lin, C.-Y., Hovy, E.: Automatic evaluation of summaries using N-gram co-occurrence statistics. In: Proceedings of the 2003 Conference of the North American Chapter of the Association for Computational Linguistics on Human Language Technology (2003)
5. Chen, J., Diekema, A., Taffett, M.D., McCracken, N., Ozgencil, N.E., Yilmazel, O., Liddy, E.D.: Question Answering: CNLP at the TREC 10 Question Answering Track. In: Proceedings of the 10th Text REtrieval Conference (2001)
6. Yang, H., Chua, T.-S.: Web-based list question answering. In: Proceedings of the 20th International Conference on Computational Linguistics (2004)
7. Greenwood, M.A., Stevenson, M., Gaizauskas, R.: The University of Sheffield's TREC 2006 Q&A Experiments. In: Proceedings of the 15th Text REtrieval Conference (2006)
8. Kazama, J., Torisawa, K.: Exploiting Wikipedia as External Knowledge for Named Entity Recognition. In: Proceedings of the 2007 Joint Conference on Empirical Methods in Natural Language Processing and Computational Natural Language Learning (EMNLP-CoNLL) (2007)
9. Hermjakob, U., Hovy, E.H., Lin, C.-Y.: Knowledge-Based Question Answering. In: Proceedings of the 6th World Multiconference on Systems, Cybernatics and Informatics (SCI-2002), Orlando, FL, U.S.A, July 14-18 (2002)

Structured Document Retrieval, Multimedia Retrieval, and Entity Ranking Using PF/Tijah

Theodora Tsikrika[1], Pavel Serdyukov[2], Henning Rode[2], Thijs Westerveld[3,*],
Robin Aly[2], Djoerd Hiemstra[2], and Arjen P. de Vries[1]

[1] CWI, Amsterdam, The Netherlands
[2] University of Twente, Enschede, The Netherlands
[3] Teezir Search Solutions, Ede, The Netherlands

Abstract. CWI and University of Twente used PF/Tijah, a flexible XML retrieval system, to evaluate structured document retrieval, multimedia retrieval, and entity ranking tasks in the context of INEX 2007. For the retrieval of textual and multimedia elements in the Wikipedia data, we investigated various length priors and found that biasing towards longer elements than the ones retrieved by our language modelling approach can be useful. For retrieving images in isolation, we found that their associated text is a very good source of evidence in the Wikipedia collection. For the entity ranking task, we used random walks to model multi-step relevance propagation from the articles describing entities to all related entities and further, and obtained promising results.

1 Introduction

In INEX 2007, CWI and the University of Twente participated in the Ad Hoc, Multimedia, and Entity Ranking tracks. In all three tracks, we used PF/Tijah [5], a flexible system for retrieval from structured document collections, that integrates NEXI-based IR functionality and full XQuery support.

In the Ad Hoc track, we participated in all three subtasks for element retrieval, and mainly investigated the effect of various length priors within a language modelling framework. We also took part in both Multimedia tasks, where we examined the value of textual and context-based evidence without considering any of the available visual evidence. For Entity Ranking, we exploited the associations between entities; entities are ranked by constructing a query-dependent entity link graph and applying relevance propagation schemes modelled by random walks.

The remainder of this paper is organised as follows. Section 2 introduces PF/Tijah. Next, Sections 3, 4, and 5 respectively discuss our participation in each of the Ad Hoc, Multimedia, and Entity Ranking tracks. Section 6 concludes this paper by highlighting our main contributions.

---

* This work was carried out when the author was at CWI, Amsterdam, The Netherlands.

N. Fuhr et al. (Eds.): INEX 2007, LNCS 4862, pp. 306–320, 2008.

2 The PF/Tijah System

PF/Tijah, a research project run by the University of Twente, aims at creating a flexible environment for setting up search systems. It achieves that by including out-of-the-box solutions for common retrieval tasks, such as index creation (that also supports stemming and stopword removal) and retrieval in response to structured queries (where the ranking can be generated according to any of several retrieval models). Moreover, it maintains its versatility by being open to adaptations and extensions.

PF/Tijah is part of the open source release of MonetDB/XQuery (available at http://www.sourceforge.net/projects/monetdb/), which is being developed in cooperation with CWI, Amsterdam and the University of München. PF/Tijah combines database and information retrieval technologies by integrating the PathFinder (PF) XQuery compiler [1] with the Tijah XML information retrieval system [11]. This provides PF/Tijah with a number of unique features that distinguish it from most other open source information retrieval systems:

– It supports retrieval of arbitrary parts of XML documents, without requiring a definition at indexing time of what constitutes a document (or document field). A query can simply ask for any XML tag-name as the unit of retrieval without the need to re-index the collection.
– It allows complex scoring and ranking of the retrieved results by directly supporting the NEXI query language.
– It embeds NEXI queries as functions in the XQuery language, leading to ad hoc result presentation by means of its query language.
– It supports text search combined with traditional database querying.

The above characteristics also make PF/Tijah particularly suited for environments like INEX, where search systems need to handle highly structured XML collections with heterogenous content. Information on PF/Tijah, including usage examples, can be found at: http://dbappl.cs.utwente.nl/pftijah/

3 Ad Hoc Track

The granularity at which to return information to the user has always been an important aspect of the INEX benchmarks. The element and passage retrieval tasks aim to study ways of pointing users to the most specific relevant parts of documents. Various characteristics of the document parts or elements are of potential value in identifying the most relevant retrieval bits. Obviously the element content is a valuable indicator, but also more superficial features like the element type, the structural relation to other elements and the depth of the XML tree may play a role.

We studied the influence of a very basic feature: element size. Size priors have played an important role in information retrieval [14,4,8]. Kamps et al. [6] studied length normalization in the context of XML retrieval and INEX collections and found that the size distribution of relevant elements differed significantly from the general size distribution of elements. Emphasizing longer elements by introducing, linear, quadratic or even cubic length priors improved the retrieval results significantly on the IEEE collection.

For this paper, we experimented with biasing towards longer elements (similarly to Kamps et al. [6]), but in the setting of the Wikipedia collection. We use a language modelling framework where document priors are incorporated as a priori probabilities of relevance based on document characteristics that are independent of a query (element size in our case). The probability of a document D given a query Q can be factored as the probability of drawing the query from the document ($P(Q|D)$: the document's language model) and the prior probability of the document $P(D)$ (the prior probability of the query $P(Q)$ does not influence the ranking and can be ignored):

$$P(D|Q) = \frac{P(Q|D)P(D)}{P(Q)} \propto P(Q|D)P(D) \tag{1}$$

where $P(Q|D)$ is estimated using a unigram language model smoothed by a Jelinek-Mercer parameter [4]. We also performed a retrospective study on the Wikipedia collection, where we analysed the size distributions of elements in the collection, in the relevant elements for the INEX 2006 Focused task, and in the elements retrieved by our baseline language model run.

3.1 Experiments with Length Priors

In our runs for INEX 2007, we experimented with different priors. We submitted runs with priors that are linear in the log of the element size (star_logLP) and runs with a normally distributed log size prior (star_lognormal). Each of the prior runs is submitted for the Focused task and in addition filtered for the Relevant in Context task (runIDs with _Ric affix); for relevant in context we grouped the results in a top 1500 baseline run by article and ordered the articles based on their top scoring element. In addition we submitted an article only baseline run, i.e. a run in which we only return full articles. This article run was submitted to both the Focused (article) and Best in Context tasks (article_BiC). Tables 1, 2, and 3 show the results for these official submissions.

Table 1. Results for the CWI/UTwente submissions to the Ad Hoc Focused task. The table shows the rank of the run among official submissions, the run identifier and the interpolated precision at 0.01 recall.

rank	runID	iP[0.01]
56	star_logLP	0.3890
59	article	0.3701
78	star_lognormal	0.0381

Table 2. Results for the CWI/UTwente submissions to the Ad Hoc Relevant in Context task. The table shows the rank of the run among official submissions, the run identifier and Mean Average generalized precision.

rank	runID	MAgP
15	star_logLP_RinC	0.1233
64	star_lognormal_RinC	0.0075

Table 3. Results for the CWI/UTwente submissions to the Ad Hoc Best in Context task. The table shows the rank of the run among official submissions, the run identifier and the Mean Average generalized precision.

rank	runID	MAgP
31	articleBic	0.1339

3.2   Analysis of Element Size

The disappointing results with the two size priors warrant a study of the distribution of element size in relevant and non-relevant elements. We studied INEX 2006 data to gain some insight. Figure 1 shows the distribution of element sizes in the Wikipedia collection as a whole and in the relevant elements. While the collection contains many small elements, these are rarely relevant. If we would not pay attention to element length and use a retrieval model that does not have a bias for elements of any size we would retrieve too many small elements. Simply giving a bias towards longer elements could improve retrieval results.

As previously mentioned, one way of compensating for this emphasis on small elements that nicely fits in the language modeling framework that we use is to incorporate document priors. The probability of relevance given a certain size can be estimated by comparing the distributions of relevant elements to those of the collection: $P_{size}(D) = P(relevant|size(D))$. This leads to the prior visualized in

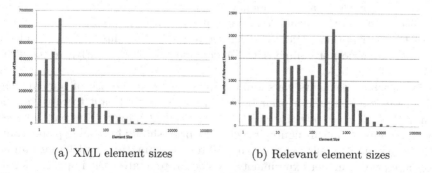

(a) XML element sizes   (b) Relevant element sizes

Fig. 1. Size distribution of collection elements and elements relevant to 2006 topics

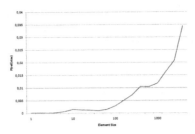

Fig. 2. Size prior estimated from INEX 2006 statistics for relevant and collection elements

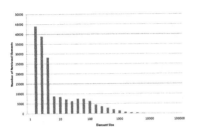

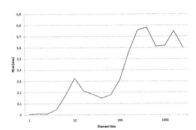

(a) Size distribution of elements retrieved in Language Modeling framework

(b) Size prior estimated from the fraction of the number of relevant and retrieved elements

Fig. 3. Size distribution of retrieved elements and prior based on comparing this distribution with size distribution of relevant elements (Figure 1b)

Figure 2. A quadratic prior as found by Kamps et al. [6] for the IEEE collection seems appropriate.

However, in reality, a retrieval model does not retrieve elements of all sizes uniformly. For example, the language model we use interpolates document and collection probabilities in a standard manner and computes the document probability based on the relative frequency of query terms in documents [4]. This has the effect that short elements containing query terms get a high score. Figure 3a shows the distribution of elements that we retrieve using this language modeling approach if we do not compensate for document length. Clearly, we retrieve a lot of small elements.

To see which elements we should emphasize given the use of our language model, we also compute a prior based on comparing relevant to retrieved elements: $P_{size}(D) = P(relevant|size(D), retrieved(D))$. Figure 3b visualises these priors. Judging from this figure, it seems the prior should have a big peak around 1000 terms and a smaller peak around 10 terms. A mixture model seems an appropriate prior. Further experiments are needed to analyse the impact of such a prior on retrieval effectiveness.

4 Multimedia Track

CWI/Utwente participated in both MMfragments and MMimages tasks of the Multimedia track. Our overall aim is to investigate the value of textual and contextual evidence given information needs (and queries) with clear multimedia character. As a result, we only submitted text-based runs without taking into account any of the provided visual evidence. Below, we discuss our approaches and experimental results for both tasks.

4.1 MMfragments Task

For MMfragments, the objective is to find relevant XML fragments (i.e., elements or passages) in the (Ad Hoc) Wikipedia XML collection given a multimedia information need. MMfragments is actually very similar to the Ad Hoc retrieval task, with the difference being that MMfragments has a multimedia character and, therefore, requires the retrieved fragments to contain at least one relevant image, together with relevant text. Furthemore, additional visual evidence, such as concepts and image similarity examples, can be provided as part of a topic. Given these similarities, MMfragments was run in conjunction with the Ad Hoc track, with MMfragments topics forming a subset of the Ad Hoc ones. In addition, MMfragments contains the same three substasks as the Ad Hoc task. This gives us the opportunity to compare the effectiveness of MMfragments runs (i.e., runs with a clear multimedia character) against Ad Hoc runs on the same topic subset.

We only participated in the Focused MMfragments task. Given the similarities with the Ad Hoc task, we decided to (i) use only the title field of the topics, (ii) apply the same three element runs as the ones submitted for the Focused Ad Hoc task (i.e., `article`, `star_logLP` and `star_lognormal`), and (iii) realise the multimedia character by filtering our results, so that we only return fragments that contain at least one image. Not all <image> tags in the (Ad Hoc) Wikipedia XML collection correspond to images that are actually part of the Wikipedia image XML collection; images that are not part of this collection will not be visible to users during assessments. Therefore, we also removed all results that contained references to images that are not in the Wikipedia image XML

Table 4. Results for the CWI/UTwente official MMfragments Focused submissions and Ad Hoc Focused runs on the MMfragments topic subset. The table shows the rank of the run among official submissions, the run identifier and the interpolated precision at 0.01 recall.

rank	runID	iP[0.01]
1	article_MM	0.3389
4	star_loglength_MM	0.2467
5	star_lognormal_MM	0.0595
-	star_loglength	0.2325
-	star_lognormal	0.1045

Table 5. Results for the CWI/UTwente official submissions and additional runs to the MMimages task. The table shows the rank of the run among official submissions, the run identifier and Mean Average Precision.

rank	runID	MAP
1	title_MMim	0.2998
3	article_MMim	0.2240
5	figure_MMim	0.1551
-	title_MMim_lengthPrior	0.3094
-	title_MMim_logLengthPrior	0.3066

collection. This way, we made sure all our returned fragments contain at least one *visible* image.

The results of our official submissions are presented in Table 4. Given our analysis of priors in Section 3.2, further experimentation is needed to determine whether other priors (e.g., quadratic and mixed priors) would lead to better performace. Finally, Table 4 also presents the results of our Ad Hoc Focused runs on the MMfragments topic subset, which indicate the usefulness of our filtering approach in the context of topics with clear multimedia character.

4.2 MMimages Task

For MMimages, the aim is to retrieve documents (images + their metadata) from the Wikipedia image XML collection. Similarly to the Ad Hoc and MMfragments tasks, our submitted runs are based on the language modelling approach. Each image is represented either by its textual metadata in the Wikipedia image XML collection, or by its textual context when that image appears as part of a document in the (Ad Hoc) Wikipedia XML collection.

To be more specific, we submitted the following three runs:

title_MMim. Create a stemmed index using the metadata accompanying the images in the Wikipedia image XML collection, and perform an article run using only the topics' title field: `//article[about(.,$title)]`.

article_MMim. Rank the articles in the (Ad Hoc) Wikipedia XML collection using each topic's title field and retrieve the images that these articles contain. Filter the results, so that only images that are part of the Wikipedia image XML collection are returned.

figure_MMim. Rank the figures with captions in the (Ad Hoc) Wikipedia XML collection using each topic's title field (`//figure[about(.,$title)]`) and return the images of these figures (ensuring that these images are part of the Wikipedia image XML collection).

Table 5 presents the Mean Average Precision (MAP) of these runs, whereas Figure 4 compares them against all the runs submitted to the MMimages task. Our experimental results indicate that these text-based runs give a highly competitive performance on the MMimages task.

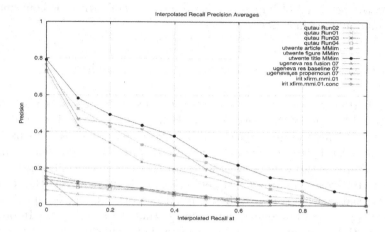

Fig. 4. MMimages results: CWI/Utwente runs compared to all submitted runs

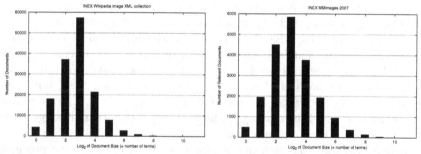

(a) Wikipedia image XML documents   (b) MMimages 2007 relevant documents

Fig. 5. Size distribution of metadata in Wikipedia image XML collection and of metadata of images relevant to MMimages 2007 topics

We incorporated a document prior based on length (defined as the number of terms in the metadata), **title_MMim_lengthPrior**, and the log of this length, **title_MMim_logLengthPrior**. By defining the priors in that manner, we are able to apply them without performing any training. Our results in Table 5 indicate that both priors improve over the corresponding baseline, with the length prior improving the most.

These runs are based on the assumption that the distribution of document size is different for relevant and non-relevant images. We perform a retrospective analysis of the distribution of length in the MMimages collection (Figure (5a)), and the relevant documents for 2007 (Figure (5b)). While the collection contains many small documents, these are rarely relevant. If we would not pay attention to document length and just use a retrieval model that does not have a bias for documents of any size, we would retrieve too many small documents. Simply giving a bias towards longer documents in the context of the INEX MMimages task has the potential of improving the retrieval result, which is confirmed by our evaluation experiments.

5   Entity Ranking by Relevance Propagation

We also participated in this year's entity ranking task. The queries here ask for a ranked list of entities, e.g., movies, flags, or diseases. Entities are usually identified by their name and type. An entity of type movie would be identified by its title. In general, the entity ranking task clearly differs from document ranking, since it requires to estimate the relevance of items that do not have text content [12,15]. In this case, the ranking can only be done by propagating the relevance from retrieved text fragments to their contained entities. Using Wikipedia as the corpus for entity ranking experiments, the setting changes slightly. In order to use the existing mark-up of the corpus – instead of employing taggers for named entity recognition – the only entities considered were those that have their own Wikipedia article. An entity is contained in an article when it is linked by that article. In consequence, the distinction of articles and entities is abandoned. Since entities have their own article, they can also be ranked directly by their content.

In the context of Wikipedia, the type of an entity is defined by the categories assigned to the entity's article. An entity can thus have several types. Furthermore, Wikipedia categories are hierarchically organized. We can thus assume that an entity does not only belong the categories assigned to it, but also to ancestor categories. However, the hierarchy of Wikipedia's categories does not form a strict tree, and thus moving too far away from the original categories can lead to unexpected type assignments.

Our entity ranking approach can be summarized by the following processing steps: (1) initial retrieval of articles, (2) building of an entity graph, (3) relevance propagation within the graph, and (4) filtering articles by the requested type. The notion of *entity graph* stands here for a query-dependent link graph, consisting of all articles (entities) returned by the initial retrieval as vertices and the link-structure among them forming the edges. Links to other articles not returned in the initial ranking are not considered in the entity graph. The entity graph can later be used for the propagation of relevance to neighbouring nodes. Starting with web retrieval [10,7,13], graph based ranking techniques have been recently used in several fields of IR [3,9,2].

5.1   Baseline: Entity Retrieval by Description Ranking

The simplest and most obvious method for entity retrieval is the ranking of their textual descriptions with some classic document retrieval method. In our experiments, we rank Wikipedia articles representing entities using a language model based retrieval method:

$$P(Q|e) = \prod_{t \in Q} P(t|e), \tag{2}$$

$$P(t|e) = (1 - \lambda_C)\frac{tf(t,e)}{|e|} + \lambda_C \frac{\sum_{e'} tf(t,e')}{\sum_{e'} |e'|} \tag{3}$$

where $tf(q,e)$ is a term frequency of q in the entity description e, $|e|$ is the description length and λ_C is a Jelinek-Mercer smoothing parameter - the probability of a term to be generated from the global language model. In all our experiments, λ_C is set to 0.8, which is standard in retrieval tasks.

However, due to several reasons this approach may produce unsatisfactory results. First, many entities have too short or empty descriptions, especially those that appear in novel and evolving domains that are just becoming known. Thus, many entities may get scores close to zero and not appear in the top. Second, many entities are described by showing the associations with other entities and in terms of other entities. This means that query terms have lesser chance in appearing in the content of a relevant description, since some concepts mentioned in its text are not explained because explanations can be found in their own descriptions.

5.2 Entity Retrieval Based on K-Step Random Walk

In our follow-up methods, we consider that relevance propagation from initially retrieved entities to the related ones is important. We imagine and model the process in which the user, after seeing initial list of retrieved entities:

- selects one document and reads its description,
- follows links connecting entities and reads descriptions of related entities.

Since we consider this random walk as finite, we assume that at some step a user finds the relevant entity and stops the search process. So, we iteratively calculate the probability that a random surfer will end up with a certain entity after K steps of walk started at one of the initially ranked entities. In order to emphasize the importance of entities to be in proximity to the most relevant ones according to the initial ranking, we consider that both (1) the probability to start the walk from certain entity and (2) the probability to stay at the entity node are equal to the probability of relevance of its description.

$$P_0(e) = P(Q|e) \tag{4}$$

$$P_i(e) = P(Q|e)P_{i-1}(e) + \sum_{e' \to e} (1 - P(Q|e'))P(e|e')P_{i-1}(e'), \tag{5}$$

The probabilities $P(e|e')$ are uniformly distributed among links outgoing from the same entity. Finally, we rank entities by their $P_K(e)$.

Linear Combination of Step Probabilities It is also possible to estimate entity relevance using several finite walks of different lengths at once. In the following modification of the above described method, we rank entities considering a weighted sum of probabilities to appear in the entity node at different steps:

$$P(e) = \mu_0 P_0(e) + (1 - \mu_0) \sum_{i=1}^{K} \mu_i P_i(e) \tag{6}$$

In our experiments we set μ_0 to 0.5 and distribute $\mu_1 \ldots \mu_K$ uniformly.

5.3 Entity Retrieval Based on Infinite Random Walk

In our second approach, we assume that the walk in search for relevant entities consists of countless number of steps. The stationary probability of ending up in a certain entity is considered to be proportional to its relevance. Since the stationary distribution of a described discrete Markov process does not depend on the initial distribution over entities, the relevance flow becomes unfocused. The probability to appear in a certain entity node becomes dependent only on its centrality, but not on its closeness to the sources of relevance. To solve this issue, we introduce regular jumps to entity nodes from any node of the entity graph after which the walk restarts and the user follows inter-entity links again. We consider that the probability of jumping to a specific entity equals to the probability of relevance of its description. This makes a random walker visit entities which are situated closer to the initially highly ranked ones more often during normal walk steps. The following formula is used for iterations until convergence:

$$P_i(e) = \lambda_J P(Q|e) + (1 - \lambda_J) \sum_{e \to e'} P(e|e') P_{i-1}(e') \tag{7}$$

where λ_J is the probability that, at any step, the user decides to make a jump and not to follow outgoing links anymore. The described discrete Markov process is stochastic and irreducible, since each entity is reachable due to the introduced jumps, and hence has a stationary distribution. Consequently, we rank entities by their stationary probabilities $P_\infty(e)$.

5.4 Experiments

We trained our models using the 28 queries from the Ad-Hoc XML Retrieval task that are also suitable for the entity ranking task. All our algorithms start from the retrieval of articles from the collection using a language modelling approach to IR for scoring documents. Then, we extract entities mentioned in these articles and build entity graphs. For the initial article retrieval, as well as for the graph generation, the PF/Tijah retrieval system was employed. For this experiment, we generated XQueries that directly produce entity graphs in *graphml* format given a title-only query. We tuned our parameters by maximization of the MAP measure for 100 initially retrieved articles.

For the following methods, we discuss their performance first on the training and then on the test data:

- **Baseline:** the baseline method which ranks entities by the relevance of their Wikipedia-articles (see Equations 2, 3),
- **K-Step RW:** the K-step Random Walk method which uses multi-step relevance propagation with K steps (see Equations 4, 5),
- **K-Step RWLin:** the K-step Random Walk method which uses the linear combination of entity relevance probabilities at different steps up to K (see Equation 6),

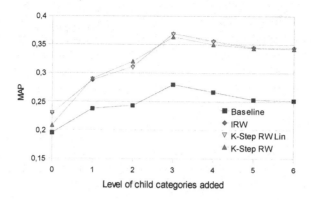

Fig. 6. MAP performance of all methods for different levels of child categories added

 – **IRW:** the Infinite Random Walk method which ranks entities by the probability of reaching them in infinity during non-stop walks (see Equation 7).

For the Entity Retrieval task, we have a query and the list of entity categories as input. However, according to the track guidelines and our own intuition, relevant entities could be found to be out of the scope of given categories.
Preliminary experiments have shown that using parent categories of any level
spoiled the performance of the Baseline method. However, it was very important to include child categories up to 3rd level both for our Baseline method
and for our remaining methods which require tuned parameters (see Figure 6).
This probably means that queries were created with an assumption that given
categories should be the greatest common super-types for the relevant entities.
It must be mentioned that we used entities of all categories for the graph construction and relevance propagation, and filtered out entities using the list of
allowed categories only at the stage of result output.

In all our methods, except the Baseline, we had to tune one specific parameter. For the K-step RW and K-step RWLin methods, we experimented with
the number of walking steps. As we see in Figure 7, both methods reach their
maximum performance after 3 steps only. The K-step RW Lin method seems
to be more robust to the parameter setup. This probably happens because it
smooths the probability to appear in a certain entity after K steps, with probabilities of visiting it earlier. The rapid decrease of performance for even steps
for the K-step RW method can be explained in the following way. A lot of relevant entities are only mentioned in the top ranked entity descriptions and do
not have their own descriptions in this top, due to their low relevance probability or due to their absence in the collection. The relevance probability of these
"outsider" entities entirely depends on the relevance of related entities, which
are not relevant entities themselves (for example, do not match the requested
entity type), but tell a lot about the ranked entity. So, all "outsider" entities
have direct (backward) links only to the entities with descriptions in the top.

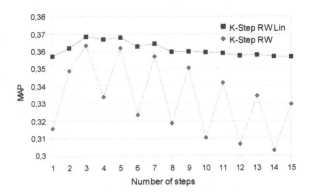

Fig. 7. MAP performance for two methods and different numbers of steps

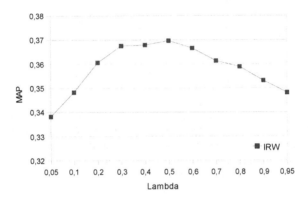

Fig. 8. MAP performance of IRW method for different values of jumping probability

Table 6. Final results for all methods for the Entity Ranking task

runID	MAP
Baseline	0.291
K-Step RW	0.281
K-Step RWLin	0.306
IRW	0.301

Since we always start walking only from the latter entities, the probability to appear in "outsider" entities at every even step is close to zero.

We also experimented with the probability to restart the walk from initially ranked entities for the IRW method. According to results shown in Figure 8, values between 0.3 and 0.5 seem to be optimal. This actually means that making only 2-3 steps (before the next restart) is the best strategy, which is also the case for the finite random walk methods.

To sum the things up, our experiments with the training data showed that all our three methods significantly outperform the **Baseline** method. However, the **K-Step RW** method produced a bit worse results than the other two.

As we see in Table 6, our final results on the test data show that both the **K-Step RWLin** and the **IRW** methods are equally more effective than the **Baseline** method. The fact that the **K-Step RW** could not outperform the **Baseline** method in our final experiments confirms its lower robustness with respect to the proper parameter setup.

6   Conclusions

This is the second year that CWI and University of Twente used PF/Tijah in INEX. The flexibility of this system is clearly demonstrated through its application in INEX tracks as diverse as ad hoc structured document retrieval, retrieval of multimedia documents and document fragments, and entity ranking.

The unigram language modelling approach we have previously applied in Ad Hoc element retrieval tasks retrieves short elements. Given that our analysis of last year's results indicates that the relevant elements tend to be longer than the ones our approach retrieves, the incorporation of length priors would be beneficial. For the Focused subtask, further experimentation is needed to determine whether the priors indicated by our recent analysis would yield better performance, whereas for the Best in Context and Relevant in Context subtasks, we need to examine in more detail our filtering strategies.

Our text only approach to Multimedia retrieval was very successful on the MMimages task. Further experimentation on the MMfragments task would reveal whether more appropriate filtering techniques or alternative priors would improve our results.

The experiments with our approaches for entity ranking demonstrated the advantage of multi-step relevance propagation from textual descriptions to related entities over the simple ranking of entity textual descriptions. The further improvement seems especially challenging because all our three methods showed quite similar effectiveness.

Acknowledgements

This material is based on work funded in part by the European Union via the European Commission project VITALAS (contract no. 045389), the Dutch National project MultimediaN, and the Netherlands Organisation for Scientific Research (NWO).

References

1. Boncz, P., Grust, T., van Keulen, M., Manegold, S., Rittinger, J., Teubner, J.: MonetDB/XQuery: A fast XQuery processor powered by a relational engine. In: Proceedings of the 25th ACM SIGMOD International Conference on Management of Data, pp. 479–490 (2006)

2. Chirita, P.-A., Diederich, J., Nejdl, W.: Mailrank: using ranking for spam detection. In: Proceedings of the 14th ACM CIKM International Conference on Information and Knowledge Management, pp. 373–380 (2005)
3. Craswell, N., Szummer, M.: Random walks on the click graph. In: Proceedings of the 30th ACM SIGIR Annual International Conference on Research and Development in Information Retrieval, pp. 239–246 (2007)
4. Hiemstra, D.: A linguistically motivated probabilistic model of information retrieval. In: ECDL 1991. LNCS, vol. 513, pp. 569–584. Springer, Heidelberg (1991)
5. Hiemstra, D., Rode, H., van Os, R., Flokstra, J.: PF/Tijah: text search in an XML database system. In: Proceedings of the 2nd International Workshop on Open Source Information Retrieval (OSIR) (2006)
6. Kamps, J., de Rijke, M., Sigurbjörnsson, B.: Length normalization in XML retrieval. In: Proceedings of the 27th ACM SIGIR Annual International Conference on Research and Development in Information Retrieval, pp. 80–87 (2004)
7. Kleinberg, J.M.: Authoritative sources in a hyperlinked environment. Journal of the ACM 46(5), 604–632 (1999)
8. Kraaij, W., Westerveld, T., Hiemstra, D.: The importance of prior probabilities for entry page search. In: Proceedings of the 25th ACM SIGIR Annual International Conference on Research and Development in Information Retrieval, pp. 27–34 (2002)
9. Kritikopoulos, A., Sideri, M., Varlamis, I.: Blogrank: ranking weblogs based on connectivity and similarity features. In: Proceedings of the 2nd International Workshop on Advanced Architectures and Algorithms for Internet Delivery and Applications, p. 8 (2006)
10. Lawrence, P., Sergey, B., Motwani, R., Winograd, T.: The PageRank citation ranking: Bringing order to the web. Technical report, Stanford University (1998)
11. List, J., Mihajlovic, V., Ramirez, G., de Vries, A., Hiemstra, D., Blok, H.: Tijah: Embracing IR methods in XML databases. Information Retrieval 8(4), 547–570 (2005)
12. Serdyukov, P., Rode, H., Hiemstra, D.: University of Twente at the TREC 2007 Enterprise Track: Modeling relevance propagation for the expert search task. In: Proceedings of the 16th Text REtrieval Conference (2007)
13. Shakery, A., Zhai, C.: A probabilistic relevance propagation model for hypertext retrieval. In: Proceedings of the 15th ACM CIKM International Conference on Information and Knowledge Management, pp. 550–558 (2006)
14. Singhal, A., Buckley, C., Mitra, M.: Pivoted document length normalization. In: Proceedings of the 19th ACM SIGIR Annual International Conference on Research and Development in Information Retrieval, pp. 21–29 (1996)
15. Zaragoza, H., Rode, H., Mika, P., Atserias, J., Ciaramita, M., Attardi, G.: Ranking very many typed entities on wikipedia. In: Proceedings of the 16th ACM CIKM International Conference on Information and Knowledge Management, Lisbon, Portugal, pp. 1015–1018 (2007)

Using Wikipedia Categories and Links in Entity Ranking

Anne-Marie Vercoustre[1], Jovan Pehcevski[1], and James A. Thom[2]

[1] INRIA Rocquencourt, France
{anne-marie.vercoustre,jovan.pehcevski}@inria.fr
[2] RMIT University, Melbourne, Australia
james.thom@rmit.edu.au

Abstract. This paper describes the participation of the INRIA group in the INEX 2007 XML entity ranking and ad hoc tracks. We developed a system for ranking Wikipedia entities in answer to a query. Our approach utilises the known categories, the link structure of Wikipedia, as well as the link co-occurrences with the examples (when provided) to improve the effectiveness of entity ranking. Our experiments on both the training and the testing data sets demonstrate that the use of categories and the link structure of Wikipedia can significantly improve entity retrieval effectiveness. We also use our system for the ad hoc tasks by inferring target categories from the title of the query. The results were worse than when using a full-text search engine, which confirms our hypothesis that ad hoc retrieval and entity retrieval are two different tasks.

1 Introduction

Entity ranking has recently emerged as a research field that aims at retrieving entities as answers to a query [7,10,12,14]. Here, unlike in the related field of entity extraction, the goal is not to tag the names of the entities in documents but rather to get back a list of the relevant entity names. It is a generalisation of the expert search task explored by the TREC Enterprise track [11], except that instead of ranking people who are experts in the given topic, other types of entities such as organisations, countries, or locations can also be ranked.

The Initiative for the Evaluation of XML retrieval (INEX) ran a new track on entity ranking in 2007, using Wikipedia as its document collection [5]. There were two tasks in the INEX 2007 XML entity ranking (XER) track: task 1 (*entity ranking*), with the aim of retrieving entities of a given category that satisfy a topic described in natural language text; and task 2 (*list completion*), where given a topic text and a small number of entity examples, the aim was to complete this partial list of answers. Two data sets were used by the participants of the XER track: a *training* data set, comprising 28 XER topics which were adapted from the INEX 2006 ad hoc topics; and a *testing* data set, comprising 46 XER topics that were proposed and assessed by the track participants.

In the XER track, the expected entities correspond to Wikipedia articles that are likely to be referred to by links in other articles. As an example, the query

N. Fuhr et al. (Eds.): INEX 2007, LNCS 4862, pp. 321–335, 2008.

"European countries where I can pay with Euros" [5] should only return a list of entities (or pages) representing relevant countries, and not include entities representing non-relevant countries nor other entities found in pages about the Euro and similar currencies.

In this paper, we describe our approach to ranking entities from the Wikipedia XML document collection. Our approach is based on the following hypotheses:

1. A good entity page is a page that answers the query, or a query extended with names of target categories (task 1) or entity examples (task 2).
2. A good entity page is a page associated with a category close to the target category (task 1) or to the categories of the entity examples (task 2).
3. A good entity page is referred to by a page answering the query; this is an adaptation of the HITS [9] algorithm to the problem of entity ranking.
4. A good entity page is referred to by contexts with many occurrences of the entity examples (task 2). A broad context could be the full page that contains the entity examples, while smaller and more narrow contexts could be elements such as paragraphs, lists, or tables.

This paper is organised as follows. After a short review of the related work and a brief presentation of the INEX Wikipedia XML collection used for entity ranking, we provide a detailed description of our entity ranking approach and the runs we submitted for evaluation to the INEX 2007 XER track. We also report on our run submissions to the INEX 2007 ad hoc track that are based on our entity ranking approach. For both tracks we submitted a run based on a full-text retrieval approach. By analysing and comparing the performances of runs based on these two approaches, we address the following research question: Are ad hoc retrieval and entity retrieval two different tasks?

2   Related Work

Entity ranking has attracted a lot of research recently. It can be seen as a generalisation of expert search where the entities of interest are not only people. For example, Craswell et al. [4] use the co-occurrence of people's names and query words in documents as evidence to rank experts. Zhu et al. [15] have extended their expert search system to allow for entity search. Their approach involves an association model based on co-occurrence of entities with query terms in documents mentioning the entity. The association can be made at multiple levels: phrase, sentence, paragraph and up to a document level, with associated weights that decrease for larger contexts. Entities are filtered by comparing their categories with the target category and its child and parent categories.

ESTER [2] was recently proposed as a system for searching text, entities and relations. ESTER relies on the Wikipedia links to identify the entities and on the context of the links for disambiguation (using 20 words around the anchor text instead of just the anchor text). Hu et al. [8] propose a linear model that uses a number of features to weight passages containing entity names. They first determine top k passages and extract the top n entities from these passages.

Features include term frequency, distance to the entity name and co-occurrences in the same section as the entity. Tsikrika et al. [13] build a graph of the initial set of documents returned in answer to the query, but do not extend the graph to linked documents outside the initial set; this graph is then used to propagate relevance in entity retrieval based on k-step or infinite random walk. Entities are filtered by using the target category and its child categories (up to a third level).

3   INEX Wikipedia XML Collection

Wikipedia is a well known web-based, multilingual, free content encyclopedia written collaboratively by contributors from around the world. Denoyer and Gallinari [6] have developed an XML-based corpus based on a snapshot of the Wikipedia, which has been used by various INEX tracks in 2006 and 2007. It differs from the real Wikipedia in some respects (size, document format, category tables), but it is a very realistic approximation.

3.1   Entities in Wikipedia

The entities have a name (the name of the corresponding page) and a unique ID in the collection. When mentioning such an entity in a new Wikipedia article, authors are encouraged to link occurrences of the entity name to the page describing this entity. This is an important feature as it makes it easy to locate potential entities, which is a major issue in entity extraction from plain text.

However in this collection, not all potential entities have been associated with corresponding pages. The INEX 2007 XER topics have been carefully designed to make sure there is a sufficient number of answer entities. For example, in the Euro page (see Fig. 1), all the underlined hypertext links can be seen as occurrences of entities that are each linked to their corresponding pages. In this figure, there are 18 entity references of which 15 are country names; specifically, these countries are all "European Union member states", which brings us to the notion of categories in Wikipedia.

3.2   Categories in Wikipedia

Wikipedia also offers categories that authors can associate with Wikipedia pages. There are 113,483 categories in the INEX Wikipedia XML collection, which are organised in a graph of categories. Each page can be associated with many categories (2.28 as an average).

"The **euro** ... is the official currency of the Eurozone (also known as the Euro Area), which consists of the European states of Austria, Belgium, Finland, France, Germany, Greece, Ireland, Italy, Luxembourg, the Netherlands, Portugal, Slovenia and Spain, and will extend to include Cyprus and Malta from 1 January 2008."

Fig. 1. Extract from the Euro Wikipedia page

Wikipedia categories have unique names (e.g. "France", "European Countries", "Countries"). New categories can also be created by authors, although they have to follow Wikipedia recommendations in both creating new categories and associating them with pages. For example, the Spain page is associated with the following categories: "Spain", "European Union member states", "Spanish-speaking countries", "Constitutional monarchies" (and some other Wikipedia administrative categories).

When searching for entities it is natural to take advantage of the Wikipedia categories since they would give a hint on whether the retrieved entities are of the expected type. For example, when looking for entities of type "authors", pages associated with the category "Novelist" are more likely to be relevant than pages associated with the category "Book".

4   Our Entity Ranking Approach

Our approach to identifying and ranking entities combines: (i) the full-text similarity of the answer entity page with the query; (ii) the similarity of the page's categories to the target categories (task 1) or to the categories attached to the entity examples (task 2); and (iii) the contexts around entity examples found in the top ranked pages returned by a search engine for the query.

We have built a system based on the above ideas, and a framework to tune and evaluate a set of different entity ranking algorithms.

4.1   Architecture

The system involves several modules and functions that are used for processing a query, submitting it to the search engine, applying our entity ranking algorithms, and finally returning a ranked list of entities. We use Zettair[1] as our choice for a full-text search engine. Zettair is a full-text information retrieval (IR) system developed by RMIT University, which returns pages ranked by their similarity score to the query. We used the Okapi BM25 similarity measure that has proved to work well on the INEX 2006 Wikipedia test collection [1].

Our system involves the following modules and functions:

- the topic module takes an INEX topic as input and generates the corresponding Zettair query and the list of target categories and entity examples; as an option, the names of target categories (task 1) or example entities (task 2) may be added to the query;
- the search module sends the query to Zettair and returns a list of ranked Wikipedia pages (typically 1500);
- the link extraction module extracts the links from a selected number of highly ranked pages,[2] together with the information concerning the paths of the links (using an XPath notation);

---

[1] http://www.seg.rmit.edu.au/zettair/
[2] We discarded external links and some internal collection links that do not refer to existing pages in the INEX Wikipedia collection.

- the category similarity module calculates a weight for a page based on the similarity of the page categories to target categories or to those attached to entity examples (see sub-section 4.2);
- the linkrank module calculates a weight for a page based (among other things) on the number of links to this page (see sub-section 4.3); and
- the full-text IR module calculates a weight for a page based on its initial Zettair score.

The global score for a page is calculated as a linear combination of three normalised scores coming out of the last three modules (see sub-section 4.4).

The architecture provides a general framework for evaluating entity ranking which allows for some modules to be replaced by more advanced modules, or by providing a more efficient implementation of a module. It also uses an evaluation module to assist in tuning the system by varying the parameters and to globally evaluate our entity ranking approach.

The major cost in running our system lies in extracting the links from the selected number of pages retrieved by the search engine. Although we only extract links once by topic and store them in a database for reuse in later runs, an online system would require extracting and storing all the links at indexing time.

4.2   Using Wikipedia Categories

To make use of the Wikipedia categories in entity ranking, we define similarity functions between the categories of answer entities and the target categories (task 1), or between the categories of answer entities and a set of categories attached to the entity examples (task 2).

Similarity measures between concepts of the same ontology, such as tree-based similarities [3], cannot be applied directly to Wikipedia categories, mostly because the notion of sub-categories in Wikipedia is not a pure subsumption relationship. Another reason is that categories in Wikipedia do not form a hierarchy (or a set of hierarchies) but form a graph with potential cycles.

Task 1. We first define a similarity function that computes the ratio of common categories between the set of categories, $\mathsf{cat}(t)$, associated to an answer entity page t, and the set $\mathsf{cat}(C) = C$, where C is the set of provided target categories:

$$S_C(t) = \frac{|\mathsf{cat}(t) \cap \mathsf{cat}(C)|}{|\mathsf{cat}(C)|} \tag{1}$$

The target categories will be generally very broad, so it is to be expected that the answer entities would not be directly attached to these broad categories. Accordingly, we experimented with several extensions of the set of categories, both for the target categories and the categories attached to answer entities.

We first experimented with extensions based on using sub-categories and parent categories in the graph of Wikipedia categories. However, on the training data set, we found that these category extensions overall do not result in an improved performance [12], and so they were not used in our INEX 2007 runs.

Another approach is to use lexical similarity between the category names. For example, "european countries" is lexically similar to "countries" since they both contain the word "countries" in their names. We use an information retrieval approach to retrieve similar categories, by constructing a separate index with Zettair of all the category names (using the names as documents). By sending both the title of the topic T and the category names C as a query to Zettair, we then retrieve all the categories that are lexically similar to C. We keep the top M ranked categories and add them to C to form the set $\mathsf{TCcat}(C)$. On the training data set, we found that the value M=5 is the optimal parameter value to retrieve the likely relevant categories for this task [12]. We then use the same similarity function as before, but where $\mathsf{cat}(C) = \mathsf{TCcat}(C)$.

We also experimented with two alternative approaches: by sending the category names C as a query to Zettair (denoted as $\mathsf{Ccat}(C)$); and by sending the title of the topic T as a query to Zettair (denoted as $\mathsf{Tcat}(C)$). On the training data set, we found that these two approaches were less effective than the $\mathsf{TCcat}(C)$ approach [12]. However, we use $\mathsf{cat}(C) = \mathsf{Tcat}(C)$ in our ad-hoc runs since no target categories are provided.

Task 2. Here, the categories attached to entity examples are likely to correspond to very specific categories, just like those attached to the answer entities. We define a similarity function that computes the ratio of common categories between the set of categories attached to an answer entity page $\mathsf{cat}(t)$ and the set of the union of the categories attached to entity examples $\mathsf{cat}(E)$:

$$S_C(t) = \frac{|\mathsf{cat}(t) \cap \mathsf{cat}(E)|}{|\mathsf{cat}(E)|}. \tag{2}$$

4.3 Exploiting Locality of Links

For task 2, exploiting locality of links around entity examples can significantly improve the effectiveness of entity ranking [10]. The idea is that entity references (links) that are located in close proximity to the entity examples, especially in list-like elements, are likely to refer to more relevant entities than those referred to by links in other parts of the page.

Consider the example of the Euro page shown in Fig. 1, where France, Germany and Spain are the three entity examples. We see that the 15 countries that are members of the Eurozone are all listed in the same paragraph with the three entity examples. In fact, there are other contexts in this page where those 15 countries also co-occur together. By contrast, although there are a few references to the United Kingdom in the Euro page, it does not occur in the same context as the three examples (except for the page itself).

We have identified in the Wikipedia collections three types of elements that correspond to the notion of lists: paragraphs (tag p); lists (tags normallist,

Table 1. List of links referring to entity examples (France, Germany, and Spain), extracted from the Wikipedia page 9272.xml

Page		Links		
ID	Name	XPath	ID	Name
9472	Euro	/article[1]/body[1]/p[1]/collectionlink[7]	10581	France
9472	Euro	/article[1]/body[1]/p[1]/collectionlink[8]	11867	Germany
9472	Euro	/article[1]/body[1]/p[1]/collectionlink[15]	26667	Spain
9472	Euro	/article[1]/body[1]/p[3]/p[5]/collectionlink[6]	11867	Germany
9472	Euro	/article[1]/body[1]/normallist[1]/item[4]/collectionlink[1]	10581	France
9472	Euro	/article[1]/body[1]/normallist[1]/item[5]/collectionlink[2]	11867	Germany
9472	Euro	/article[1]/body[1]/normallist[1]/item[7]/collectionlink[1]	26667	Spain
9472	Euro	/article[1]/body[1]/normallist[1]/item[8]/collectionlink[1]	26667	Spain

numberlist, and definitionlist); and tables (tag table). We use an algorithm for identifying the (static) element contexts on the basis of the leftmost occurrence of any of the pre-defined tags in the absolute XPaths of the entity examples. The resulting list of element contexts is sorted in a descending order according to the number of distinct entity examples contained by the element. If two elements contain the same number of distinct entity examples, the one that has a longer XPath length is ranked higher. Finally, starting from the highest ranked element, we filter all the elements in the list that either contain or are contained by that element. We end up with a final list of (one or more) non-overlapping elements that represent the statically defined contexts for the page.[3]

Consider Table 1, where the links to entity examples are identified by their absolute XPath notations. The three static contexts that will be identified by the above algorithm are the elements p[1], normallist[1] and p[3]. The first two element contexts contain three (distinct) examples, while the last one contains only one entity example.

The drawback of this approach is that it requires a predefined list of static elements that is dependent on the collection. The advantage is that the contexts are fast to identify. We have also experimented with an alternative algorithm that dynamically identifies the link contexts. On the training data set, we found that this algorithm does not significantly improve the entity ranking performance over the algorithm that uses the static contexts [10].

4.4   Score Functions and Parameters

The core of our entity ranking approach is based on combining different scoring functions for an answer entity page, which we now describe in more detail.

---

[3] In the case when there are no occurrences of the pre-defined tags in the XPath of an entity example, the document element (article[1]) is chosen to represent the element context.

LinkRank score. The linkrank function calculates a score for a page, based on the number of links to this page, from the first N pages returned by the search engine in response to the query. The number N has been kept to a relatively small value mainly for performance purposes, since Wikipedia pages contain many links that would need to be extracted. We carried out some experiments with different values of N and found that N=20 was a good compromise between achieving efficient performance and discovering more potentially good entities [14].

The linkrank function can be implemented in a variety of ways. We have implemented a linkrank function that, for an answer entity page t, takes into account the Zettair score of the referring page $z(p)$, the number of distinct entity examples in the referring page $\#ent(p)$, and the locality of links around the entity examples:

$$S_L(t) = \sum_{r=1}^{N} \left(z(p_r) \cdot g(\#ent(p_r)) \cdot \sum_{l_t \in L(p_r,t)} f(l_t, c_r | c_r \in C(p_r)) \right) \qquad (3)$$

where $g(x) = x + 0.5$ (we use 0.5 to allow for cases where there are no entity examples in the referring page); l_t is a link that belongs to the set of links $L(p_r, t)$ that point from the page p_r to the answer entity t; c_r belongs to the set of contexts $C(p_r)$ around entity examples found for the page p_r; and $f(l_t, c_r)$ represents the weight associated to the link l_t that belongs to the context c_r.

The weighting function $f(l_t, c_r)$ is represented as follows:

$$f(l_t, c_r) = \begin{cases} 1 & \text{if } c_r = p_r \text{ (the context is the full page)} \\ 1 + \#ent(c_r) & \text{if } c_r = e_r \text{ (the context is an XML element)} \end{cases}$$

A simple way of defining the context of a link is to use its full embedding page [14]. In this work we use smaller contexts using predefined types of elements such as paragraphs, lists and tables (as described in sub-section 4.3).

Category similarity score. The category score $S_C(t)$ is calculated using equation (1) for task 1 and equation (2) for task 2 (as described in sub-section 4.2).

For task 1, we consider variations on the category score $S_C(t)$ based on lexical similarities of category names (see sub-section 4.2), by replacing cat(C) with TCcat(C).

For task 2 we do not use any category extensions since, on the training data set, we found that extending the set of categories attached to both entity examples and answer entities did not increase the entity ranking performance [12].

Z score. The Z score assigns the initial Zettair score to an answer entity page. If the answer page does not appear among the initial ranked list of pages returned by Zettair, then its Z score is zero:

$$S_Z(t) = \begin{cases} z(t) & \text{if page } t \text{ was returned by Zettair} \\ 0 & \text{otherwise} \end{cases} \qquad (4)$$

The Z score is not the same as the plain Zettair score, since our system extracts new entities (pages) from the links contained in the highest N pages returned by Zettair; these new pages may or may not be included in the initial 1500 pages retrieved by Zettair.

Global score. The global score $S(t)$ for an answer entity page is calculated as a linear combination of three normalised scores: the normalised linkrank score $nS_L(t)$, the category similarity score $nS_C(t)$ and the Z score $nS_Z(t)$:

$$S(t) = \alpha \cdot nS_L(t) + \beta \cdot nS_C(t) + (1 - \alpha - \beta) \cdot nS_Z(t) \qquad (5)$$

where α and β are two parameters that can be tuned differently depending on the entity retrieval task.

We consider some special cases that allow us to evaluate the effectiveness of each module in our system: $\alpha = 1, \beta = 0$, which uses only the linkrank score; $\alpha = 0, \beta = 1$, which uses only the category score; and $\alpha = 0, \beta = 0$, which uses only the Z score. More combinations for the two parameters are explored in the training phase of our system. The optimal combination is then used on the testing data set.

5 Experimental Results

In this section, we present results that investigate the effectiveness of our entity ranking approach when applied to both the INEX 2007 XER and ad hoc tracks.

For the XER track, we submitted three runs for task 1 (entity ranking) and three runs for task 2 (list completion). For this track, we aim at investigating the impact of using various category and linkrank similarity techniques on the entity ranking performance; we also compare the run performances with a full-text retrieval run as a baseline. For the ad hoc track, we submitted three entity ranking runs that correspond to the three individual modules of our system and compare their performances to the performance of the full-text Zettair run submitted by RMIT.

5.1 Runs Description

Table 2 lists the six XER and four ad hoc runs that we submitted to INEX 2007. With the exception of the plain Zettair run, all the runs were created by using our entity ranking system. However, as seen in the table the runs use various parameters whose values are mainly dependent on the task. Specifically, runs differ depending on whether (or which) Zettair category index is used, which of the two types of link contexts is used, whether categories or example entities are used from the topic, and which combination of values is assigned to the α and β parameters.

For example, the run "run 3" for XER task 1 can be interpreted as follows: the Zettair index of category names is used to extract the top five ranked categories, using both the title and the category names (TC) from the INEX topic as a

Table 2. List of six XER and four ad hoc runs submitted for evaluation. "Cat-sim" stands for category similarity, "Ctx" for context, "Cat" for categories, "Ent" for entities, "T" for title, "TC" for title and categories, "C" for category names, "CE" for category and entity names, "FC" for full page context, and "EC" for element context.

| Run ID | cat-sim | α | β | Category index | | | Topic | | |
				Query	Type	M	Ctx	Cat	Ent
Zettair		–	–	–	–	–	–	–	–
XER task 1									
run 1	cat(C)-cat(t)	0.0	1.0	–	–	–	FC	Yes	No
run 2	TCcat(C)-cat(t)	0.0	1.0	TC	C	5	FC	Yes	No
run 3	TCcat(C)-cat(t)	0.1	0.8	TC	C	5	FC	Yes	No
XER task 2									
run 1	cat(E)-cat(t)	1.0	0.0	–	–	–	EC	No	Yes
run 2	cat(E)-cat(t)	0.0	1.0	–	–	–	EC	No	Yes
run 3	cat(E)-cat(t)	0.2	0.6	–	–	–	EC	No	Yes
Ad hoc retrieval task									
run 1	Tcat(C)-cat(t)	0.0	0.0	T	CE	10	FC	No	No
run 2	Tcat(C)-cat(t)	1.0	0.0	T	CE	10	FC	No	No
run 3	Tcat(C)-cat(t)	0.0	1.0	T	CE	10	FC	No	No

query. This set of five categories is used as an input in the category similarity function (TCcat(C)). The full page context (FC) is used to calculate the scores in the linkrank module. The final scores for answer entities are calculated by combining the scores coming out of the three modules ($\alpha = 0.1$, $\beta = 0.8$).

5.2   XER Track

Two data sets were used by the participants of the XER track: a training data set and a testing data set. The training data set is based on a selection of topics from the INEX 2006 ad hoc track, resulting in total of 28 topics with corresponding relevance assessments. The testing data set consists of two subsets: a subset of topics based on a selection of topics from the INEX 2007 ad hoc track, and a subset of topics specifically developed by participants for the purposes of the XER track. The complete testing data set results in total of 46 topics with corresponding relevance assessments.

We use mean average precision (MAP) as our primary method of evaluation, but also report results using several alternative measures that are typically used to evaluate the retrieval performance: mean of P[5] and P[10] (mean precision at top 5 or 10 entities returned), and mean R-precision (R-precision for a topic is the P[R], where R is the number of entities that have been judged relevant for the topic). For task 1 all the relevant entities in the relevance assessments are used to generate the scores, while for task 2 we remove the entity examples both from the list of returned answers and from the relevance assessments, as the task is to find entities other than the provided examples.

Table 3. Performance scores for Zettair and our three XER submitted runs on the training data set (28 topics) and testing data set (46 topics), obtained with different evaluation measures for INEX 2007 XER task 1: entity ranking. For each data set, the best performing score under each measure is shown in bold.

Run ID	cat-sim	α	β	P[r] 5	10	R-prec	MAP
Training data set							
Zettair		–	–	0.229	0.232	0.208	0.172
run 1	cat(C)-cat(t)	0.0	1.0	0.229	0.250	0.215	0.196
run 2	TCcat(C)-cat(t)	0.0	1.0	0.307	0.318	0.263	0.242
run 3	TCcat(C)-cat(t)	0.1	0.8	**0.379**	**0.361**	**0.338**	**0.287**
Testing data set							
Zettair		–	–	0.230	0.211	0.208	0.186
run 1	cat(C)-cat(t)	0.0	1.0	0.283	0.243	0.235	0.199
run 2	TCcat(C)-cat(t)	0.0	1.0	0.322	0.296	0.300	0.243
run 3	TCcat(C)-cat(t)	0.1	0.8	**0.378**	**0.339**	**0.346**	**0.294**

Task 1: Entity ranking. Table 3 shows the performance scores on both the training and the testing data sets for task 1, obtained for Zettair and our three submitted XER runs. Runs 1 and 2 use scores coming out from the category module only ($\alpha = 0.0$, $\beta = 1.0$) while run 3 uses a combination of linkrank, category, and Z scores ($\alpha = 0.1$, $\beta = 0.8$). Runs 2 and 3 use lexical similarity for extending the set of target categories.

When comparing the performances of runs that use the category module only (runs 1 and 2), we observe that run 2 that uses lexical similarity between category names (TCcat(C)) is more effective than the run that uses the topic-provided target categories (cat(C)). With MAP, the difference in performance between the two runs is statistically significant ($p < 0.05$). We also observe that the third run, which uses combined scores from the three modules, performs the best among the three. To find the optimal values for the two combining parameters for this run, we calculated MAP over the 28 topics in the training data set as we varied α from 0 to 1 in steps of 0.1. For each value of α, we also varied β from 0 to $(1 - \alpha)$ in steps of 0.1. We found that the highest MAP score (0.287) is achieved for $\alpha = 0.1$ and $\beta = 0.8$ [12]. This is a statistically significant 19% relative performance improvement over the best score achieved by using only the category module (α0.0–β1.0). The same performance behaviour among the three XER runs is also observed on the testing data set.

From Table 3 we also observe that, irrespective of the data set used, the three entity ranking runs outperform the plain Zettair run. This suggests that using full-text retrieval alone is not an effective retrieval strategy for this task. The differences in performance between each of the three runs and Zettair are statistically significant ($p < 0.05$) only for the two entity ranking runs that use lexical similarity between category names (runs 2 and 3 in Table 3).

Table 4. Performance scores for Zettair and our three XER submitted runs on the training data set (28 topics) and testing data set (46 topics), obtained with different evaluation measures for INEX 2007 XER task 2: list completion. For each data set, the best performing score under each measure is shown in bold.

Run	cat-sim	α	β	P[r] 5	P[r] 10	R-prec	MAP
Training data set							
Zettair	–	–	–	0.229	0.232	0.208	0.172
run 1	cat(E)-cat(t)	1.0	0.0	0.214	0.225	0.229	0.190
run 2	cat(E)-cat(t)	0.0	1.0	0.371	0.325	0.319	0.318
run 3	cat(E)-cat(t)	0.2	0.6	**0.500**	**0.404**	**0.397**	**0.377**
Testing data set							
Zettair	–	–	–	0.183	0.170	0.173	0.155
run 1	cat(E)-cat(t)	1.0	0.0	0.157	0.150	0.163	0.141
run 2	cat(E)-cat(t)	0.0	1.0	0.370	0.298	0.292	0.263
run 3	cat(E)-cat(t)	0.2	0.6	**0.409**	**0.330**	**0.336**	**0.309**

When comparing the MAP scores obtained for runs submitted by all XER track participants, our INRIA run 3 was ranked as the third best performing run among the 20 submitted runs for INEX 2007 XER task 1.

Task 2: List completion. Table 4 shows the performance scores on both the training and testing data sets for task 2, obtained for Zettair and our three submitted XER runs. With the first two runs, we want to compare two entity ranking approaches: the first that uses scores from the linkrank module only (run 1), and the second that uses scores from the category module only (run 2). We observe that using categories is substantially more effective than using the linkrank scores. With MAP, the difference in performance between the two runs is statistically significant ($p < 0.05$) on both data sets.

Run 3 combines the scores from the three modules. To find the optimal values for the two combining parameters for this run, we again used the training data set and varied the values for parameters α and β. We found that the highest MAP score (0.377) was achieved for $\alpha = 0.2$ and $\beta = 0.6$ [10]. This is a statistically significant 19% relative performance improvement over the best score achieved by using only the category module. From Table 4 we see that the same performance behaviour among the three XER runs is also observed on the testing data set.

When the three XER runs are compared with the plain Zettair run, we observe a slightly different performance behaviour depending on the data set used. Specifically, on the training data set the three XER runs outperform the plain Zettair run, while on the testing data set only runs 2 and 3 outperform Zettair which in turn outperforms run 1 (the run that uses linkrank scores only). A more detailed per-topic analysis of this behaviour revealed that this is a result of the different "nature" of the two subsets used in the testing data set. Specifically, Zettair outperformed run 1 only on the 21 topics comprising the ad hoc testing topic subset, while run 1 outperformed Zettair on the 25 topics comprising the testing topic subset developed by the XER participants. This indicates that the ad hoc topic subset may need to be further revised and adapted if it is to be reliably used for XER-specific retrieval tasks.

When comparing the MAP scores obtained for runs submitted by all XER track participants, our INRIA run 3 was ranked as the best performing run among the 10 submitted runs for INEX 2007 XER task 2.

5.3   Ad Hoc Track

There are no target categories and example entities provided for the retrieval tasks of the INEX 2007 ad hoc track. However, we wanted to apply our algorithms to test 1) whether some indication of page categories would improve the ad hoc retrieval performance, and 2) whether extracting new entities from the pages returned by Zettair would be beneficial for ad hoc retrieval.

We submitted four runs for the INEX 2007 ad hoc track: Zettair, representing a full-text retrieval run, and three entity ranking runs. As shown in Table 2, run 1 uses only the Z module for ranking the answer entities, run 2 uses only the linkrank module, while run 3 uses only the category module. For each of the 99 topics with relevance assessments used in the INEX 2007 ad hoc track, we created the set of target categories by sending the title T of the query to the Zettair index of categories that has been created by using the names of the categories and the names of all their attached entities as corresponding documents.

Table 5 shows the performance scores on the INEX 2007 ad hoc data set, obtained for Zettair and our three submitted entity ranking runs. Two retrieval scenarios are distinguished in the table: a *document retrieval* scenario (the first four result columns in Table 5), where we compare how well the runs retrieve relevant documents; and a *focused retrieval* scenario (the last three result columns in Table 5), where we compare how well the runs retrieve relevant information within documents.

For the document retrieval scenario, we observe that Zettair outperforms the other three XER runs. The differences in performance between Zettair and any of these three runs are statistically significant ($p < 0.05$). Among the three

Table 5. Performance scores for Zettair and our three XER submitted runs on the ad hoc data set (99 topics), obtained with different evaluation measures for the INEX 2007 ad hoc track. For each measure, the best performing score is shown in bold.

Run	α	β	P[r] 5	10	R-prec	MAP	Foc iP[0.01R]	RiC MAgP	BiC MAgP
Zettair	–	–	**0.513**	**0.469**	**0.326**	**0.292**	**0.483**	**0.136**	**0.192**
run 1	0.0	0.0	**0.513**	**0.469**	0.303	0.247	**0.483**	0.115	0.163
run 2	1.0	0.0	0.339	0.289	0.170	0.121	0.289	0.045	0.068
run 3	0.0	1.0	0.406	0.368	0.208	0.157	0.380	0.078	0.113

XER runs, the run that only uses the Z scores performs significantly better than either of the other two runs, followed by the run that only uses the category scores which in turn performs significantly better than the worst performing run that only uses the linkrank scores.

The same trend among the four runs is observed across the three sub-tasks of the focused retrieval scenario, where again Zettair is able to better identify and retrieve the relevant information compared to the other three XER runs.

The obvious conclusion of our ad hoc experiments is that Zettair, which is specifically designed for full-text retrieval, performs better than our entity ranking system specifically designed for entity retrieval.

6   Conclusion and Future Work

We have presented our entity ranking approach for the INEX Wikipedia XML document collection which is based on exploiting the interesting structural and semantic properties of the collection.

On both the training and the testing data sets, we have shown that our entity ranking system outperforms the full-text search engine in the task of ranking entities. On the other hand, using our entity ranking system for ad-hoc retrieval did not result in any improvement over the full-text search engine. This confirms our hypothesis that the tasks of ad hoc retrieval and entity retrieval are two very different tasks.

Our entity ranking system was one of the best performing systems when comparing the entity ranking performances of all the participating systems in the INEX 2007 XER track. In the future, we aim at further developing our entity ranking algorithms by incorporating natural language processing techniques that we expect would reveal more potentially relevant entities.

Acknowledgements

Part of this work was completed while James Thom was visiting INRIA in 2007.

References

1. AwangIskandar, D., Pehcevski, J., Thom, J.A., Tahaghoghi, S.M.M.: Social media retrieval using image features and structured text. In: Fuhr, N., Lalmas, M., Trotman, A. (eds.) INEX 2006. LNCS, vol. 4518, pp. 358–372. Springer, Heidelberg (2007)
2. Bast, H., Chitea, A., Suchanek, F., Weber, I.: ESTER: efficient search on text, entities, and relations. In: Proceedings of the 30th ACM International Conference on Research and Development in Information Retrieval, Amsterdam, The Netherlands, pp. 671–678 (2007)
3. Blanchard, E., Kuntz, P., Harzallah, M., Briand, H.: A tree-based similarity for evaluating concept proximities in an ontology. In: Proceedings of 10th conference of the International Fedederation of Classification Societies, Ljubljana, Slovenia, pp. 3–11 (2006)

4. Craswell, N., Hawking, D., Vercoustre, A.-M., Wilkins, P.: P@noptic expert: searching for experts not just for documents. In: Proceedings of the Australasian Web Conference (Ausweb 2001), Coffs Harbour, Australia (2001)
5. de Vries, A.P., Vercoustre, A.-M., Thom, J.A., Craswell, N., Lalmas, M.: Overview of the INEX 2007 Entity ranking track. In: Fuhr, N., et al. (eds.) INEX 2006. LNCS, vol. 4862, pp. 138–147. Springer, Heidelberg (2007)
6. Denoyer, L., Gallinari, P.: The Wikipedia XML corpus. SIGIR Forum. 40(1), 64–69 (2006)
7. FissahaAdafre, S., de Rijke, M., Sang, E.T.K.: Entity retrieval. In: Proceedings of International Conference on Recent Advances in Natural Language Processing (RANLP - 2007), Borovets, Bulgaria, September 27-29 (2007)
8. Hu, G., Liu, J., Li, H., Cao, Y., Nie, J.-Y., Gao, J.: A supervised learning approach to entity search. In: Ng, H.T., Leong, M.-K., Kan, M.-Y., Ji, D. (eds.) AIRS 2006. LNCS, vol. 4182, pp. 54–66. Springer, Heidelberg (2006)
9. Kleinberg, J.M.: Authoritative sources in hyperlinked environment. Journal of the ACM 46(5), 604–632 (1999)
10. Pehcevski, J., Vercoustre, A.-M., Thom, J.A.: Exploiting locality of Wikipedia links in entity ranking. In: Macdonald, C., Ounis, I., Plachouras, V., Ruthven, I., White, R.W. (eds.) ECIR 2008. LNCS, vol. 4956, pp. 258–269. Springer, Heidelberg (2008)
11. Soboroff, I., de Vries, A.P., Craswell, N.: Overview of the TREC 2006 Enterprise track. In: Proceedings of the Fifteenth Text REtrieval Conference (TREC 2006), pp. 32–51 (2006)
12. Thom, J.A., Pehcevski, J., Vercoustre, A.-M.: Use of Wikipedia categories in entity ranking. In: Proceedings of the 12th Australasian Document Computing Symposium, Melbourne, Australia, pp. 56–63 (2007)
13. Tsikrika, T., Serdyukov, P., Rode, H., Westerveld, T., Aly, R., Hiemstra, D., de Vries, A.P.: Structured document retrieval, multimedia retrieval, and entity ranking using PF/Tijah. In: Fuhr, N., et al. (eds.) INEX 2006. LNCS, vol. 4862, pp. 306–320. Springer, Heidelberg (2008)
14. Vercoustre, A.-M., Thom, J.A., Pehcevski, J.: Entity ranking in Wikipedia. In: Proceedings of the 23rd Annual ACM Symposium on Applied Computing (SAC 2008), Fortaleza, Brazil, pp. 1101–1106 (2008)
15. Zhu, J., Song, D., Rueger, S.: Integrating document features for entity ranking. In: Fuhr, N., et al. (eds.) INEX 2006. LNCS, vol. 4862, pp. 336–347. Springer, Heidelberg (2008)

Integrating Document Features for Entity Ranking

Jianhan Zhu, Dawei Song, and Stefan Rüger

Knowledge Media Institute
The Open University, United Kingdom
{j.zhu,d.song,s.rueger}@open.ac.uk

Abstract. The Knowledge Media Institute of the Open University participated in the entity ranking and entity list completion tasks of the Entity Ranking Track in INEX 2007. In both the entity ranking and entity list completion tasks, we have considered document features in addition to a basic document content based relevance model. These document features include categorizations of documents, relevance of category names to the query, and hierarchical relations between categories. Furthermore, based on our TREC2006 and 2007 expert search approach, we applied a co-occurrence based entity association discovery model to the two tasks based on the assumption that relevant entities often co-occur with query terms or given relevant entities in documents. Our initial experimental results show that, by considering the predefined category, its children and grandchildren in the document content based relevance model, the performance of our entity ranking approach can be significantly improved. Consideration of the predefined category's parents, a category name based relevance model, and the co-occurrence model is not shown to be helpful in entity ranking and list completion, respectively.

Keywords: entity ranking, list completion, entity retrieval, categories.

1 Introduction

In this year's Entity Ranking Track, there are two related tasks, i.e., entity ranking and entity list completion, on the Wikipedia dataset. A special feature of the Wikipedia dataset is that each document corresponds to an entity. Given a query topic, the aim of entity ranking is to find a list of entities that are relevant to the query topic. A category as part of the query topic specifies the type of entities that should be returned. Some entities have been labeled with certain categories in the dataset. Since entity labeling has been done collaboratively and voluntarily by users, there is no guarantee that all entities are labeled, and all entities are correctly labeled. Therefore, we assume that the categories for each entity can only be used as a guideline. We identify four types of entities that are potentially relevant to a query topic in terms of their categorization. First, the entities are labeled with the specified category. Second, the entities are labeled with categories related to the specified category. Third, the entities are labeled with neither the specified categories nor any category related to the specified category. Fourth, the entities are not labeled.

N. Fuhr et al. (Eds.): INEX 2007, LNCS 4862, pp. 336–347, 2008.

The Entity Ranking Track is related to the Expert Search task in the TREC (Text REtrieval Conference) 2005, 2006, and 2007 Enterprise Search tracks [1][2][3]. Given a query topic, the aim of expert search is to find a ranked list of experts from a list of candidates in an organization or domain. We successfully used a two-stage model in expert search in TREC2006 and 2007 Expert Search tasks. The two-stage model consists of a document relevance model where a number of documents relevant to the query topic are discovered, and a co-occurrence model where experts' relevance to the query topic are measured by their co-occurrences with query terms in a text window in these relevant documents. The two-stage model is also compatible with how users search for experts on the web, i.e., they find relevant documents on a topic through a search engine, and then read these documents in order to find out experts in these documents. Based on the similarity between entity ranking and expert search, we use the two-stage model as one component in entity ranking.

Entity ranking is more general than expert search since in entity ranking, entities of any types can be retrieved for a topic. The nature of Wikipedia dataset makes the entity ranking track different from expert search task, since in entity ranking each document corresponds to an entity while in expert search expert names are mentioned in documents and named entity recognition tools need to be employed in identifying these occurrences of expert names.

Entity list completion can be seen as a special case of entity ranking task. In entity list completion, a few entities relevant to a query topic are given. These entities can be used as relevance feedback information for finding other relevant entities. We think there are mainly two ways for using this relevance feedback information. First, use these entities and their corresponding documents as relevance feedback information. Second, based on the observation that these entities may often co-occur with other entities that are also relevant to the query topic, we propose to use a co-occurrence model for measuring the relevance between new entities and these given entities.

We think that entity ranking is sensitive to multiple document features that need to be taken into account in finding relevant entities on the Wikipedia dataset. Therefore, a number of components considering these document features in our approach include: 1. Document content based relevance to the query topic, 2. Specified category in the query topic, 3. Sub-categories and parents of the specified category, 4. The content based relevance of category names of each document to the query topic, and 5. a novel multiple-window based co-occurrence model.

We proposed the multiple-window based co-occurrence model in TREC 2006 and 2007 for expert search [4][5]. Similarly, we have applied the multiple-window based approach to entity ranking. Entities are mentioned in other documents. The contexts of these occurrences of entities often include query terms in the query topic. We assume that there are associations between an entity and query terms on multiple levels, i.e., from phrase, sentence, paragraph, etc., up to document levels. All these levels of associations need to be considered in the co-occurrence model. Increased window sizes often lead to more coverage of associations while introducing noise. We propose a novel weighted multiple window size based approach as opposed to a single fixed window size based approach in previous association discovery research [6]. In entity list completion, we have considered the co-occurrences of given relevant entities and new entities.

The rest of this paper is organized as follows. In Section 2, we introduce our entity ranking approach. We extend our entity ranking approach for entity list completion in Section 3. We report our experimental results, and submitted runs on Wikipedia dataset in Section 4 and 5, respectively. Finally, we conclude and discuss future work in Section 6.

2 Entity Ranking

For each document, which corresponds to an entity, we use its content based relevance to the query topic as the baseline model. We enhance the baseline model by taking into account multiple document features, i.e., the entity's categories' relations with the specified category, the entity's categories' content based relevance to the query topic, and the entity's co-occurrences with the query terms in other documents.

2.1 Content Based Relevance

If an entity is relevant to a topic, the content of the document representing the entity is likely to contain terms in the query topic. We used three standard relevance models, i.e., Boolean, BM25, and Lucene's span relevance models, for judging the relevance of the document content to the topic.

BM25 is a probabilistic IR model. We used the BM25 equation of Okapi [7] for the relevance model. Given a query q and document d, we get

$$p(d \mid q) \propto \sum_{T \in q} w \frac{(k_1 + 1)tf}{K + tf} \frac{(k_3 + 1)qtf}{k_3 + qtf} + k_2 |q_i| \frac{avdl - dl}{avdl + dl} \tag{1}$$

where $w = log((N - n + 0.5)/(n + 0.5))$ is the IDF of T; N is the number of documents in the dataset; n is the number of documents where T appears; K is $k_1((1-b) + b*dl/avdl)$; k_1, b, k_2 and k_3 are parameters; tf is the frequency of T in d; qtf is the frequency of T in q; dl is the length of d; and $avdl$ is the average document length. Based on the suggested parameter values in Okapi [7], we set the values of k_1, b, k_2 and k_3 as 1.4, 0.6, 0.0, and 8.0, respectively.

Boolean query model specifies that all query terms must occur in a document.

$$p(d \mid q) \propto coord \cdot \frac{1}{\sqrt{\sum_{T \in q} idf^2}} \sum_{T \in q} tf \cdot idf \cdot \frac{avdl}{dl} \tag{2}$$

where $coord$ is the number of query terms that are found in d divided by the total number of terms in the query, and idf is $1 + log(N/(n+1))$.

Span query model is based on co-occurrences of all query terms in text windows. The score for a matching span s is as follows:

$$p(s \mid q) \propto coord \cdot sloppyFreq(s) \cdot \frac{1}{\sqrt{\sum_{T \in q} idf^2}} \sum_{T \in q} idf \cdot \frac{avdl}{dl} \tag{3}$$

where *sloppyFreq(s)=1/(slop+1)* is a factor that decreases as the sloppiness of the matching span increases[1]; the effect is to favor more exact matches.

In order to help equalize the scoring between documents with many matching spans and those with few matches, the score for all matching spans in a document *d* is taken as the square root of the total score of all matches as follows:

$$p(d \mid q) \propto \sqrt{\sum_{s \; in \; d} p(s \mid q)} \tag{4}$$

We apply our relevance model to entity ranking in Section 4.

2.2 Entity's Categories

An entity's category information can help entity retrieval in mainly three aspects.

First, since a preferred category is specified as part of a query topic, if there is a match between an entity's categories and the preferred category, the relevance of the entity to the query topic should be largely boosted.

Second, since the categorization of entities is incomplete and some relevant entities may not be labeled with the preferred category, we need to find categories which are relevant to the preferred category and were used to label other relevant entities. We propose to find the sub-categories and parents of the preferred category. If there is a match between these categories and an entity's categories, the relevance of the entity to the query topic will still be boosted although the entity's categories may not contain the preferred category.

In the hierarchy of categories for the Wikipedia dataset, the links between categories do not always represent an "is-a" relationship between two categories, i.e., sometimes the child may not be a sub-class of the parent. In order to avoid the "concept drift" in the hierarchy, categories related to the preferred category are only limited to its parents and children in our approach, although we will investigate the effect of incorporating more distantly linked categories in future work.

Third, if an entity is relevant to a query topic, the names of the entity's categories can often contain terms in the query topic. We propose to create a metadata field for an entity by joining its' categories' names together, and use a standard relevance model, such as BM25, Boolean, or span models, to measure the relevance between this metadata field and the query topic.

We envisage that the categorization information associated with the Wikipedia dataset can significantly assist entity retrieval. The assumption can be tested based on an anatomy of our entity retrieval system studying the effect of multiple document features in entity retrieval that will be shown in Section 4.

2.3 Entity's Co-occurrences with Query Terms

So far, entity ranking task is similar to a document ranking problem, i.e., judging the relevance between a number of documents and a query topic for producing a ranked

---

[1] "The maximum allowable positional distance between terms to be considered a match is called slop. Distance is the number of positional moves of terms to reconstruct the phrase in order" [8].

list of documents. Categorization information associated with documents can be used to filter or weight search results based on the similarity between their categories and the predefined category. However, we propose an entity co-occurrence model, which is based on the context information of each entity in documents, to further enhanced entity ranking.

The proposed entity co-occurrence model is very similar to our co-occurrence model, which takes into account contextual information of experts in documents, in the TREC expert search task. Similarly, each entity occurs in a number of documents, and the contexts of these occurrences can help us estimate the relevance between the query and the entity.

In TREC2006 and 2007, we have successfully employed a novel two-stage multiple window based approach for expert search. Now we propose to apply the two-stage model to the entity ranking task. Given a query q, an entity e's relevance to the query in a co-occurrence model is $p_{co-occ}(e|q)$. We get:

$$p_{co-occ}(e \mid q) = \sum_d p(e,d \mid q) = \sum_d p(d \mid q)p(e \mid d, q) \tag{5}$$

where d is a document, $p(d|q)$ is the document relevance model, and $p(e|d,q)$ is the co-occurrence model.

We use one of the three relevance models presented in Section 2.1 for the first stage. In the second stage, an entity's relevance to the query topic is judged based on the co-occurrences of the entity and query topic terms in documents.

Since entity's association with a query topic can be of multiple levels, from phrase, sentence, paragraph, up to document levels, we propose a novel multiple window based approach to capture all these levels of associations. We assume that smaller text windows lead to more accurate associations and larger windows may introduce noise thus leading to less accurate associations. Therefore, we take a weighted sum of the relevance between an entity and a topic based on a number of text windows, where smaller windows are given higher weights and larger windows are given lower weights.

Suppose that, in a document d, there are M occurrences of an entity e as $\{e_k\}$ ($k=1,..., M$). We use L windows with incremental sizes, i.e., $\{W_j\}$ ($j=1,..., L$), for associating each entity occurrence e_k with query terms in d. For e_k, the smallest window in $\{W_j\}$, SW_k, which can enable e_k to co-occur with all query terms in SW_k, is used to measure the association between e_k and the query; if such a window does not exist, the association score between e_k and the query is zero. For example, suppose that we use three windows $\{20, 40, 80\}$. If one occurrence of an expert, e_k, does not co-occur with all query terms within the 20-sized window but does co-occur with all of the query terms within the 40-sized window, then we use the window size 40 to measure their associations. Therefore, for different occurrences of experts, different window sizes may be used for association discovery. This gives us more flexibility than the use of one fixed sized window only. Thus, in d, the association between e and the query is a weighted sum of the association scores between all the occurrences of e with the query, respectively, as follows:

$$p(e \mid d, q_i) \propto \sum_{\substack{k=1,...,M \\ e_k \text{ and } q_i \text{ co-occur} \\ SW_k \text{ is the smallest}}} f(SW_k) \cdot P(e_k \mid d, q_i, SW_k) \tag{6}$$

$f(SW_k)$, as a function of the window size, is the weight for the association score between e_k and the query in d. Generally, the smaller the window size, the higher the weight, thus the weight is inversely proportional to the window size.

We extend the co-occurrence model proposed by Cao et al. [9] to our multiple-window-based co-occurrence model and define $P(e_k|d,q_i, SW_k)$ as:

$$p(e_k \mid d, q_i, SW_k) \propto \mu \frac{pf(e_k, SW_k)}{pf_{total}(SW_k)} + \frac{1-\mu}{df_e} \sum_{d_i : e \in d_i} \frac{1}{n_e} \sum_{\substack{e_j : e_j \in d_i \\ e_j \text{ and } q_i \text{ co-occur} \\ SW_j \text{ is the smallest}}} \frac{pf(e_j, SW_j)}{pf_{total}(SW_j)} \tag{7}$$

where $pf(e_k, SW_k)$ is the frequency of e_k in window SW_k, $pf_{total}(SW_k)$ is the total frequency of entities in SW_k, df_e is document frequency of e, n_e is the number of occurrences of e in d_i. We use a Dirichlet prior to smooth parameter μ:

$$\mu = \frac{pf_{total}(SW_k)}{pf_{total}(SW_k) + \kappa}$$

Here κ is the average of term frequency of all occurrences of all entities inside all windows in the dataset.

We test the effectiveness of our co-occurrence model in Section 4.

2.4 Our Combined Entity Ranking Approach

Our overall entity ranking approach integrates the document relevance model in Section 2.1, weighting function based on entity's categories and relevance of an entity's category names in Section 2.2, and innovative co-occurrence model in Section 2.3.

Given an entity e, a predefined category c, and a query q, therefore, the overall relevance of e given q is:

$$P_{overall}(e \mid q, c) = w_c(w_{content} P(d_e \mid q) + w_{name} p(name_e \mid q) + w_{co-occ} P_{co-occ}(e, q)) \tag{8}$$

where c is the predefined category, d_e is the document representing entity e, $name_e$ is the joint category names of e, w_c is a weight based on the relation between the the entity's categories and the predefined category, $w_{content}$, w_{name}, and w_{co-occ} are the weights for the document relevance model, category names based relevance model, and co-occurrence model, respectively. By adjusting $w_{content}$, w_{name}, and w_{co-occ}, we can tune the effect of the three models in entity ranking, and by adjusting w_c, we can tune the effect of predefined category, its parents, and its children in entity ranking. Finally, the overall relevance of e given q is used to rank entities.

3 Entity List Completion

Entity list completion can be seen as a special case of entity ranking where a few given relevant entities can be used as relevance feedback information. We have incorporated the given relevant entities in our two-stage approach. We assume that entities relevant to the query topic tend to co-occur often with the given entities in documents.

Again, we adopted the novel multiple-window based approach for integrating multiple levels of associations between an entity and any of the given entity. Based on Equation 10, we get

$$p_{overall}(e \mid q, c) = w_c (w_{content} p(d_e \mid q) + w_{name} p(name_e \mid q) + w_{co-occ} p_{co-occ}(e, q)$$
$$+ w_{co-occ, given} \sum_j p_{co-occ}(e, e_j))$$

(9)

where w_{co-occ} is the weight for the co-occurrence model for the entity and given entities, and e_j is a given entity.

4   Experimental Results

The aim of our experiments is to test the effect of the basic document relevance model and different document features, i.e., categorizations of documents, relevance of category names to the query, and hierarchical relations between categories in entity ranking and list completion.

We pre-processed the dataset by removing HTML tags. We indexed and searched the dataset using Lucene. We used a pure document content based Boolean relevance model as the baseline shown in Table 1, i.e., in Equation 8, w_c is set as 1.0, $w_{content}$ is set as 1.0, w_{name} is set as 0, and w_{co-occ} is set as 0. We improve the baseline by adding categorization information and/or the co-occurrence model in getting other runs shown in Table 1.

Table 1. Experimental results for entity ranking

Runs	MAP	R-Prec	Bpref	P@10	Num_rel_ret
baseline	0.1943	0.2239	0.2697	0.2174	623
Cat1	0.2712	0.3036	0.3530	0.3000	596
Cat2	0.2609	0.2763	0.3796	0.2804	600
Cat3	0.3116	0.3351	0.3907	0.3543	655
Cat4	0.3306	0.3584	0.4156	0.3652	669
Cat5	0.3206	0.3457	0.4047	0.3478	654
Cat6	0.2475	0.2799	0.3331	0.2870	594
Cat-CoOcc1	0.3069	0.3447	0.3995	0.3457	687
Cat-background1	0.2827	0.3357	0.3882	0.3450	452
Cat-background2	0.3298	0.3532	0.4160	0.3587	668
Cat-background3	0.3313	0.3639	0.4151	0.3652	671
Cat-background4	0.3308	0.3650	0.4132	0.3652	672

Cat 1: We assume that entities labeled with the predefined category should be given higher weight, and set w_c as 1.0 for entities labeled with the predefined category, and 0.3 otherwise for run Cat1 in Table 1. We can see that MAP, R-Prec, Bpref and P@10 are all significantly improved compared with the baseline showing that categorization information is very helpful in entity ranking. However, the number of relevant entities discovered decreases from 623 to 596, showing that the integration of categorization

information helps put relevant entities near the top of ranked lists at the expense that some entities not labeled with the predefined category are put lower down ranked lists.

Cat 2: We take into account the relevance of entity's category names to query topics in the model by setting w_{name} as 0.4 for run Cat2 in Table 1. We can see that the results degrade due to the combination. The reason might be that terms in an entity's category names may often be mentioned in the entity's document already and simply combining the relevance scores linearly may not be very helpful in entity ranking.

Cat 3: We assume that entities labeled with the children of the predefined category should also be considered. Therefore, we set w_c as 1.0 for entities labeled with the predefined category, 0.8 for entities labeled with the children of the predefined category, and 0.3 otherwise for run Cat3 in Table 1. We can see that MAP, R-Prec, Bpref and P@10 are all significantly improved compared with Cat1. It is worth noting that the number of relevant entities retrieved also significantly improves compared with that for Cat1. This proves that some entities are not labeled with the predefined category directly but labeled with the children of the predefined category. By taking into account children of the predefined category, the retrieval performance significantly improves.

Cat 4: We assume that entities labeled with the grandchildren of the predefined category should be considered. Therefore, we set w_c as 1.0 for entities labeled with the predefined category, 0.8 for entities labeled with the children of the predefined category, 0.65 for entities labeled with the grandchildren of the predefined category, and 0.3 otherwise for run Cat4 in Table 1. We can see that all performance measures improve compared with those for Cat3, showing that grandchildren of the predefined category are helpful in entity ranking.

Cat 5: We assume that entities labeled with the grand-grandchildren of the predefined category should be considered. Therefore, we set w_c as 1.0 for entities labeled with the predefined category, 0.8 for entities labeled with the children of the predefined category, 0.65 for entities labeled with the grandchildren of the predefined category, 0.55 for entities labeled with the grand-grandchildren of the predefined category, and 0.3 otherwise for run Cat5 in Table 1. We can see that all performance measures degrade compared with those for Cat4, showing that the introduction of grand-grandchildren of the predefined category may introduce more noise than helpful information in entity ranking. This is also consistent with the observation of concept drift in the categorization hierarchy.

Cat 6: We assume that entities labeled with the parents of the predefined category should be considered. Therefore, we set given higher weight, and set w_c as 1.0 for entities labeled with the predefined category, 0.7 for entities labeled with the parents of the predefined category, and 0.3 otherwise for run Cat6 in Table 1. We can see that all performance measures degrade compared with those for Cat1, showing that parents of the predefined category are not very helpful in entity ranking due to the reason that they are probably too general.

We further improve the categorization information enhanced baseline by integrating with the co-occurrence model. We trained our co-occurrence model on the TREC2006 expert search test collection. On the basis of run Cat4, we get:

Cat-CoOcc 1: We set w_{co-occ} as 0.3 for run CatCoOcc1 in Table 1. We can see that MAP, R-Prec, Bpref, and P@10 all degrade compared with those for Cat4. However, the number of relevant entities retrieved improves from 669 to 687, showing that the integration of the co-occurrence model helps find more relevant entities at the expense of putting many relevant entities lower down the ranked lists than Cat4 does.

Furthermore, we study the effect of the background document relevance model in entity ranking. On the basis of the best performing run, Cat4, we get:

Cat-background 1: We assume that entities not labeled with the predefined category, its children, or grandchildren are not relevant, i.e., we set w_c as 1.0 for entities labeled with the predefined category, 0.8 for entities labeled with the children of the predefined category, 0.65 for entities labeled with the grandchildren of the predefined category, and 0 otherwise for run Cat-background1 in Table 1. We can see that all performance measures degrade compared with those for Cat4. Especially, the number of relevant entities retrieved decreases sharply from 669 to 452, showing that entities in the dataset are not completely labeled, and it is necessary to include a background model for more successful entity ranking.

We study the effect of the weight for the background model in the following runs:

Cat-background 2: On the basis of Cat4, we set w_c as 0.2 for the background model in run Cat-background2 in Table 1. We can see that the MAP degrades slightly compared with that for Cat4.

Cat-background 3: On the basis of Cat4, we set w_c as 0.35 for the background model in run Cat-background3 in Table 1. We can see that the MAP improves slightly compared with that for Cat4.

Cat-background 4: On the basis of Cat4, we set w_c as 0.4 for the background model in run Cat-background4 in Table 1. We can see that the MAP degrades slightly compared with that for Cat4.

We further study how to use the given relevant entities for entity list completion using the co-occurrence model proposed in Section 3.

Table 2. Experimental results for entity list completion

Runs	MAP	R-Prec	Bpref	P@10	Num_rel_ret
Cat-CoOcc-feedback1	0.2725	0.3005	0.3471	0.2935	558
Cat4	0.2727	0.3005	0.3473	0.2935	558

In Table 2, on the basis of run Cat4, we integrate the co-occurrence model for the following run:

Cat-CoOcc-feedback 1: In Equation 9, we set w_c as 1.0 for entities labeled with the predefined category, 0.8 for entities labeled with the children of the predefined category, 0.65 for entities labeled with the grandchildren of the predefined category, and 0.3 otherwise, w_{co-occ} as 0, and $w_{co-occ, given}$ as 0.4.

For comparison purpose, we remove given relevant entities from run Cat4, which does not use any relevance feedback information, and get the results for Cat4 in Table 2. We can see that the introduction of the co-occurrence model does not help improve the performance of entity ranking. We think this may be due to the reason that each entity's document already contain detailed and complete information, therefore, the co-occurrence model introduce information that is already covered in the entity's document. We will study more effective use of relevance feedback information in future work.

5 Our Submitted Runs

We submitted three entity ranking runs and one list completion run to the Entity Ranking track, and their results are shown in Table 3. The descriptions of our four runs are as follows.

ou_er01: Boolean model, w_c is set as 1.0 for entities labeled with the predefined category, 0.8 for entities labeled with the children of the predefined category, 0.65 for entities labeled with the parent of the predefined category, or 0.3 otherwise, $w_{content}$ is set as 1.0, w_{name} is set as 0.4, and w_{co-occ} is set as 0.2.

ou_er02: Boolean model, w_c is set as 1.0 for entities labeled with the predefined category, 0.8 for entities labeled with the children of the predefined category, 0.65 for entities labeled with the parent of the predefined category, or 0.3 otherwise, $w_{content}$ is set as 1.0, w_{name} is set as 0.6, and w_{co-occ} is set as 0.4.

ou_er03: Boolean model, w_c is set as 1.0 for entities labeled with the predefined category, 0.8 for entities labeled with the children of the predefined category, 0.8 for entities labeled with the parent of the predefined category, or 0.5 otherwise, $w_{content}$ is set as 1.0, w_{name} is set as 0.4, and w_{co-occ} is set as 0.2.

ou_lc01: Boolean model, w_c is set as 1.0 for entities labeled with the predefined category, 0.8 for entities labeled with the children of the predefined category, 0.65 for entities labeled with the parent of the predefined category, or 0.3 otherwise, $w_{content}$ is set as 1.0, w_{name} is set as 0.4, w_{co-occ} is set as 0.2, and $w_{co-occ,given}$ is set as 0.5.

Table 3. Submitted runs for entity ranking and list completion

Runs	MAP	R-Prec	Bpref	P@10
ou_er01	0.2582	0.2958	0.3855	0.2913
ou_er02	0.2306	0.2583	0.3639	0.2630
ou_er03	0.2306	0.2583	0.3639	0.2630
ou_lc01	0.2072	0.2213	0.2384	0.2389

6 Conclusions

We have participated in both entity ranking and list completion tasks in INEX2007. Based on the assumption that entity ranking is sensitive to multiple document features, we propose a novel approach for integrating multiple document features for effective entity ranking. In our approach, we have considered the content of the document describing an entity, matching between the entity's categories and the preferred category, the effect of hierarchical relations between categories, and the content of categories. In addition, we integrate a co-occurrence model, which considers multiple levels of associations between an entity and a query topic, in entity ranking.

We treat entity list completion as a special case of entity ranking by using the given relevant entities as relevance feedback information for incorporation into our co-occurrence model, which considers multiple levels of associations between an entity and each given relevant entity.

Our experimental results show that a document content based relevance model can be significantly improved by considering the categorization information of documents. In particular, consideration of the predefined category, its children, and grandchildren is helpful in entity ranking, while consideration of the predefined category's grand-grandchildren seems not very helpful. Consideration of the predefined category's parents is not help in entity ranking. We think the reason may be due to "concept drift" in the category hierarchy.

On the other hand, entity ranking based purely on documents labeled with the predefined category, its children, and grandchildren can be significantly improved by integrating with the baseline, showing that there are still a number of entities which are not labeled with the predefined category or its children and grandchildren which still are relevant to the query topic.

Interestingly, the incorporation of both category name based relevance model and our co-occurrence model is not helpful in both entity and list completion, respectively. We think this may be due to the reason that each entity's document already contain detailed and complete information, therefore, both the category name based relevance model and our co-occurrence model introduce information that is already covered in the entity's document. We will carry out more systematic research to reconfirm our findings in our experimental results. We will also study more effective approach of using relevance feedback information in the form of given relevant entities in entity list completion.

Acknowledgements

The work reported in this paper is funded in part by an IBM 2007 UIMA innovation award and the JISC (Joint Information Systems Committee) funded DYNIQX (Metadata-based DYNamIc Query Interface for Cross(X)-searching content resources) project.

References

[1] Bailey, P., Craswell, N., de Vries, A.P., Soboroff, I.: Overview of the TREC 2007 Enterprise Track (DRAFT). In: Proc. of The Sixteenth Text REtrieval Conference (TREC 2007), Gaithersburg, Maryland USA (2007)

[2] Craswell, N., de Vries, A.P., Soboroff, I.: Overview of the TREC-2005 Enterprise Track. In: Proc. of The Fourteenth Text REtrieval Conference (TREC 2005) (2005)

[3] Soboroff, I., de Vries, A.P., Craswell, N.: Overview of the TREC 2006 Enterprise Track. In: Proc. of The Fifteenth Text REtrieval Conference (TREC 2006), Gaithers-burg, Maryland, USA (2007)

[4] Zhu, J., Song, D., Rüger, S., Eisenstadt, M., Motta, E.: The Open University at TREC 2006 Enterprise Track Expert Search Task. In: Proc. of The Fifteenth Text REtrieval Conference (TREC 2006) (2007)

[5] Zhu, J., Song, D., Rüger, S., Eisenstadt, M., Motta, E.: The Open University at TREC 2006 Enterprise Track Expert Search Task. In: Proc. of The Sixteenth Text REtrieval Conference (TREC 2007), Notebook (2007)

[6] Conrad, J.G., Utt, M.H.: A System for Discovering Relationships by Feature Extraction from Text Databases. In: Proc. of SIGIR 1994, pp. 260–270 (1994)

[7] Robertson, S.E., Walker, S., Beaulieu, M.M., Gatford, M., Payne, A.: Okapi at TREC-4. In: NIST Special Publication 500-236: The Fourth Text REtrieval Conference (TREC-04), pp. 73–96 (1995)

[8] Hatcher, E., Gospodnetic, O.: Lucene in Action. Manning Publications Co. (2004) ISBN: 1932394281

[9] Cao, Y., Liu, J., Bao, S., Li, H.: Research on Expert Search at Enterprise Track of TREC 2005. In: Proc. of TREC (2005)

A Comparison of Interactive and Ad-Hoc Relevance Assessments

Birger Larsen[1], Saadia Malik[2], and Anastasios Tombros[3]

[1] Royal School of Library and Information Science, Denmark
blar@db.dk
[2] University Duisburg-Essen, Germany
malik@is.informatik.uni-duisburg.de
[3] Queen Mary University of London, UK
tassos@dcs.qmul.ac.uk

Abstract. In this paper we report an initial comparison of relevance assessments made as part of the INEX 2006 Interactive Track (itrack'06) to those made for the topic assessment phase of the INEX 2007 ad-hoc track. The results indicate that that there are important differences in what information was assessed under the two different conditions, but it also suggests a certain level of agreement in what constitutes relevant and non-relevant information. In addition, there are indications that the task type has an influence on the distribution of relevance assessments.

1 Introduction

In this paper, we report on a comparison of relevance assessments made as part of the INEX 2006 interactive track [7] (itrack'06) and those made as part of the topic assessment phase for the INEX 2007 ad-hoc track. Our analysis is based on eight topics that were assessed as part of both tracks.

The conditions under which the eight topics were assessed were significantly different, with searchers in itrack'06 assessing the usefulness of elements in addressing information seeking tasks, while topic assessors for the ad-hoc track focused on providing comprehensive assessments for each retrieved document. These different conditions provide the main motivation for carrying out this research. More specifically, we are primarily interested in investigating:

- The extent to which the different conditions affect the relevance of document elements, as perceived by itrack'06 searchers and ad-hoc topic assessors.
- The overlap of the assessed information, i.e. to what extent the information that searchers and assessors perceived as being useful in their respective tasks was similar.

In addition, the eight topics used in the study are classified into different task types [7,12], providing thus the opportunity to also study the effect of different topic types. Further, in itrack'06 two versions of an XML IR system were used (more details in section 2.1 and in [7]), allowing us to also study the effect of system type perception of document element relevance.

N. Fuhr et al. (Eds.): INEX 2007, LNCS 4862, pp. 348–358, 2008.

There has been significant work on the study of relevance assessments and agreement between assessors in the context of the Text Retrieval Conferences - TREC [1, 9, 13, 14, 15]. The main emphasis has been on binary relevance assessments, since this has been the basis for evaluation in TREC. In one of the few studies that have used multi-scale relevance assessments, Voorhees [14] used the TREC-9 web track data and a three-point relevance scale (not relevant, relevant, highly relevant) in order to examine the effect in evaluation stability of considering only highly relevant documents. Voorhees found that there is a negative effect on stability by the consideration of only highly relevant documents.

Most of the past work on relevance assessments in the context of TREC has also focused relevance judgements made by the TREC assessors, not by online searchers. Some exceptions involve interactive searching and judgement are the work by Cormack et al. [1], and Sanderson and Joho [9], interactive searching, judging and query reformulation are used for forming relevance assessments. In the study by Cormack et al., it was reported that an agreement level of 40% existed between relevance assessments made by interactive searching and by TREC assessors. Voorhees [13] has also examined inter-assessor agreement for a subset of the TREC-4 data (only between TREC assessors), and found similar levels of agreement. Inter-assessor agreement has generally been considered a problem area in IR evaluation in the context of TREC.

In the remaining of this paper, we first describe some methodological issues in section 2, we then present some initial results and analysis in section 3, and we conclude and outline our further plans for analysis in section 4.

2 Methodology

In this section we describe the methodology of our study. First in sections 2.1 and 2.2 we briefly summarise the frameworks under which relevance assessments were made for itrack'06 and the INEX 2007 ad-hoc track, respectively, and in section 2.3 we discuss the methodology by which the assessments in the two tracks were compared.

2.1 Interactive Track 2006

In the INEX 2006 interactive track (itrack'06) searchers from various participating institutions were asked to find information for addressing information seeking tasks by using two interactive retrieval systems: one based on a Passage retrieval backend[1] and one on an Element retrieval backend[2]. Both versions had similar search interfaces but differed in the returned retrieval entities: The passage retrieval backend returned non-overlapping passages derived by splitting the documents linearly. The element retrieval system returned elements of varying granularity based on the hierarchical document structure. The frontend was a modified version of the Daffodil system [3],

---

[1] The Passage retrieval backend was based on CSIRO's Panoptic™/Funnelback™ platform. See http://www.csiro.au/csiro/content/standard/pps6f,,.html for more information.

[2] The Element retrieval backend was based on Max Planck Institute for Informatics' TopX platform. See [11] for more information.

and the document collection used was the INEX Wikipedia corpus [2]. For a full description of the systems used in itrack'06 the reader can refer to [6].

Twelve search tasks of three different types [12] (*Decision making, Fact finding* and *Information gathering*), further split into two structural kinds (*Hierarchical* and *Parallel*), were used in the track [7]. The tasks were split into different categories allowing the searchers a choice between at least two tasks in each category, and at the same time ensuring that each searcher will perform at least one of each type and structure.

An important aspect of the study was to collect the searcher's assessments of the relevance of the information presented by the system. We chose to use a relevance scale based on work by Pehcevski et al. [8]. Searchers were asked to select an assessment score *for each viewed piece of information* that reflected the usefulness of the seen information in solving the task. Five different scores were available, expressing two aspects, or dimensions, in relation to solving the task: How much **relevant information** does the part of the document contain, and how much **context is needed** to understand the element? This was combined into five scores as follows:

- **Not relevant (NR).** The element does not contain any information that is useful in solving the task
- **Relevant, but too broad (TB).** The element contains relevant information, but also a substantial amount of other information
- **Relevant, but too narrow (TN).** The element contains relevant information, but needs more context to be understood
- **Partial Relevant answer (PR).** The element has enough context to be understandable, but contains only partially relevant information
- **Relevant answer (R).** The element contains highly relevant information, and is just right in size to be understandable.

In the interactive track, the intention is that each viewed element should be assessed with regard to its relevance to the topic by the searcher. This was, however, not enforced by the system as it may be regarded as intrusive by the searchers [6]. Note that in contrast to the assessments made for the ad-hoc track, there is no requirement for searchers to view each retrieved element as independent from other components viewed. Experiences from user studies clearly show that users learn from what they see during a search session. To impose a requirement for searchers to discard this knowledge would create an artificial situation and will restrain the searchers from interacting with the retrieved elements in a natural way.

Overall, 88 interactive track searchers made 2170 relevance assessments for the eight tasks analysed in this paper. Table 1 in Section 3 gives a detailed account of this data.

2.2 INEX 2007 Ad-Hoc Assessments

The purpose of the INEX 2007 ad-hoc track is to create a test collection consisting of a corpus of documents, a set of questions directed at the documents (called topics) and a set of relevance assessments specifying which documents (or the elements that are part thereof) that are relevant to each topic [4]. The elements to be assessed were identified by pooling the output of multiple retrieval systems following the method first proposed in [10]; the pool of retrieved elements for each topic was then assessed by the topic author.

In INEX 2007 the assessment process focussed on the notion of specificity, that is, the extent to which the element focuses on the information need expressed in the topic [4]. A highlighting approach was taken, where the assessor first skims the document and then highlights any parts that contain only relevant information. From this, the specificity of any element with highlighted content can be calculated automatically. This may be done by computing the ratio of relevant content (rsize) to all content (size), measured in the number of characters.

All twelve topics that were used in itrack'06 were also submitted as topics for the ad-hoc track. Up to the point of writing this paper, full assessments for eight of these topics were available – we use these as the basis of our result presentation and analysis in section 3.

2.3 Mapping Ad-Hoc and Interactive Track Assessments

Whereas the interactive track assessments are given in terms of one of the five categories in section 2.1, the ad-hoc assessments are of a continuous nature. Thus a mapping between them is needed for comparisons. As mentioned above, there was a difference in the scope of the two types of assessments: where the ad-hoc track aimed at getting comprehensive assessments for each retrieved document, the interactive track searchers were free to assess as much or as little information as they saw fit. In addition, no attempt was made to control learning effects across a search session in the interactive track, while ad-hoc assessors were explicitly asked to assess each element on its own merit.

In the interactive track, non-relevant elements could be specified explicitly (by selecting the NR assessment), as well as implicitly (by searchers viewing an element but not giving any assessment). As such, there is a good correspondence with the ad-hoc track, where only relevant information was highlighted and the rest ignored.

The notion of relevant information (R) in the interactive track would correspond in the ad-hoc assessments to elements that are either fully highlighted or have a large ratio of highlighted content, for example elements with more than 75% relevant content might be considered as being relevant. Following the same line of argument, the interactive track notion of Too Broad (TB) would correspond to elements that in the ad-hoc assessments have a relatively small amount of highlighted content, for example, elements with less than 25% relevant content might be considered as being Too Broad.

It is, however, more difficult to identify a direct parallel to the notion of Too Narrow (TN) in the ad-hoc assessment data. It might be argued though that it is unlikely that small elements would have been relevant to the itrack'06 topics. Pragmatically, such small elements can be filtered out by excluding elements smaller than a given absolute size, e.g., 125 characters[3]. A similar reasoning based on absolute size could

---

[3] Based on that a typical sentence length in English text is around 125 characters (http://hearle. nahoo.net/Academic/Maths/Sentence.html).

be applied as a supplemental criterion to the notion of Relevant (R): elements that contain, e.g., 500 characters of highlighted content could be deemed Relevant, regardless of the ratio of highlighted content.

The notion of Partial Relevant Answer (PR) is also difficult to translate to the ad-hoc assessments, because only relevant information was highlighted in the assessment process.

3   Results and Analysis

In the interactive track 88 searchers were recruited by 8 research groups, and overall they completed 334 search sessions[4]. Table 1 presents some basic statistics for the assessments provided as part of itrack'06. For the eight topics analysed in the present paper, 2170 elements were assessed. As different searchers would often assess the same elements for the same topic, the number of unique assessed elements was 1039 (an average of 2.1 assessments per element). For 177 of these uniquely assessed elements, two or more different assessments (e.g. R, TB and TB) were given by searchers. These present a particular challenge in our study, because we need to arrive at a single assessment for each element in order to compare it to the ad-hoc assessments.

Table 1. Basic statistics on the relevance assessments provided by the INEX 2006 interactive track searchers (including elements that were viewed, but not assessed)

Total number of assessments (including elements assessed more than once)	2170
Unique elements assessed	1039
Unique elements with two or more different assessments	177

In Table 2, we provide details about how these different assessments are distributed among the 1039 uniquely assessed elements. Both rows and columns list the relevance categories and the table shows how many elements have been assessed under both categories by any number of different searchers. There are for instance 57 elements that have been assessed both as Relevant and as Too Broad.

The distribution of values in Table 2 is fairly uniform, with the maximum value being the 10% of the elements marked as NA and R. This largest value corresponds to searchers viewing, but not assessing (NA), elements that other searchers had assessed as relevant. Overall, elements that were not assessed by some searchers but were assessed by other searchers (i.e. the NA row) correspond to the largest percentage in Table 2. Elements assessed as non-relevant (NR) are noteworthy as they correspond to cases where searchers have explicitly indicated that the elements are particularly ill-fitted to the topic. Elements assessed as non-relevant overlap with relevant of any category in 3-5% of the cases. In the heuristics applied to derive a single assessment for the 177 elements, special weight is given to those that were explicitly assessed as non-relevant.

---

[4] Due to system problems, logs of some search sessions had to be excluded.

Table 2. Details of how different assessments are distributed among document elements in raw counts (left) and percentages over the 1039 unique assessed elements (right)

	R	NA	NR	PR	TB	TN		R	NA	NR	PR	TB	TN
R	-	103	52	68	57	36	R	-	9.9%	5.0%	6.5%	5.5%	3.5%
NA	103	-	77	75	59	34	NA	9.9%	-	7.4%	7.2%	5.7%	3.3%
NR	52	77	-	47	32	19	NR	5.0%	7.4%	-	4.5%	3.1%	1.8%
PR	68	75	47	-	35	20	PR	6.5%	7.2%	4.5%	-	3.4%	1.9%
TB	57	59	32	35	-	18	TB	5.5%	5.7%	3.1%	3.4%	-	1.7%
TN	36	34	19	20	18	-	TN	3.5%	3.3%	1.8%	1.9%	1.7%	-

We applied the following heuristics to arrive at a single category of relevance for each of the 177 elements that were assessed differently by different searchers:

1. For elements that were viewed, but not-assessed, the explicit assessments are given priority.
2. If there was a majority vote, the majority category was chosen regardless of the difference.
3. If there was a tie with an element assessed as non-relevant, NR was chosen.
4. In remaining ties, any elements assessed as Relevant were categorised as relevant.
5. Any outstanding ties (i.e., between PR, TB and TN in any combination) were left as ties (indicated as -tie- below).

Table 3 shows the resulting distribution of the interactive track assessments in total and over the eight topics. Less than 25% were Partially Relevant, Narrow or Broad including only 10 ties. The rest are roughly divided into three equally sized groups of relevant, non-relevant and non-assessed elements, each of around 25%.

Table 3. Distribution of interactive track assessments over topics after application of heuristics on elements with two or more different assessments

Topic	T1	T3	T4	T5	T7	T8	T9	T12	Total
R	15	52	11	26	21	37	67	50	279
NA	21	31	27	23	16	55	60	35	268
NR	13	31	16	60	20	71	42	10	263
PR	4	16	9	14	11	25	15	11	105
TB	5	16	4	7	6	5	18	3	64
-tie-	1	5		1	1	1	1		10
TN	9	7	3	5	11	5	4	6	50
Total	68	158	70	136	86	199	207	115	1039

In order to compare the interactive assessments to those of the ad-hoc track, we applied the mapping heuristics discussed in Section 2.3 to the ad-hoc assessments. We regard any element with 75% or more highlighted content as relevant (R), and any with less than 25% as Too Broad. We thus arrive at a set of inferred assessments where the ad-hoc assessments are mapped to the interactive track Relevant and Too Broad relevance categories as shown in Table 4. 801 elements that were assessed in the interactive track but not assessed in the ad-hoc track are also shown (the NA column). In addition, the 39 assessments that fall outside the range defined by the inferred R and TB categories are shown distributed over 5 intermediate bins according to the rsize/size ratio. Excluding 23 elements that were viewed but not assessed in the interactive track (NA, second row) leaves only 215 elements that were assessed in both tracks.

The data from Table 4 suggest that there is little agreement in what kind of information interactive and ad-hoc assessors deem as useful for the same information-seeking tasks, since there is relatively small overlap in the 215 common elements assessed. A further observation from the data is that, with regards to the commonly assessed elements, there is a certain degree of agreement on relevant and not relevant information, as demonstrated by the level of agreement in the R and NR[5] rows. For instance, of the 129 elements assessed as relevant in the interactive track, 75 were relevant in the ad-hoc assessments and 12 more had between 50% - 75% relevant content as measured by the rsize to size ratio. In addition, looking at marginal cases such as TB and TN in the interactive assessments, we notice that relatively few of these are Relevant in the ad-hoc data.

Table 4. Distribution of inferred relevance categories (Relevant and Too Broad) of ad-hoc assessments as well as non-assessed ad-hoc elements over interactive track assessments

Ad-hoc data: Inferred relevance categories & non-assessed elements

rsize/size	0.3-0.4	0.4-0.5	0.5-0.6	0.6-0.7	0.7-0.8	R	TB	Total	NA	Grand total
R	7	2	3	5	4	**75**	**33**	129	150	279
NA						**13**	**10**	23	245	268
NR	2	3			1	**6**	**13**	25	238	263
PR	1	1	2	1	2	**11**	**5**	23	82	105
TB			2		1	**9**	**12**	24	40	64
-tie-						**1**	**1**	2	8	10
TN	1		1			**4**	**6**	12	38	50
Total	11	6	8	6	8	119	80	238	801	1039

(Interactive track data)

The rather small overlap between the two sets of assessments indicates that each set contains significant numbers of elements not assessed in the other set. To investigate the nature of the unique contribution by the interactive track, we have checked

---

[5] Especially so given that non-assessed (NA) elements in the ad-hoc track are an explicit indication of non relevance.

how many of the 801 elements not assessed in the ad-hoc track were actually present in the ad-hoc assessment pools. Table 5 shows that 510 of the interactive track elements were not even included in the ad-hoc track pools, that is, they were not found by any of the systems of the ad-hoc track participants. In slightly more than half of these cases, the interactive track searchers found these elements either non-relevant or not worth assessing. However, in 117 cases (23%) they did find the elements fully relevant and in another 114 cases (22%) relevant to some degree (i.e., PR,TB, TN or a tie between these). Thus, at least from the perspective of interactive track searchers, there were much more relevant information to be found for these 8 tasks than identified in the ad-hoc track.

Table 5. Distribution of non-assessed elements from the ad-hoc track over interactive track assessments, including and excluding elements in the ad-hoc pools

	NA	NA, not in ad-hoc pool
R	150	117
NA	245	150
NR	238	129
PR	82	49
TB	40	25
-tie-	8	7
TN	38	33
Total	801	510

Finally, we investigate if there were any differences in the perceived relevance depending on the task type, and depending on the type of backend used. As the number of mutually judged elements in the ad-hoc and interactive track is quite small the full set of interactive track assessments are used for this analysis. Figure 1 shows the distribution of inferred relevance categories over the three tasks types used in the study. Comparing across task types there are indications that the searchers found a larger proportion of Relevant and a smaller proportion of Non-relevant elements for the Information gathering tasks. For the Fact finding tasks, the trend is the opposite. The Decision making tasks lie in the middle of these two extremes, with a relative low proportion of Non-relevant and a slightly lager proportion of Too broad and Too narrow than either of the other two task types. The element and passage backend systems thus performed better for the more general Information gathering tasks, and somewhat poorer for the more specific Fact finding tasks. This may seem counter intuitive, bearing in mind that the goal of XML element retrieval is to support more focused retrieval. It may, however, be partially explained by the fact that keyword only queries with no structural hints were used in the study.

Figure 2 below shows the distribution of the inferred relevance categories on the two backend systems. The distribution is quite similar, with only a slight tendency for more Relevant and less non-assessed elements in the passage system.

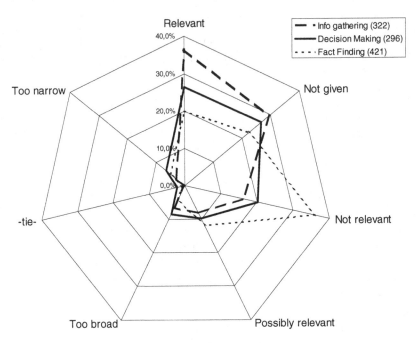

Fig. 1. Distribution of inferred relevance categories over the three task types in the study: Information gathering, Decision making and Fact finding

Fig. 2. Distribution of inferred relevance categories over backend systems: element and passage

4 Concluding Remarks and Future Work

We reported an initial comparison of relevance assessments made as part of the INEX 2006 Interactive Track (itrack'06) to those made for the topic assessment phase of the INEX 2007 ad-hoc track. The data that we presented suggest that there are significant differences in what information was assessed under the two different conditions, but it also suggests a certain level of agreement in what constitutes relevant and non-relevant information for those elements that were assessed in both tracks. In addition, a noteworthy amount of additional relevant elements were identified by interactive track searchers. There are also indications that the task type has an influence on the distribution of relevance assessments, and that there were not great differences between the assessments given in the element and passage backend retrieval systems. For future work, we plan to investigate the effect of different relevance schemes (e.g. by removing the 'partially relevant' level), and we also plan to further investigate the effect that specific differences in the assessment conditions might have had in the relevance assessments. Kazai's initial analysis indicate [5], based on video recordings, that the process of giving assessments in itself may have affected the data, leading to increased interaction.

References

1. Cormack, G.V., Palmer, C.R., Clarke, C.L.A.: Efficient construction of large test collections. In: Proceedings of the 21st ACM SIGIR Conference, pp. 282–289 (1998)
2. Denoyer, L., Gallinari, P.: The Wikipedia XML corpus. SIGIR Forum. 40(1), 64–69 (2006)
3. Fuhr, N., Klas, C.P., Schaefer, A., Mutschke, P.: Daffodil: An integrated desktop for supporting high-level search activities in federated digital libraries. In: Agosti, M., Thanos, C. (eds.) ECDL 2002. LNCS, vol. 2458, pp. 597–612. Springer, Heidelberg (2002)
4. Fuhr, N., Lalmas, M., Trotman, A., Kamps, J. (eds.): Focused access to XML documents. In: INEX 2006. LNCS, vol. 4518. Springer, Heidelberg (2007) (to appear)
5. Kazai, G.: Search and navigation in structured document retrieval: Comparison of user behaviour in search on document passages and XML elements. In: Proceedings of the 12th Australasian Document Computing Symposium (ADCS 2007) (2007)
6. Larsen, B., Tombros, A., Malik, S.: Obtrusiveness and relevance assessment in interactive XML IR experiments. In: Trotman, A., Lalmas, M., Fuhr, N. (eds.) INEX 2005, pp. 39–42 (2005)
7. Malik, S., Tombros, A., Larsen, B.: The interactive track at INEX 2006. In: Fuhr, N., Lalmas, M., Trotman, A. (eds.) Proceedings of the 5th International Workshop of the Initiative for the Evaluation of XML Retrieval, pp. 387–399 (2006)
8. Pehcevski, J., Thom, J.A., Vercoustre, A.M.: Users and assessors in the context of INEX: Are relevance dimensions relevant? In: Trotman, A., Lalmas, M., Fuhr, N. (eds.) INEX 2005, pp. 47–62 (2005)
9. Sanderson, M., Joho, H.: Forming test collections with no system pooling. In: Proceedings of the 27th ACM SIGIR Conference, pp. 33–40 (2004)
10. Spärck Jones, K., van Rijsbergen, C.J.:: Report on the need for and provision of an 'ideal' information retrieval test collection. British Library Research and Development Report 5266, University Computer Laboratory, Cambridge (1975)

11. Theobald, M., Schenkel, R., Weikum, G.: An efficient and versatile query engine for TopX search. In: Proceedings of the 31st International Conference on Very Large Data Bases (VLDB), pp. 625–636 (2005)
12. Toms, E.G., O'Brien, H., MacKenzie, T., Jordan, C., Freund, L., Toze, S., Dawe, E., MacNutt, A.: Task effects on interactive search – The Query Factor. In: INEX 2006. Springer, Heidelberg (to appear, 2008)
13. Voorhees, E.M.: Variations in relevance judgments and the measurement of retrieval effectiveness. In: Proceedings of the 21st ACM SIGIR Conference, pp. 315–323 (1998)
14. Voorhees, E.M.: Evaluation by highly relevant documents. In: Proceedings of the 24th ACM SIGIR Conference, pp. 74–82 (2001)
15. Zobel, J.: How reliable are the results of large-scale retrieval experiments? In: Proceedings of the 21st ACM SIGIR Conference, pp. 307–314 (1998)

Task Effects on Interactive Search: The Query Factor

Elaine G. Toms[1], Heather O'Brien[1], Tayze Mackenzie[1], Chris Jordan[1],
Luanne Freund[2], Sandra Toze[1], Emilie Dawe[1], and Alexandra MacNutt[1]

[1] Centre for Management Informatics, Faculty of Management
Dalhousie University, Halifax, NS, Canada
[2] SLAIS, University of British Columbia
Vancouver, BC, Canada
{etoms,hlobrien,tayze.mackenzie,sandra.toze,emilie.dawe,
AL512723}@dal.ca, luanne.freund@ubc.ca, chris.jordan@acm.org

Abstract. The purpose of this research is to examine how search differs according to selected task variables. Three types of task information goals and two types of task structures were explored. This mixed within- and between-subjects designed study had 96 participants complete three of 12 tasks in a laboratory setting using a specialized search system based on Lucene. Using a combination metrics (user perception collected by questionnaires, transaction log data, and characteristics of relevant documents), we assessed the effect of goals and structure on search as demonstrated through queries and their use in interactive searching.

Keywords: Information search, task, query analysis, Wikipedia.

1 Introduction

The concept of task is central to our understanding of work and play within information environments. Within knowledge work processes, work tasks have evolved from activities that are primarily skill- or rule-based to those that are knowledge-based. Although treated in an ad hoc manner within information retrieval research, in real-world settings, work tasks establish parameters that influence the conduct of information search. By studying the impact of variation in task characteristics upon searching, it may be possible to develop search systems that are tailored to the needs of different types of situations and user groups. To examine the effect of task, we isolated two attributes of tasks: structure and information goal. We speculate that differences in these task attributes will be reflected in variations in interactive search behaviour. In the research reported here, we focus on queries and querying behaviour as evidence of the search process and the primary tool used by searchers to carry out their search tasks. A discussion of results based on performance and outcome is contained elsewhere.

2 Previous Research

2.1 Tasks

The concept of task has multiple definitions and has been analyzed and operationalized in research in many different ways. Bystrom and Hansen [7] characterize a task

N. Fuhr et al. (Eds.): INEX 2007, LNCS 4862, pp. 359–372, 2008.
© Springer-Verlag Berlin Heidelberg 2008

as an activity that has a beginning and an end, has requirements (may be conditional or unconditional), and has both a goal/result and reason/purpose. Task is an inherently hierarchical concept in that tasks can be decomposed into various levels of activity. In general, we distinguish a work task (choose and purchase a yacht) from an information task (find which models of yacht are available). Another important distinction can be made between the static conception of task as an assignment description that lays out the goals, requirements, methods and constraints of the task prior to it being carried out, and the post-processing of the task, once the outcomes or products are known. A task exist as an objective task description communicated by whoever assigns the task, or as a subjective perception of the assigned task on the part of the task performer. Gill and Hicks [11] suggest that a task is a set of assigned: a) goals to be achieved, b) instructions to be performed, or c) a mix of the two. The performed task is the set of actions taken to perform a task, or the task performer's subjective perception of the task as performed. Gwizdka and Spence [11] defined task as "a sequence of actions performed by the searcher in the process of looking for information to satisfy current information need."

One of the challenges in studying task effect is the large number of task attributes that have been identified. For example, Li & Belkin [20] suggested that tasks have the following facets: Origin, Performer, Time, Topic, Process, Product, and Goal. Our work with tasks builds on a similar framework of facets: Motivation, Requirements and Constraints, Goals, Domain and Topic, Process, Structure and Outcome. Notably a task may have a differing set of characteristics if considered once the task is assigned, that differs from the set that would be available after the task was accomplished.

One of the most common task characteristics to be studied is complexity, a characteristic related to the topic and the process. Campbell [8] defined complex tasks as one in which the doer makes decisions or judgments, or solves a problem in situations with varying degrees of information and uncertainty. Bystrom and Jarvelin [6] found that as task complexity increases, the complexity of the information need increases, the need for domain and problem solving information increases, more internal sources are used and the number of sources in general increases. However, it can be difficult to establish the complexity of a task in isolation from the doer. Bell and Ruthven [3], for example, state that the complexity of an information task may be affected by the searchers ability to articulate the information goal and interpret the relevancy of the results.

As such, investigations of task must take into consideration a range of factors, such as characteristics of the doers, the doers' perception of their tasks, the nature of the product or task goal, the constraints around the task (e.g. time), the accessibility of information that will enable the successful completion of the task, and the usability, and interactivity of the medium for locating that information [14,19]. This is a complex set of variables that is difficult to test and isolate in experimental settings.

Since tasks are usually goal-directed, the nature and type of goal is another important characteristic. Common types of search task goals are fact-finding, learning or information gathering, decision-making, know-item searching and problem solving [10, 17]. Evidence suggests that difference by task may influence search behaviour but has not been confirmed.

2.2 Queries

Queries are the primary tool available to searchers carrying out search tasks, and as such provide evidence of how search tasks are performed. Queries are viewed as an articulation of a searcher's information need [25]. Yet, Kelly and Fu [18] point out users may "censor" their queries based on how they think the system's automatic processing functions, which is one among a number of reasons that search queries tend to be very brief [25]. Lau and Goh [19], for example, found that the most common query length in their study was two terms (34.1%), followed by one (23.6%) and three (17.3%). Other researchers have reported average query lengths of 2.35 [13, 22]. For a majority of information retrieval systems, the size of the search box may encourage 'cryptic' querying, though the relationship between query length and search performance has not been statistically significant, although, query length has been associated with user satisfaction [1]. In addition, users have indicated a preference for human rather than system assisted query generation; the former tend to contain more terms (5.87 terms on average) and are more grammatically complex [25].

In addition to query length, the intent or goal has also been examined. Broder [4] proposed three types of queries: navigational, where the user plots a course to a particular website or piece of information; informational, where the intention is to peruse documents or lists of links; and transactional, where the purpose is to interact with a website through actions such as downloading files, purchasing a product, etc. His set has been applied and modified [13, 28]. The challenge with this work is the level of accuracy that can be achieved: how can a query be classified given the limited information that is available, e.g., the one to five words in a query?.

Research has also examined the composition of queries. More specifically, analyses of search logs have shown that users pose queries in the form of questions or complete sentences [24, 25]. However, Toms and Freund [25] found that, when asked to use a question format to query a search engine, searchers found it more difficult than typing in key words, and that this task did not result in improved search results.

Complimentary to research pertaining to how 'real' users query for information, a great deal of focus has been placed on the retrieval system. Kang and Kim [16] argue that it is not users who need to employ different strategies, but search engines. Some search engines have added features such as: term suggestions for query reformulation (Alta Vista, Kazoo, Surfwax, InfoNetWare); automatic clustering or categorization of results (Northern Light, Teoma); and visualization of search results (Kazoo, Vivismo). Other information retrieval systems have attempted to improve search through query expansion, "the process of supplementing the original query with additional terms, and it can be considered as a method for improving retrieval performance" [9], and sometimes the user is active in the process. Lists of keywords and phrases [5] or sets of retrieved documents [27] are presented to the user; their decision to click on these constitutes relevance feedback to the system.

To date, task has emerged as an element of context that has the potential to affect how search is conducted and affect how systems might match the task, and not just the query terms. In the research reported here, we identified two characteristics of task, and tested the differences in search.

3   Methods

We conducted an experiment in a university lab setting enabling the efficient collection of data from 5-10 participants in a single session (for a total of 96 participants from multiple sessions). To conduct the research, we created a search system using open source software, and with a specialized interface using open source software and the Wikipedia document collection.

3.1   Tasks

Initially we considered using INEX 2006's original task pool developed according to a multitude of attributes (e.g., domain, level of specificity or abstraction, named objects, etc.). But those tasks were too simple for human searchers, e.g., a single keyword search would likely net the most relevant page for most of the topics, and thus not useful in the study of interactive IR

The 12 tasks were developed for this research according to a set of principles: 1) no search could be answered in a single page; 2) the task required searchers to actively make a decision about what information was truly relevant to complete a task. In addition, we considered that tasks have semantic content that requires interpretation, and also have syntactic content – structure – that physically represents the task. It is from these two elements that attributes of task were derived for testing: task type and task structure. Tasks were constructed according to Borlund's [3] Situated Work Task Situations (SWTS). SWTSs include a work task and context in addition to the search topic, which prompts for more natural search behaviour and provides a basis for assessing relevance. SWTSs are now used quite commonly in interactive search studies as a means of providing context and operationalizing differences in contextual variables, such as tasks or domains. The resulting tasks are instances of Broder's informational task type [4].

3.2   Variables

Task Type. This characteristic of task contained three levels of task goal:

a) Fact Finding: The objective is to find specific facts or pieces of information;
b) Information Gathering: The objective is to collect information about a topic, often from more than one source;
c) Decision Making: The objective is to select a course of action from among multiple alternatives.

Task Structure. The tasks were also split into two categories, depending on the "structure" of the task:

a) Parallel: The search uses multiple concepts that exist on the same level in a conceptual hierarchy; this is a breadth search (and, in a traditional Boolean system likely was a series of OR relationships).
b) Hierarchical: The search uses a single concept for which multiple attributes or characteristics are sought; this is a depth search, that is, a single topic explored more widely.

The 12 tasks are too lengthy to include in this paper, but some examples follow:

Decision making/Hierarchical Task: Your friends who have an interest in art have been debating the French Impressionism exhibit at a local art gallery. One claims that Renoir is the best impressionist ever, while the other argues for another. You decide to do some research first so you can enter the debate. You consider Degas, Monet and Renoir to construct an argument for the one that best represents the spirit of the impressionist movement. Who will you choose and why?

Information Gathering/Parallel Task: Friends are planning to build a new house and have heard that using solar energy panels for heating can save a lot of money. Since they do not know anything about home heating and the issues involved, they have asked for your help. You are uncertain as well, and do some research to identify some issues that need to be considered in deciding between more conventional methods of home heating and solar panels.

Fact Finding/Hierarchical Task: A friend has just sent an email from an Internet café in the southern USA where she is on a hiking trip. She tells you that she has just stepped into an anthill of small red ants and has a large number of painful bites on her leg. She wants to know what species of ants they are likely to be, how dangerous they are and what she can do about the bites. What will you tell her?

Metrics. Each independent variable was assessed using the following metrics: Number of [modified] queries; Time in search segment; Length of query; Number of keywords not present in assigned search task; Number of [unique] webpages viewed; Number of webpages added to Bookbag; Number of [unique] 'objects' used.

3.3 System – WikiSearch

WikiSearch runs on Lucene 2.2, an open source search engine using the vector space model. We indexed the Wikipedia XML documents with the Lucene standard analyzer, using its default stemming and stop word filtering. Each resulting 'document' is composed of two fields, one holding the title and the other handling the rest of the contents. WikiSearch contains a customized interface with special features written using a combination of server-side PHP, and client-side Javascript. First, to eliminate the labyrinth effect of layering multiple pages that may result in constant backtracking, wikiSearch is a single interface divided into three logical frames:

The **Page Display** contains a scrollable wiki page that can be selected from the results (or History or Bookbag). Each page contains two types of links: ordinary hypertext links that connect among the wiki pages, and link that acts as a search to the wiki. Thus, when links are discovered that might serve as a search term, that capability is supplied. Within the Page Display, a further list of Suggested Pages is provided. This set was created by entering the entire first paragraph of that page as a search string. This set can serve two purposes: providing more specific pages about the topic, or by providing distractions.

The **Search** section contains the omnipresent searchbox. But, to conserve space, the results section contains only titles, while a mouseover provides a snippet containing a word-in-context summary. Below the search results is a History section that contains a reminder of both past searches and past articles that were viewed.

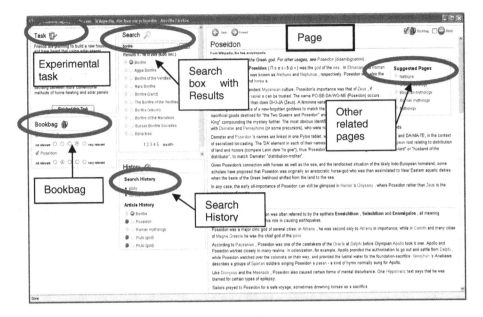

Fig. 1. WikiSearch Interface

The **Task** section contains the contents of the experimental task and serves as a reminder. Below the task is a **Bookbag** that is used to collect pages that are useful to the task. This idea is similar to the shopping cart used in the online shopping environment. In the Bookbag, each page is rated by the participant before the task is considered finished. Pages can be removed from the bookbag from within it, but are added to the Bookbag from the Page Display, the Search Results, and the History sections. In addition, from the Page Display, the Search Results, or the History sections, a participant can drop a page in the Garbage can so that it never need to be viewed again for this search task.

3.4 Participants

The 96 participants (M=49, F=47) were primarily students from the university community, and from mixed disciplines. This convenient sample who was recruited via listservs and recruitment posters placed around the campus received a $10 honorarium. 90% were students: 25% held undergraduate degrees and 12% graduate or other degrees. 84.4% were under 27. They were an experienced search group with 86.5% searching for something one or more times a day, and also relatively frequent users of the Wikipedia (54% use it at least weekly).

3.5 Procedure

Participants interacted with wikiSearch via an enhanced version of WiiRE (Web Interactive Information Retrieval Experimentation) [26]. WiiRE was re-written in PHP and led the participant through the experimental process using a series of webpages.

Responses to questionnaires and the contents of a customized logfile were stored in a mySQL database.

Data collection took place in a laboratory setting that ran 5 to 7 people at a time. Participants were presented with the following steps: 1) Introduction, 2) Consent Form, 3) Demographics and Use Questionnnaire, 4) Tutorial and practice time using the wikiSearch system, 5) Pre-Task Questionnaire, 6) Assigned Task, 7) Post-Task Questionnaire, 8) Steps 5 to 7 were repeated for the other two tasks, 9) Post-Session Questionnaire, 10) SUS Questionnaire, 11) Thank-you for participating page.

After completing the demographic information, each participant performed three randomly assigned search tasks. For each task, participants were introduced to search task and the pre-task questions. Upon completing each task, participants completed a post-task questionnaire.

3.6 Data Analysis

Data was extracted from the log files for this analysis and analyzed primarily using SPSS' univariate and multivariate analysis of variance. In addition, the queries were isolated for additional content analysis.

4 Results

The 96 participants created a total of 2245 queries for the 12 tasks, producing on average 7.8 queries per task. Of those queries, 971 were created for Hierarchically structured tasks, and 1274 for Parallel tasks. When examined by task type, 752, 798, and 694 tasks were used with Decision Making, Fact Finding and Information Gathering tasks. A Chi-Square test of independence found no relationship between Task Structure and Task Type (Chi square=31.777, df=2, p<.0001), indicating that the two attributes of task are independent.

Table 1. Number of Tasks by Structure and Type

		Task Structure		
		Hierarchical	**Parallel**	Mean
Task Type	**Decision Making**	321	432	753
	Fact Finding	294	504	798
	Information Gathering	356	338	694
	Mean	971	1274	2245

Queries were assessed by time as illustrated in Table 2. This is the amount of time that elapsed from the point that a query was submitted to the point when the next query was submitted or the task was declared finished. This relates to the amount of time the user spent with selecting from results, and reading pages associated with that query. There is an interaction effect ($F(1,5)=5.896$, $p<.0001$). The time is different by structure for Hierarchical and Parallel tasks ($F(2,5)=5.826$, $p<.003$); more time was spent on the result of a query for Hierarchical tasks, than for Parallel tasks $(F(2,5)=15.378, p<.0001)$, but the same effect was not present by Task type $(F(2,5)=0.080, p=.924)$.

Table 2. Time per Queries by Task Structure and Type

		Task Structure		Mean
		Hierarchical	**Parallel**	
Task Type	**Decision Making**	57.8	45.6	50.8
	Fact Finding	61.8	40.0	48.0
	Information Gathering	51.6	53.0	52.3
	Mean	56.8	45.3	50.3

Each query was assessed according to length – the number of keywords that each contains. Overall, query length varied from 1 to 15 keywords with a mean length of 2.52 keywords. As illustrated in Table 3, queries varied significantly in length according to task type: on average Information Gathering tasks had the largest queries, and Decision Making the smallest $(F(2,5)=30.022, p<.0001)$, and a Bonferroni post hoc test confirmed the differences among the three types. In addition, Hierarchical tasks tended to have larger queries than Parallel $(F(2,5)=14.681, p<.0001)$. Fact Finding and Information Gathering tasks with a Hierarchical structure tended to be larger than those types with Parallel structure. The same could not be said for the Decision Making tasks $(F(2,5)=25.785, p<.0001)$.

Table 3. Length of Query by Task Structure and Type

		Task Structure		Mean
		Hierarchical	**Parallel**	
Task Type	**Decision Making**	2.12	2.40	2.28
	Fact Finding	2.96	2.26	2.52
	Information Gathering	2.90	2.67	2.79
	Mean	2.66	2.42	2.52

We tested the amount of overlap between words in a query and words that may have been extracted from the assigned tasks. As illustrated in Table 4, human-generated queries contained more keywords that those present in the assigned task statements. All queries had on average at least 1.5 more keywords than those present in the task statement.   While there were no differences by task structure $(F(2,5)=0.413, p=.520)$, there were significant differences by task type $(F(2,5)=38.432, p<.0001)$. Those in Information Gathering tasks were more likely to add additional, unprompted terminology than those in Decision Making, and likewise among the other types, a finding confirmed in post hoc Bonferroni tests.

Table 4. Query overlap with Task statement by Type and Structure

		Task Structure		
		Hierarchical	Parallel	Mean
Task Type	Decision Making	1.31	1.25	1.28
	Fact Finding	1.52	1.56	1.55
	Information Gathering	1.89	1.81	1.85
	Mean	1.59	1.52	1.55

For each query submitted, we examined the number of pages selected from the results pages and viewed, as well as the unique pages viewed per query. Not all queries resulted in page views, and not all pages viewed were unique. As illustrated in Table 5, the total number of pages viewed per query is the first number, and the second number after the "/" is the unique number of pages viewed per query issued. Participants examined the same number of pages ($F(2,5)=.277$, $p=.598$) and the same number of unique pages ($F(2,5)=.074$, $p=.785$).regardless of the task structure. However, participants varied in the number of pages viewed ($F(2,5)=4.406$, $p=.012$).and number of unique pages viewed by task type ($F(2,5)=6.269$, $p=.002$). Post hoc pairwise comparison showed that Decision Making and Information Gathering tasks do not differ by page views or unique pages views, but all other pairs showed differences. There was no interaction effect of structure and type.

As described in Methods, participants were asked to add pages to the Bookbag that were relevant for responding to the task. As illustrated in Table 6, queries generated differing numbers of possible relevant pages according to task structure and type, but there was no interaction effect of either variable. Queries for Parallel tasks generated fewer items for the Bookbag than Hierarchical tasks ($F(2,5)=4.005$, $p=.045$). Similarly, differences were observed by Task type ($F(2,5)=3.249$, $p=.039$), although post hoc pairwise comparisions limited those difference to Decision Making and Fact Finding.

Table 5. Number of Pages viewed from Queries by Task Type and Structure

		Task Structure		
		Hierarchical	Parallel	Mean
		Actual / Unique	Actual / Unique	Actual / Unique
Task Type	Decision Making	0.91 / 0.84	0.83 / 0.77	0.87 / 0.80
	Fact Finding	0.73 / 0.66	0.74 / 0.68	0.74 / 0.67
	Information Gathering	0.94 / 0.87	0.93 / 0.88	0.94 / 0.87
	Mean	0.87 / 0.80	0.82 / 0.76	0.84 / 0.78

Participants had several tools available for use (see Figure 1) including History, Bookbag, Results page, wiki Pages etc. We examined the number of tools used per query, and of those the number of unique ones used per query. There was no interaction effect of type and structure. However, there was main effects of both structure

(F(2,5)=4.783, p=.029) and type (F(2,5)=4.529, p=.011) for total number of objects used (and similarly for unique objects used). The differences by levels of task type were assessed using post hoc pairwise comparisons; Decision Making tasks used more tools than Fact Finding for total number used and unique tools used..

Table 6. Number of Pages added to Bookbag per Query by Task Type and Structure

		Task Structure		
		Hierarchical	Parallel	Mean
Task Type	Decision Making	0.61	0.50	0.54
	Fact Finding	0.50	0.36	0.41
	Information Gathering	0.48	0.51	0.49
	Mean	0.53	0.44	0.48

Table 7. Number of Interface Objects Used per Query by Task Type and Structure

		Task Structure		
		Hierarchical	Parallel	Mean
		Actual/Unique	Actual/Unique	Actual/Unique
Task Type	Decision Making	2.06 / 1.26	1.84 / 1.10	1.93 / 1.17
	Fact Finding	1.53 / 1.08	1.44 / 0.95	1.47 / 1.00
	Information Gathering	1.95 / 1.26	1.44 / 1.11	1.70 / 1.19
	Mean	1.86 / 1.20	1.57 / 1.04	1.70 / 1.11

In addition, the location of query words in the documents was assessed. We wondered if the query words in different types of tasks were more likely to be found in different parts of documents. As illustrated in Table 8, we examined the location of query words in titles, in the first 150 works in the document and in the body of the document.

Table 8. Location of Query words in Document by Task Type and Structure

	Titles		First 150 words		Body	
	Hier	Par	Hier	Par	Hier	Par
Decision Making	2.9	4.4	8.0	7.8	42.4	34.8
Fact Finding	4.1	3.1	9.2	5.3	34.1	22.1
Information Gathering	5.0	5.4	9.9	9.4	31.8	66.9

While the variation in numbers is to be expected across these three content segments due to the difference in size of each, there were effects by task type on all three measures. More words for Information Gathering tasks were found in all three elements than for the other two tasks. When compared by structure, differences appears on in first 150 words: in general Hierarchical tasks had more words in that element than Parallel tasks.

6 Discussion

Results indicated differences by both task type and structure, and selected interaction effects between the two variables:

Differences by Task Structure: Users formulated fewer queries for Hierarchical tasks but the result of a query took more time to process than those for Parallel tasks. Parallel tasks contained multiple concepts and may therefore have required more queries than Hierarchical tasks, which consisted of a single albeit complex concept. Hierarchical tasks used more interface tools, and resulted in more items being added to the Bookbag. Overall it would appear that Hierarchical tasks required more effort considering most metrics other than number of queries.

Differences by Task Type: There were notable differences among task types for number of queries and time to construct queries. Decision Making and Fact Finding tasks contained more queries than Information Gathering. In the case of Information Gathering tasks, the searchers objective was to collect information but, unlike Fact Finding or Decision Making tasks, make no value judgments on the information (e.g. the "right" answer) or use it to select a course of action, and thus one would expect these to require more queries. However, time did not factor into those differences. As expected, Fact Finding tasks used few pages than the other two; one would expect to find the answer for this type of task on a single page. Fact Finding tasks used longer queries than Decision making, but not Information Gathering. Perhaps the concept embedded in a Fact Finding task was so crystal clear that the essence was represented as a holist concept. Queries for Fact Finding tasks generated fewer relevant results pages than either Decision Making or Information Gathering tasks, based on users' relevance assessment.

Users in this study generated, on average, queries that were 2.52 terms in length, which is comparable with ten year old data, and significantly less that the approximately six words found by Wacholder and Liu [26]. Our findings indicated that, even when searchers have a written directive of what they are searching for, they still generated their own keywords. Of note is that Decision Making tasks had the least amount of user-formulated keywords, while Information Gathering had the most. It may be that the nature of Information Gathering tasks "provokes" other information needs during the course of searching [18]. This finding, however, has ramifications for non-user-centric experimentation. When systems are tuned to a 'designer' set of queries, the evaluation may not be pertinent to the real world.

An interactive feature of the interface was a "bookbag" – akin to the online "shopping cart" – that allowed searchers to save relevant results pages. More items were saved for Hierarchical tasks than for Parallel tasks which is somewhat surprising since parallel tasks were expected to include more concepts. Potentially in Wikipedia, information is densely packed in fewer pages such that for . As expected, for Decision Making tasks, users placed more pages (per query) in the book bag than other tasks, particularly Fact Finding. Perhaps this has to do with the essence of the task – making a decision, which was not evident among those doing the Information Gathering task.

Clearly from this research is an effect of task predicted by so many before us (see for example [15]). This work selected but two aspects from an almost endless set of potential attributes. The question remains as to which characteristics have the most discriminatory power and thus likely to have value implicitly in filter results.

Future Research

Eye tracking methods have been used to address scan path patterns of Google results pages. Findings indicated how many abstracts searchers skimmed on results pages, but also demonstrated that queries comprised of one or two keyword resulted in linear scanpaths [20]. Given our findings pertaining to difference in queries across task structures and types, specifically with regard to the interface features and results pages viewed, eye tracking is a promising method for understanding users' behaviour during the search process based on task.

In addition, task itself is a complex problem. While we isolated two aspects of task, there are many others, that my have more significant differences with potential as a differentiator of task. In tandem with that issue is concept of measurement. We used a wide range of metrics for this study, but we have yet to fully understand which metrics are appropriate for measuring interactive IR in the context of task.

7 Conclusions

In the past, users have been shown to place minimal effort in generating queries and examining results [14, 28]. In this study we collected metrics for query length, time taken to create queries, use of self-generated terms, pages viewed, and items declared relevant. Overall, our results demonstrate different levels of effort expended by participants relative to task types and structures. This underscores the need to understand the effects of task on search behaviour and the ways in which information retrieval interfaces may enhance search performance across various tasks, and indeed, how search algorithm might be customized according to task.

Acknowledgments. The authors acknowledge the support of the National Science and Engineering Research Council of Canada, the Canada Research Chairs Program, and the Canada Foundation for Innovation, and the 96 anonymous participants.

References

1. Belkin, N.J., et al.: Query length in interactive information retrieval. In: Proceedings of ACM SIGIR, Toronto, CA, pp. 205–212 (2003)
2. Bell, D.J., Ruthven, I.: Searcher's Assessments of Task Complexity for Web Searching. In: McDonald, S., Tait, J.I. (eds.) ECIR 2004. LNCS, vol. 2997, pp. 57–71. Springer, Heidelberg (2004)

3. Borlund, P.: The IIR evaluation model: a framework for evaluation of interactive information retrieval systems. Inf. Res. 8(3) (2003)
   http://informationr.net/ir/8-3/paper152.html
4. Broder, A.: A taxonomy of web search. SIGIR Forum. 36(2) (2002)
5. Bruza, P., McArthur, R., Dennis, S.: Interactive Internet search: Keyword, directory and query reformulation mechanisms compared. In: ACM SIGIR, Athens, Greece, pp. 280–287 (2000)
6. Bystrom, K., Jarvelin, K.: Task complexity affects information seeking and use. Information Processing and Management 31(2), 191–213 (1995)
7. Bystrom, K., Hansen, P.: Conceptual framework for tasks in information studies. Journal of the American Society for Information Science and Technology 56(10), 1050–1061 (2005)
8. Campbell, D.J.: Task complexity: A review and analysis. The Academy of Management Review 13(1), 40–52 (1988)
9. Efthimiadis, E.N.: Query expansion. Annual Review of Information Science and Technology 31, 121–187 (1996)
10. Freund, L., Toms, E.G., Waterhouse, J.: Modeling the information behaviour of software engineers using a work task framework. In: Proceedings of the ASIST Annual Meeting, Charlotte, NC, November 1-3, vol. 42 (2005)
11. Gill, T.G., Hicks, R.C.: Task Complexity and Informing Science: A Synthesis. Informing Science 9 (2006)
12. Gwizdka, J., Spence, I.: What can searching Behavior tell us about the difficulty of information tasks? A study of Web navigation. In: Proceedings of the Annual Meeting of ASIS&T 43, Austin, TX, USA (2006)
13. Jansen, B.J., Booth, D.L., Spink, A.: Determining the informational, navigational, and transactional intent of Web queries. Information Processing & Management. Corrected Proof 1618 (in Press)
14. Jansen, B.J., Spink, A., Bateman, J., Saracevic, T.: Real life information retrieval: Study of user queries on the Web. SIGIR Forum. 32(1), 5–17 (1998)
15. Jarvelin, K., Ingwersen, P.: Information seeking research needs extension towards tasks and technology. Information Research 10(1) (2004),
   http://informationr.net/ir/10-11/paper212.html
16. Kang, I.-H., Kim, G.C.: Integration of multiple evidences based on a query type for web search. Information Processing & Management 40(3), 459–478 (2004)
17. Kellar, M., Watters, C., Sheppard, M.: A field study characterizing Web-based information-seeking tasks. Journal of the American Society for Information Science and Technology 58, 999–1018 (2007)
18. Kelly, D., Fu, X.: Eliciting better information need descriptions from users of information search systems. Information Processing & Management 43(1), 30–46 (2007)
19. Lau, E.P., Goh, D.H.-L.: In search of query patterns: A case study of a university OPAC. Information Processing & Management 42(5), 1316–1329 (2006)
20. Li, Y.: Task Type and a Faceted Classification of Tasks. In: Proceedings of the Annual Meeting of ASIS&T, Providence, Rhode Island [poster] (2004)
21. Lorigo, L., Pan, B., Hembrooke, H., Joachims, T., Granka, L., Gay, G.: The influence of task and gender on search and evaluation behavior using Google. Information Processing & Management 42(4), 1123–1131 (2006)
22. Rieh, S.Y., Xie, H.: Analysis of multiple query reformulations on the web: The interactive information retrieval context. Information Processing & Management 42(3), 751–768 (2006)

23. Silverstein, C., Henzinger, M., Marais, H., Moricz, M.: Analysis of a very large Alta Vista query log (Technical Report 1998-014). Palo Alto, CA COMPAQ System Research Center (1998)
24. Spink, A., Gunar, O.: E-commerce Web queries: Excite and Ask Jeeves study. First Monday 6(7) (2001)
25. Toms, E.G., Freund, L.: Priming the query specification process. Proceedings of ASIS&T 40(1), 381–388 (2003)
26. Toms, E.G., Freund, L., Cara Li, C.: WiIRE: the Web interactive information retrieval experimentation system prototype. Information Processing & Management 40(4), 655–675 (2004)
27. Wacholder, N., Liu, L.: Assessing term effectiveness in the interactive information access process. Information Processing & Management (in Press) 1618
28. White, R.W., Drucker, S.M.: Investigating behavioral variability in web search. In: WWW 2007, Banff, Alberta, May 8-12, pp. 21–30 (2007)

Overview of INEX 2007 Link the Wiki Track

Darren Wei Che Huang, Yue Xu, Andrew Trotman, and Shlomo Geva

Faculty of Information Technology, Queensland University of Technology,
Brisbane Queensland Australia
w2.huang@student.qut.edu.au, yue.xu@qut.edu.au,
andrew@cs.otago.ac.nz, s.geva@qut.edu.au

Abstract. Wikipedia is becoming ever more popular. Linking between documents is typically provided in similar environments in order to achieve collaborative knowledge sharing. However, this functionality in Wikipedia is not integrated into the document creation process and the quality of automatically generated links has never been quantified. The Link the Wiki (LTW) track at INEX in 2007 aimed at producing a standard procedure, metrics and a discussion forum for the evaluation of link discovery. The tasks offered by the LTW track as well as its evaluation present considerable research challenges. This paper briefly described the LTW task and the procedure of evaluation used at LTW track in 2007. Automated link discovery methods used by participants are outlined. An overview of the evaluation results is concisely presented and further experiments are reported.

Keywords: Wikipedia, Link Discovery, Assessment, Evaluation.

1 Introduction

In 2007, Geva and Trotman suggested the Link the Wiki track that aims to provide an evaluation forum for link discovery in Wikipedia and for objectively evaluating the performance of such algorithms. Wikipedia is composed of millions of articles in English and it offers many attractive features as a corpus for information retrieval tasks. The INEX Wikipedia collection has been converted from its original wiki-markup into XML [1]. This collection is composed of a set of XML files where each file corresponds to an online article in Wikipedia.

Links between pages are essential for navigation, but most systems require authors to manually identify each link. Authors must identify both the anchor and the target page in order to build a knowledge network. This creates a heavy and often unnecessary burden on content providers [2] who would prefer to focus on the content and let the system assist in discovering the relationship between the new content and content already in the collection. Without assistance, as the size of the collection increases, link creation and maintenance can become unmanageable. The maintenance cost of keeping the entire network up to date is huge – and Wikipedia has seen faster than linear growth for many years. Authors are typically unaware of all available links, and even if they are aware of the pre-existing content they are unlikely to be aware of

N. Fuhr et al. (Eds.): INEX 2007, LNCS 4862, pp. 373–387, 2008.

newly created content to which they could link. Page maintenance, in particular linking to content added after a page is created, is a burden on content providers who often do not maintain their content (hence the collaborative nature of these information resources). Ellis et al. [3] have shown significant differences in the links assigned by different people. To eliminate the human effort required to build a highly accurate linking network, to reduce the chance of erroneous links, and to keep links up-to-date, automatic link discovery mechanisms are needed.

The user scenario for the Link the Wiki task is that of an end user who creates a new article in the Wikipedia. A Wikipedia link discovery system then automatically selects a number of prospective anchor texts, and multiple link destinations for each anchor. This is namely the discovery of *outgoing links*. A Wikipedia link discovery system also offers prospective updates to related links in other (e.g. older) wiki articles, which may point to a Best Entry Point (BEP) within this newly created article. In this way, *incoming links* are generated. Therefore, links on each article can always be up-to-date with the latest information existing within the wiki system.

At INEX 2007, the LTW task addressed only document-to-document links, in order to bootstrap the track. Systems were required to discover incoming and outgoing links for selected topics. Evaluation was based on existing Wikipedia links and performance was measured using standard IR metrics.

The remainder of this paper is organized as follows. In the next section, we review the link discovery literature, Wikipedia related research, assessment and evaluation. In section 3 we describe the LTW track at INEX, which comprises the description of task and submission. Section 4 investigates Wikipedia links while the procedure of assessment and evaluation including the result set generation and evaluation procedure is depicted in Section 5. Following that, the techniques used by all participants and the results of evaluation are depicted, and the better performing runs are discussed in some more detail. Finally, conclusions and future direction are presented.

2   Background of Link Discovery

As suggested by Wilkinson & Smeaton [2], navigation between linked documents is a great deal more than simply navigating multiple results of a single search query. Linking between digital resources is becoming an ever more important way to find information. Through hypertext navigation, users can easily understand context and realize the relationships of related information. However, since digital resources are distributed it has become a difficult task for users to maintain the quality and the consistency of links. Automatic techniques to detect the semantic structure (e.g. hierarchy) of the document collection and the relatedness and relationships of digital objects have been studied and developed [4]. Early works, in the 1990s, determined whether and when to insert links between documents by computing document similarity. Approaches such as term repetition, lexical chains, keyword weighting and so on were used to calculate the similarity between documents [5, 6, 7]. These approaches were based on a document-to-document linking scenario, rather than identifying which parts of which documents were interrelated.

Adafre and de Rijke [8] state that most links in Wikipedia are conceptual. The Wikipedia linking network offers hierarchical information and links aim to expand on the concepts in their anchors. The anchors imply the concept while the links are complementary to the concept. Since there is no strict standard of editing there are problems with *over linking* and *missing links*. They proposed a method of discovering *missing links* in Wikipedia pages by clustering topically related pages using *LTRank* and identified link candidates by matching anchor texts. Page ranks using the *LTRank* method are based on the co-citation and page title information. Experiment results show reasonable outcome.

Jenkins [9] developed a link suggestion tool, *Can We Link It*. This tool extracts a number of anchors which have not been linked within the current article and that might be linked to other pages in the Wikipedia. With this tool, the user can accept, reject, or click *"don't know"* to leave a link as undecided. Using this tool the user can add new anchors and corresponding links back to the Wikipedia article.

A collaborative knowledge management system, called *PlanetMath*, based on the Noosphere system has been developed for mathematics [10]. It is encyclopaedic, (like the Wikipedia), but mainly used for the sharing of mathematical knowledge. Since the content is considered to be a semantic network, entries should be cross-referenced (linked). An automatic linking system provided by Noosphere employs the concept of conceptual dependency to identify each entry for linking. A classification hierarchy used in online encyclopedias is used to improve the precision of automatic linking. In practice, the system looks for common anchors that are defined in multiple entries and creates links between them, once the page metadata is identified as related. Based on the Noosphere system, NNexus (Noosphere Networked Entry eXtension and Unification System) was developed to automate the process of the automatic linking procedure [11]. This was the first automatic linking system to eliminate the linking efforts required by page authors. Declarative linking priorities and clauses are specified to enhance the linking precision. An approach, called *invalidation index*, was developed to invalidate entries belonging to those concepts where there are new entries. Reputation based collaborative filtering techniques could be used to provide personalized links.

Research on the Wikipedia has been undertaken in recent years. A set of experiments, based on Markov Chains [12], for finding related pages within Wikipedia collection was undertaken using two Green-based methods [13], *Green* and *Sym-Green*, and three classical approaches, *PageRankOfLinks*, *Cosine with tf-idf weight* and *Co-citations*. The results show the Green method has better performance at finding similar nodes than only relying on the graph structure. Although page titles and category structure can be used to classify documents in Wikipedia, properties such as the internal text of the articles, the hierarchical category, and the linking structure should be used [14]. *Wikirelate* proposed by Strube and Ponzetto uses *Path*, *Information content* and *Text overlap* measures to compute the semantic relatedness of words [15]. These measures mainly rely on either the texts of the articles or the category hierarchy. Gabrilövich and Markövitch [16] introduce a new approach called Explicit Semantic Analysis (ESA), which computes relatedness by comparing two weighted vectors of Wikipedia concepts that represent words appearing within the content. Common to this research is the use of the *existing* linking structure and content (category, etc.).

Various link-based techniques based on the correlation between the link density and content have been developed for a diverse set of research problems including link discovery and relevance ranking [8]. Moreover, communities can be identified by analysing the link graph [17]. Beside co-citation used by Kumar et al. [18] to measure similarity, bibliographic coupling and SimRank based on citation patterns, and the similarity of structural context (respectively), have also been used to identify the similarity of web objects [19]. The companion algorithm derived from HITS has also been proposed for finding related pages (by exploiting links and their order on a page) [20, 21].

The assessment of IR results has been a challenge in IR experiments for many years because despite the existence of some standard procedures, relevance is hard to define and cross-assessor agreement levels are often low (so individual judgments come under dispute). Worse, it is difficult to compare IR methods which are able to retrieve highly relevant documents with those that retrieve less relevant documents because assessments are usually binary. The use of Precision-Recall (P-R) curves is typical in IR; however, Schamber [22] argues that traditional P-R based comparison using binary relevance cannot adequately capture the variability and complexity of relevance. Relevance is a multilevel circumstance where, for a user, the degree of relevance may vary from document to document.

Several studies done in [23] have examined components that influence judgments and the criteria of relevance (including graded relevance) in information seeking and retrieval. Kekalainen and Jarvelin [24] argued that evaluation methods should be flexible enough to handle different degrees of judgment scales. They proposed generalized precision and recall that can incorporate a continuous relevance scale into the traditional precision and recall measures. Their experiments demonstrate that the evaluation approach can distinguish between retrieval methods fetching highly relevant documents from those retrieving partially relevant documents.

3   The Link-the-Wiki Track

3.1   Tasks

Wikipedia is composed of millions of interlinked articles in numerous languages and offers many attractive features for retrieval tasks. The current INEX Wikipedia collection contains a snapshot of the Wikipedia English collection from 2006 and contains 660,000 documents and is about 4GB in size. In INEX 2007 the linking task used 90 topics, nominated from the existing collection by participants [25]. The task is two fold:

1. *Outgoing links*: Recommend anchor text and destinations within the Wikipedia collection.
2. *Incoming links*: Recommend incoming links from other Wikipedia documents.

Missing topics were regarded as having a score of zero for the purpose of calculating system rank when using all topics. Up to 250 outgoing links and 250 incoming links were allowed for each topic. Surplus links were discarded in computing the performance.

3.2   Submission

Each participant was encouraged to submit up to 5 runs. The submission example can be found in Appendix A-1. Finally, thirteen runs were submitted by 4 groups in 2007.

The University of Amsterdam had 5 runs, Waterloo submitted 1 run, Otago 5 runs and QUT 2 runs.

4 Wikipedia Links

The 90 LTW topics were examined to discover the relationship between anchors and page titles. The result showed that in 81 of the 90 topics at least 50% of the links match an existing page name (see Table 1). This could be because the links were generated through careful construction by a user, or automatically by matching page names, either way such links are relatively easy to find. For example, the word, *Explorer*, in the document can be manually linked using Wiki markup [[destination page title| anchor text]] to the page titles, *Explorer (Learning for Life)*, *Explorer (album)* or *Explorer (novel)*, depending on the context. According to this investigation, we can expect a method that systematically matches potential anchor strings with page names to identify most links – and to achieve a better recall (e.g. near 1) when most page names and anchors are exact matches. Although this implies that we can expect high recall from simple page-name matching strategies, it does not necessarily mean that we can expect high precision – many matching links are not relevant (e.g. polyvalent terms). As the Wikipedia is a huge repository of definitions it is quite easy to find matching page names which are not relevant.

Table 1. Ratio of matching names between anchors and links

Ratio of Match	Number of Topics
90% ~ 100%	1
80% ~ 90%	8
70% ~ 80%	26
60% ~ 70%	35
50% ~ 60%	16
40% ~ 50%	2
30% ~ 40%	2

5 Assessment and Evaluation

The main focus of the Link-the-Wiki track in 2007 was to explore an automated evaluation procedure without human assessment effort. The incoming and outgoing links were retrieved directly from the existing collection links to form the result set for evaluation. This can completely eliminate the assessment effort. Accompanying the automatically generated result set (see Appendix A-2), the proposed evaluation tool was developed to examine the performance of the link discovery methods. We notice in the result set that incoming links outnumber outgoing links. This could be because the number of outgoing links may be restricted by the length of the document

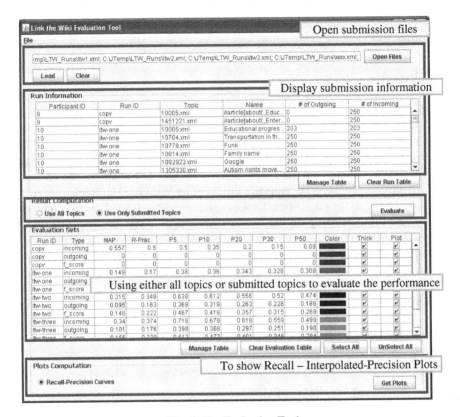

Fig. 1. The Evaluation Tool

(only proper anchors will be specified) but there can be many different pages linked to the topic page.

As we can see above, we treat the Wikipedia links as the ground-truth. However, they are obviously not perfect. Many links in Wikipedia are automatically generated but some of them do not have a clear topic relation. *Year* links, for example, are very often unrelated to the content of the document, but are easy to discover. Such links have probably no utility for IR. Problematically, when used in evaluation as the ground truth, they may also lead to optimistic evaluation results when easily identified by link-discovery systems. Many potentially good links that have not been identified by Wikipedia users are amenable to automatic discovery. Such useful returned links which are missing from the ground truth could result in poor evaluation scores for link discovery systems, hence leading to pessimistic evaluation results. So although it is not possible to quantify the absolute performance of link discovery by using automated result assessment, the procedure we used provides a trade-off between assessment effort and absolute accuracy of measurement.

It is reasonable to conjecture that *comparative evaluation* of methods and systems is still informative. Through the investigation of comparative analysis of automated linking system for the Wikipedia, it should remain possible to improve link discovery methods.

An evaluation tool, named *ltwEval*, was developed for LTW 2007 (see Figure 1). The performance measures include *Mean Average Precision (MAP)*, *precision at the point of the number of relevant documents (R-Prec)*, and *precision at varying numbers of documents retrieved (P@r)*. Plots of Interpolated Precision-Recall for incoming, outgoing and a combined score are also computed for comparison. By combined score we refer to the harmonic mean of the various values obtained for incoming and outgoing links. The *ltwEval* program was developed in Java for platform independence, but is GUI driven and provides more extensive functionality than traditional evaluation software. This assists participants by making result exploration and analysis easier.

6   Link Discovery Methods

6.1 Approaches and Evaluation Results

In this section we briefly describe the approaches that were taken by participants.

The University of Amsterdam system assumed that Wikipedia pages link to each other when articles are similar or related in content. For each of the 90 topics, the system queried the index of the entire collection, (excluding the topics). This was done by using the full topic as the query, but excluding stop words, and with important terms derived from a language model. The top 100 files (anchors) were selected for each topic. They experimented with line matching from the orphans to the anchor files. For the outgoing links, the system matched each line of a topic with the lines of the anchors until a matching line has been found. For the incoming links, the system iterated over all lines of each anchor for each line of the topic. The generated runs were based on the names of the pages, exact lines, and longest common substrings (LCSS) expanded with WordNet synonyms. The results show that the run based on restricting the line matching to the names of pages performed best. Therefore, submitted runs have a good performance on average, especially for incoming links (see Figure 2).

The University of Otago system identified terms within the document that were over represented by comparing term frequency in the document with the expected term frequency (computed as the collection frequency divided by document frequency). From the top few over-represented terms they generated queries of different lengths. A BM25 ranking search engine was used to identify potentially relevant documents. Links from the source document to the potentially relevant documents (and back) were constructed. They showed that using 4 terms per query was more effective than fewer or more.

The University of Waterloo system found the first 250 documents (in document collection order) that contain the topic titles and then generated article-to-article Incoming links. For outgoing links, they performed link analysis. The system computed the probabilities that each candidate anchor would be linked to a destination file. The

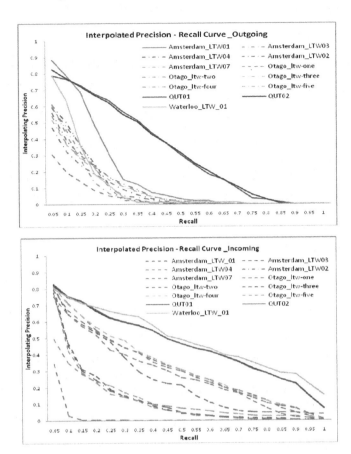

Fig. 2. Interpolated Precision-Recall of Outgoing and Incoming links

probability that a candidate anchor would be linked was computed (essentially) as the ratio of the number of times that the anchor text was actually linked in the collection, to the number of times that the anchor text appeared in the collection.

The Queensland University of Technology (QUT) system identified incoming links using a ranked search for documents that were *about* the new document title. Outgoing links were identified by running a window over the new document text and looking for matching document titles in the collection. The window size varied from 12 words down to 1 word, and included stop words. Longer page names were ranked higher than shorter page names, motivated by the observation that the system was less likely to hit on a longer page name by accident. From the official results shown in Table 2, 3 and Figure 2, QUT runs have better performance than any other run proposed by participants, utilizing page title matching. However, the Waterloo system, using link analysis, was subsequently shown to outperform this system, for outgoing link discovery, when suitably configured.

Table 2. MAP of Outgoing and Incoming Links

	MAP Outgoing Links			MAP Incoming Links	
1	QUT02	0.484	1	QUT02	0.318
2	QUT01	0.483	2	QUT01	0.314
3	Waterloo_LTW_01	0.465	3	Amsterdam_LTW_01	0.147
4	Otago_ltw-four	0.339	4	Otago_ltw-four	0.102
5	Otago_ltw-five	0.319	5	Otago_ltw-five	0.101
6	Otago_ltw-three	0.318	6	Waterloo_LTW_01	0.093
7	Otago_ltw-two	0.284	7	Otago_ltw-three	0.092
8	Amsterdam_LTW_01	0.226	8	Otago_ltw-two	0.081
9	Otago_ltw-one	0.123	9	Amsterdam_LTW04	0.080
10	Amsterdam_LTW03	0.110	10	Amsterdam_LTW02	0.080
11	Amsterdam_LTW02	0.108	11	Amsterdam_LTW03	0.073
12	Amsterdam_LTW04	0.093	12	Amsterdam_LTW07	0.067
13	Amsterdam_LTW07	0.004	13	Otago_ltw-one	0.048

Table 3. R-Precision of Outgoing and Incoming Links

	R-Prec Outgoing Links			R-Prec Incoming Links	
1	QUT01	0.415	1	Waterloo_LTW_01	0.512
2	QUT02	0.411	2	QUT02	0.505
3	Otago_ltw-four	0.183	3	QUT01	0.503
4	Otago_ltw-five	0.183	4	Otago_ltw-four	0.379
5	Amsterdam_LTW_01	0.182	5	Otago_ltw-three	0.363
6	Otago_ltw-three	0.173	6	Otago_ltw-five	0.356
7	Otago_ltw-two	0.156	7	Otago_ltw-two	0.331
8	Amsterdam_LTW02	0.154	8	Amsterdam_LTW_01	0.258
9	Amsterdam_LTW04	0.149	9	Amsterdam_LTW02	0.165
10	Amsterdam_LTW03	0.141	10	Otago_ltw-one	0.153
11	Amsterdam_LTW07	0.127	11	Amsterdam_LTW03	0.144
12	Waterloo_LTW_01	0.103	12	Amsterdam_LTW04	0.142
13	Otago_ltw-one	0.098	13	Amsterdam_LTW07	0.020

6.2 Discussion of Best Approaches

In this section we concentrate on the performance of the two most successful approaches at INEX 2007 [25, 26], the Waterloo and QUT systems.

The best performing approaches were those that used either existing anchors to predict suitable anchors (Waterloo), or matching document titles to predict suitable anchors. The performance of these 2 approaches for discovering outgoing links (note: produced *after* the INEX 2007 workshop and some implementation corrections) are depicted in Fig. 3. It can be seen that both approaches produce a very good result with high precision over a wide range of recall levels. This is precisely the kind of performance needed to satisfy a user.

There are considerable differences between the two approaches. The Waterloo approach relies on the availability of an extensive pre-existing web of anchor to document links in the collection. This pre-requisite may not always be satisfied, particularly when a new cluster of documents in a new domain is added to the collection in bulk, or when a new Wikipedia-like resource is created. However, the approach can discover links that are not solely based on a match between anchor text and a document title. If an anchor is frequently linked to a document with a different title, it will become a highly probable link. For instance, the Waterloo system was able to link *Educational Philosophy* to a document titled *The Philosophy of Education*. By contrast, the QUT approach only discovered matching document titles. In regard to the investigation described in Section 4, LTW approaches aiming at matching anchors with page titles can achieve a certain level of performance. Although the performance of QUT is somewhat lower than that of Waterloo, the approach is applicable to any collection, regardless of the pre-existing link structure. It could immediately be applied to any document collection, completely devoid of links or with pre-existing links.

Fig. 3 presents the precision-recall curves for the two systems. *Anchors 90* is the Waterloo system and *Page Titles 90* is the QUT approach. Both are shown for the 90 INEX topics. The result shows that the anchor-based approach (Waterloo) is better at almost all recall points. In order to verify the scalability and reliability of the INEX evaluation itself, the QUT system was also tested with 6600 randomly selected topics (1% of the collection) – the plot entitled *Page Titles 6600* corresponds to this experiment. It demonstrates that the approach taken by INEX LTW in 2007 is robust and that 90 topics represent an adequate number of topics for the track.

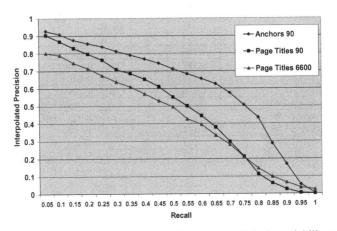

Fig. 3. System performance of discovering outgoing links in scalability test

7   Conclusions and Future Direction

As far as we are aware, the Link the Wiki task at INEX is the first to offer extensive reusable independent evaluation resources for link discovery. Although in 2007 the LTW track still operated the evaluation at the document level, the LTW track has launched a forum to discuss the performance of results for extracting link discovery. The procedure of the LTW track has been defined and an evaluation tool has also been developed to speed the exploration of submission runs. Evaluation results were analyzed in this paper and the main findings described. Using a very large set of documents (1% of the collection), extensive linking experiment has been undertaken and the result has showed that linking is feasible and effective.

It is noticed that document-to-document link discovery systems are very good exhibiting high precision levels at most points of recall, systems are scalable and that several different techniques might be used. This result motivates us to examine (and outline future work) anchor to Best-Entry-Point identification. In future INEX evaluations the task will be defined as anchor to BEP link discovery, and allow multiple links per anchor (actually, the latter is essential for manual evaluation purposed where two systems might link the same anchor to different documents, both of which are correct). Traditional performance measures such as MAP will be adapted to address the performance differences of link-discovery methods in this new scenario. Manual assessment would allow us to study more deeply the nature of link discovery, to identify those links returned by automatic systems that have not yet been identified by Wikipedia authors, and those automatic links that already exist in the Wikipedia and which are not useful (e.g. *year* links are common, yet often of no use).

References

1. Denoyer, L., Gallinari, P.: The Wikipedia XML Corpus. SIGIR Forum. 40(1), 64–69 (2006)
2. Wilkinson, R., Smeaton, A.F.: Automatic Link Generation. ACM Computing Surveys 31(4) (December 1999)
3. Ellis, D., Furner-Hines, J., Willett, P.: On the Measurement of Inter-Linker Consistency and Retrieval Effectiveness in Hypertext Database. In: Proceedings of the 17th Annual International Conference on Research and Development in Information Retrieval, Dublin, Ireland, July 1994, pp. 51–60 (1994)
4. Green, S.J.: Building Hypertext Links By Computing Semantic Similarity. IEEE Transactions on Knowledge and Data Engineering 11(5), 713–730 (1999)
5. Allan, J.: Building Hypertext using Information Retrieval. Information Processing and Management 33(2), 145–159 (1997)
6. Green, S.J.: Automated Link Generation: Can We Do Better than Term Repetition? In: Proceedings of the 7th International World Wide Web Conference, Brisbane, Australia, April 14-18, pp. 75–84 (1998)
7. Zeng, J., Bloniarz, O.A.: From Keywords to Links: an Automatic Approach. In: Proceedings of the International Conference on Information Technology: Coding and Computing (ITCC 2004), April 5-7, pp. 283–286 (2004)
8. Adafre, S.F., de Rijke, M.: Discovering missing links in Wikipedia. In: Proceedings of the SIGIR 2005 Workshop on Link Discovery: Issues, Approaches and Applications, Chicago, IL, USA, August 21-24 (2005)
9. Jenkins, N.: Can We Link It (2007),
   http://en.wikipedia.org/wiki/User:Nickj/Can_We_Link_It
10. Krowne, A.: An Architecture for Collaborative Math and Science Digital Libraries. Thesis for Master of Science Virginia Polytechnic Institute and State University (July 19, 2003)
11. Gardner, J., Krowne, A., Xiong, L.: NNexus: Towards an Automatic Linker for a Massively-Distributed Collaborative Corpus. In: Proceedings of the International Conference on Collaborative Computing: Networking, Applications and Worksharing, November 17-20, pp. 1–3 (2006)
12. Norris, J.R.: Markov chains, Cambridge Series in Statistical and Probabilistic Mathematics. Cambridge University Press, Cambridge (1999)

13. Ollivier, Y., Senellart, P.: Finding Related Pages Using Green Measures: An Illustration with Wikipedia. In: Proceedings of the 22nd National Conference on Artificial Intelligence (AAAI 2007), Vancouver, Canada, July 22-26 (2007)

14. Schönhofen, P.: Identifying decument topics using the Wikipedia category network. In: Proceedings of the 2006 IEEE/EIC/ACM International Conference on Web Intelligence (WI 2006), Hong Kong, December 18-22 (2006)

15. Strube, M., Ponzetto, S.P.: WikiRelate! Computing Semantic Relatedness Using Wikipedia. In: Proceedings of the 21th National Conference on Artificial Intelligence (AAAI 2006), Boston, Massachusetts, USA, July 16-17, pp. 16–20 (2006)

16. Gabrilovich, E., Markovitch, S.: Computing Semantic Relatedness using Wikipedia-based Explicit Semantic Analysis. In: Proceedings of the 20th International Joint Conference on Artificial Intelligence (IJCAI 2007), Hyderabad, India, January 6-12 (2007)

17. Kumar, R., Raghavan, P., Rajagopalan, S., Tomkins, A.: Trawling the Web for Emerging Cyber-Communities. Computer Networks 31(11-16), 1481–1493 (1999)

18. Jeh, G., Widom, J.: SimRank: a measure of structural-context similarity. In: Proceedings of the 8th ACM SIGKDD international conference on Knowledge discovery and data mining (KDD 2002), Edmonton, Canada, July 23-26, pp. 538–543 (2002)

19. Kessler, M.M.: Bibliographic coupling between scientific papers. American Documentation 14(10-25) (1963)

20. Dean, J., Henzinger, M.R.: Finding related pages in the World Wide Web. Computer Networks 31(11-16), 1467–1479 (1999)

21. Kleinberg, J.: Authoritative sources in a hyperlinked environment. In: Proceedings of the 9th Annual ACM–SIAM Symposium on Discrete Algorithms, San Francisco, CA, USA, January 25-27, pp. 668–677 (1998)

22. Schamber, L.: Relevance and Information Behavior, Annual review of information science and technology, vol. 29, pp. 3–48. Information Today, Medford (1994)

23. Vakkari, P., Hakala, N.: Changes in relevance criteria and problem stages in task performance. Journal of Documentation 56(5), 540–562 (2000)

24. Kekäläinen, J., Järvelin, K.: Using graded relevance assessments in IR evaluation. Journal of the American Society for Information Science and Technology 53(13), 1120–1129 (2002)

25. Huang, W.C., Xu, Y., Geva, S.: Overview of INEX 2007 Link the Wiki track. In: Pre-Proceedings of the INEX 2007 Conference, Dagstuhl, Germany (2007)

26. Itakura, K.Y., Clarke, C.L.A.: University of Waterloo at INEX2007: Ad Hoc and Link-the-Wiki Tracks. In: Pre-Proceedings of the INEX 2007 Conference, Dagstuhl, Germany, pp. 380–387 (2007)

Appendix A

A.1 Example Submission

```
<inex-submission participant-id="12" run-id="LTW_01" task="LinkTheWiki">
  <details>
    <machine>
      <cpu>Intel(R) Pentium (R) D</cpu>
      <speed>3.00GHz</speed>
      <cores>2</cores>
      <hyperthreads>None</hyperthreads>
      <memory>2GB</memory>
```

```
      </machine>
      <time>166295 seconds</time>
    </details>
    <description>Using text chunking etc.</description>
    <collections>
     <collection>wikipedia</collection>
    <collections>
    <topic file="13876.xml" name="Albert Einstein">
     <outgoing>
      <link>
       <anchor>
        <file>13876.xml</file>
        <start>/article[1]/body[1]/p[3]/text()[2].10</start>
        <end>/article[1]/body[1]/p[3]/text()[2].35</end>
       </anchor>
       <linkto>
        <file>123456.xml</file>
        <bep>/article[1]/sec[3]/p[8]<bep>
       </linkto>
      </link>
      ...
     </outgoing>
     <incoming>
      <link>
       <anchor>
        <file>654321.xml</file>
        <start>/article[1]/body[1]/p[3]/text()[2].10</start>
        <end>/article[1]/body[1]/p[3]/text()[2].35</end>
       </anchor>
       <linkto>
        <file>13876.xml</file>
        <bep>/article[1]/sec[3]/p[8]<bep>
       </linkto>
      </link>
      ...
     </incoming>
    </topic>
   </inex-submission>
```

These two sections of tags can be left as empty (*e.g.* *<bep></bep>*) since the task is operated at the document level.

A.2 Official Result Set

Topics	# of Outgoing	# of Incoming	Topics	# of Outgoing	# of Incoming
Donald Bradman (87021.xml)	72	144	Dalai Lama (8133.xml)	71	237
Unified Modeling Language (32169.xml)	62	91	Within You Without You (1451526.xml)	13	11
Sukhoi Su-33 (552810.xml)	23	15	Software engineering (27010.xml)	107	404
Funk (10778.xml)	126	755	Philately (23681.xml)	41	108
Star Trek (26717.xml)	143	1649	Marie Curie (20408.xml)	75	127

Cartilage (166945.xml)	41	166	Stockholm syndrome (90910.xml)	49	36
Organic food (177593.xml)	73	50	Pink Floyd (24370.xml)	175	718
Pope Clement V (24102.xml)	69	56	Wavelet compression (50911.xml)	21	13
David (8551.xml)	124	513	Computer science (5323.xml)	241	1606
Aranyaka (321947.xml)	10	6	Pizza (24768.xml)	189	262
Greater Tokyo Area (354951.xml)	32	28	Joshua (16121.xml)	57	136
Xorn (322085.xml)	42	17	Skin cancer (64993.xml)	18	54
Kennewick Man (92818.xml)	47	10	Prince (artist) (57317.xml)	252	475
Frank Klepacki (752559.xml)	13	1	Family name (10814.xml)	165	474
University of London (60919.xml)	193	564	Search engine (27804.xml)	64	254
Latent semantic analysis (689427.xml)	16	10	Charleston, South Carolina (61024.xml)	200	947
Use case (300006.xml)	12	16	Elf (9896.xml)	235	378
Gout (55584.xml)	95	118	Akira Kurosawa (872.xml)	95	186
Thomas Edison (29778.xml)	132	358	Database (8377.xml)	99	186
Baylor University basketball scandal (493525.xml)	44	3	Radical feminism (25998.xml)	29	49
Search engine optimization (187946.xml)	49	45	Educational progressivism (10005.xml)	6	15
Civil Constitution of the Clergy (410450.xml)	40	34	Software development process (27565.xml)	49	33
Nokia (21242.xml)	48	196	Alastair Reynolds (69168.xml)	29	40
Achilles (305.xml)	124	219	Kazi Nazrul Islam (539155.xml)	31	20
Sunscreen (294419.xml)	38	46	Muammar al-Qaddafi (53029.xml)	159	149
Experiential education (447089.xml)	16	17	Neo-Byzantine architecture (1453013.xml)	36	5
Yitzhak Rabin (43983.xml)	77	145	Waseda University (376791.xml)	67	85
Triple J's Impossible Music Festival (2542756.xml)	103	1	Text Retrieval Conference (1897206.xml)	9	2
World Wide Web Consortium (33149.xml)	23	181	Autism rights movement (1305330.xml)	86	27
Excel Saga (265496.xml)	74	73	Ballpoint pen (4519.xml)	53	55
Link popularity (210641.xml)	20	6	Digital library (8794.xml)	13	43

Coca-Cola (6690.xml)	171	506	Sloe gin (392900.xml)	13	7
Entertainment robot (1451221.xml)	17	3	Koala (17143.xml)	70	104
Indira Gandhi (15179.xml)	100	199	Billie Holiday (50420.xml)	53	196
Leukemia (18539.xml)	64	403	Softball (80763.xml)	50	368
Miss Universe (150340.xml)	159	182	Information retrieval (15271.xml)	40	45
Neuilly-sur-Seine (234647.xml)	18	80	Cheminformatics (575697.xml)	13	17
Jihad (16203.xml)	56	254	Requirement (544592.xml)	9	27
Google (1092923.xml)	192	541	Susan Haack (321979.xml)	27	10
Joseph Stalin (15641.xml)	373	1324	Math rock (221484.xml)	72	49
Seasonal energy efficiency ratio (2189642.xml)	8	0	Transportation in the Faroe Islands (10704.xml)	18	0
Sony (26989.xml)	136	965	Anthropology (569.xml)	129	808
Doctor of Philosophy (8775.xml)	64	2110	Red Bull (61123.xml)	75	74
Taiwanese aborigines (53787.xml)	68	86	Lithography (18426.xml)	32	281
Hyperlink (49547.xml)	60	118	Isaac Newton (14627.xml)	207	611

Using and Detecting Links in Wikipedia

Khairun Nisa Fachry[1], Jaap Kamps[1,2], Marijn Koolen[1], and Junte Zhang[1]

[1] Archives and Information Studies, Faculty of Humanities, University of Amsterdam
[2] ISLA, Faculty of Science, University of Amsterdam

Abstract. In this paper, we document our efforts at INEX 2007 where we participated in the Ad Hoc Track, the Link the Wiki Track, and the Interactive Track that continued from INEX 2006. Our main aims at INEX 2007 were the following. For the Ad Hoc Track, we investigated the effectiveness of incorporating link evidence into the model, and of a CAS filtering method exploiting the structural hints in the INEX topics. For the Link the Wiki Track, we investigated the relative effectiveness of link detection based on retrieving similar documents with the Vector Space Model, and then filter with the names of Wikipedia articles to establish a link. For the Interactive Track, we took part in the interactive experiment comparing an element retrieval system with a passage retrieval system. The main results are the following. For the Ad Hoc Track, we see that link priors improve most of our runs for the Relevant in Context and Best in Context Tasks, and that CAS pool filtering is effective for the Relevant in Context and Best in Context Tasks. For the Link the Wiki Track, the results show that detecting links with name matching works relatively well, though links were generally under-generated, which hurt the performance. For the Interactive Track, our test-persons showed a weak preference for the element retrieval system over the passage retrieval system.

1 Introduction

In this paper, we describe our participation in the INEX 2007 Ad Hoc and Link the Wiki tracks, and the INEX 2006 Interactive Track. For the Ad Hoc track, our aims were: a) to investigate the effectiveness of incorporating link evidence into the model, to rerank retrieval results and b) to compare several CAS filtering methods that exploit the structural hints in the INEX topics. Link structure has been used effectively in Web retrieval [9] for known-item finding tasks. Although the number of incoming links is not effective for general ad hoc topics on Web collections [5], Wikipedia links are of a different nature than Web links, and might be more effective for informational topics.

For the Link the Wiki Track, we investigated the relative effectiveness of link detection based on the Wikipedia article's name only, and on the matching arbitrary text segments of different pages. Information Retrieval methods have been employed to automatically construct hypertext on the Web [2], as well for specifically discovering missing links in Wikipedia [4]. The track is aimed at detecting missing links between a set of topics, and the remainder of the

N. Fuhr et al. (Eds.): INEX 2007, LNCS 4862, pp. 388–403, 2008.
© Springer-Verlag Berlin Heidelberg 2008

collection, specifically detecting links between an origin node and a destination node. To detect whether two nodes are implicitly connected, it is necessary to search the Wikipedia pages for some text segments that both nodes share.

For the Interactive Track, we took part in the interactive experiment comparing an element retrieval system with a passage retrieval system. The element retrieval system returns element of varying granularity based on the hierarchical document structure and passage retrieval returns non-overlapping passages derived by splitting the document linearly. Trotman and Geva [16] argued that, since INEX relevance assessments are not bound to XML element boundaries, retrieval systems should also not be bound to XML element boundaries. Their implicit assumption is that a system returning passages is at least as effective and useful as a system returning XML elements. Since the document structure may have additional use beyond retrieval effectiveness, think for example of browsing through a result article using a table of contents, the INEX 2006 Interactive Track set up a concerted experiment compare an element retrieval system to a passage retrieval system [11]. The INEX 2006 Interactive Track run well into INEX 2007, so we report our findings here.

The rest of the paper is organized as follows. First, Section 2 describes our retrieval approach. Then, in Section 3, we report the results for the Ad Hoc Track: the Focused Task in Section 3.1; the Relevant in Context Task in Section 3.2; and the Best in Context Task in Section 3.3. Followed by Section 4, which details our approach and results for the INEX 2007 Link the Wiki Track. In Section 5 we discuss our INEX 2006 Interactive Track experiments. Finally, in Section 6, we discuss our findings and draw some conclusions.

2 Experimental Setup

2.1 Collection, Topics, and Relevance Judgments

The document collection is based on the English Wikipedia [17]. The collection has been converted from the wiki-syntax to an XML format [3]. The XML collection has more than 650,000 documents and over 50,000,000 elements using 1,241 different tag names. However, of these, 779 tags occur only once, and only 120 of them occur more than 10 times in the entire collection. On average, documents have almost 80 elements, with an average depth of 4.82.

There have been 130 topics selected for the INEX 2007 Ad Hoc track, which are numbered 414-543. Table 1 shows some statistics on this years assessments. We have included the numbers from last years assessments for comparison. The number of relevant articles and passages is slightly higher than last year, while the number of assessed topics is lower. Last year, 114 topics were assessed, with 49.54 relevant articles and 79.68 relevant passages per topic. This year, 107 topics were assessed, with 60.66 relevant articles and 107.31 relevant passages per topic. The average number of relevant passages per relevant articles is 1.61 for the 2006 topics and 1.77 for the 2007 topics. On the other hand, the size of the relevant passages this year has decreased compared to last year. Both average (931) and

Table 1. Relevant passage statistics

Description	Statistics	
	2006	2007
# topics	114	107
# articles with relevance	5,648	6,491
# relevant passages	9,083	11,482
mean length relevant passage	1,090	931
median length relevant passage	297	272

median (272) size (in character length) are lower than last year (1,090 and 297 respectively).

2.2 Indexing

Our indexing approach is based on our earlier work [8, 13, 14, 15].

- *Element index*: Our main index contains all retrievable elements, where we index all textual content of the element including the textual content of their descendants. This results in the "traditional" overlapping element index in the same way as we have done in the previous years [14].
- *Contain index*: We built an index based on frequently retrieved elements. Studying the distribution of retrieved elements, we found that the `<article>`, `<body>`, `<section>`, `<p>`, `<normallist>`, `<item>`, `<row>` and `<caption>` elements are the most frequently retrieved elements. Other frequently retrieved elements are `<collectionlink>`, `<outsidelink>` and `<unknownlink>` elements. However, since these links contain only a few terms at most, and say more about the relevance of another page, we didn't add them to the index.

For all indexes, stop-words were removed, but no morphological normalization such as stemming was applied. Queries are processed similar to the documents, we use either the CO query or the CAS query, and remove query operators (if present) from the CO query and the about-functions in the CAS query.

2.3 Retrieval Model

Our retrieval system is based on the Lucene engine with a number of home-grown extensions [7, 10].

For the Ad Hoc Track, we use a language model where the score for an element e given a query q is calculated as:

$$P(e|q) = P(e) \cdot P(q|e) \tag{1}$$

where $P(q|e)$ can be viewed as a query generation process—what is the chance that the query is derived from this element—and $P(e)$ an element prior that provides an elegant way to incorporate link evidence and other query independent evidence [6, 9].

We estimate $P(q|e)$ using Jelinek-Mercer smoothing against the whole collection, i.e., for a collection D, element e and query q:

$$P(q|e) = \prod_{t \in q} \left((1 - \lambda) \cdot P(t|D) + \lambda \cdot P(t|e) \right), \tag{2}$$

where $P(t|e) = \frac{\text{freq}(t,e)}{|e|}$ and $P(t|D) = \frac{\text{freq}(t,D)}{\sum_{e' \in D} |e|}$.

Finally, we assign a prior probability to an element e relative to its length in the following manner:

$$P(e) = \frac{|e|^\beta}{\sum_e |e|^\beta}, \tag{3}$$

where $|e|$ is the size of an element e. The β parameter introduces a length bias which is proportional to the element length with $\beta = 1$ (the default setting). For a more thorough description of our retrieval approach we refer to [15]. For comprehensive experiments on the earlier INEX data, see [12].

2.4   Link Evidence as Document Priors

One of our aims for the Ad Hoc Track this year was to investigate the effectiveness of using link evidence as an indicator of relevance. We have chosen to use the link evidence priors to rerank the retrieved elements, instead of incorporating it directly into the retrieval model.

In the official runs, we have only looked at the number of incoming links (indegree) per article. Incoming links can only be considered at the article level, hence we apply all the priors at the article level, i.e., all the retrieved elements from the same article are multiplied with the same prior score. We experimented with *global* indegree, i.e., the number of incoming links from the entire collection, and *local* indegree, i.e., the number of incoming links from within the subset of articles retrieved for one topic. Although we tried global and local indegree scores separately as priors, we limit our discussion to a weighted combination of the two degrees, as this gave the best results when we tested on the 2006 topics. We compute the link degree prior $P_{\text{LocGlob}}(d)$ for an article d as:

$$P_{\text{LocGlob}}(d) \propto 1 + \frac{\text{Indegree}_{\text{Local}}(d)}{1 + \text{Indegree}_{\text{Global}}(d)} \tag{4}$$

Since the local indegree of an article is at most equal to the global indegree (when all the articles pointing to it are in the subset of retrieved articles), $P_{\text{LocGlob}}(d)$ is a number between 1 and 2. This is a much more conservative prior than using the indegree, local or global, directly. We will, for convenience, refer to the link evidence as prior, even though we do not actually transform it into a probability distribution. Note that we can turn any prior into a probability distribution by multiplying it with a constant factor $\frac{1}{\sum_{d \in D} prior(d)}$, leading to the same ranking.

3   Ad Hoc Retrieval Results

This year, there was no official Thorough task. The remaining tasks were the same as last year: Focused, Relevant in Context and Best in Context. To get CAS runs, we use a filter over the CO runs, using the pool of target elements of all topics. If a tag X is a target element for a given topic, we treat it as target element for all topics. We pool the target element tags of all topics, resulting in the following tags (by decreasing frequency): `<article>`, `<section>`, `<figure>`, `<p>`, `<image>`, `<title>`, and `<body>`. Then, we filter out all other elements from the results list of each topic. In other words, a retrieved element is only retained in the list if it is a target element for at least one of the topics.

For the Focused Task, no overlapping elements may be returned. For the Relevant in Context Task, all retrieved elements must be grouped per article, and for the Best in Context Task only one element or article offset may be returned indicating the best point to start reading. However, since both our indexes contain overlapping elements, the initials runs might contain overlapping results.

The link degrees in the official runs where erroneous, so we report on updated versions of the official runs, where only the degrees are different. We used the following Thorough runs as base runs for the various tasks:

- `element`: a standard *element* index run, with $\beta = 1$ and $\lambda = 0.15$.
- `contain`: a standard *contain* index run, with $\beta = 1$ and $\lambda = 0.15$.

where

- `+link` means the elements of the top 100 articles are reranked using the link prior.
- `+pool` means the run is filtered on the pool of target elements from the CAS queries.

3.1   Focused Task

To ensure the Focused run has no overlap, it is post-processed by a straightforward list-based removal strategy. We traverse the list top-down, and simply remove any element that is an ancestor or descendant of an element seen earlier in the list. For example, if the first result from an article is the article itself, we will not include any further element from this article.

Table 2 shows the results for the Focused Task. Looking at the two base runs first, we see that the *element* run scores better on very early precision, but loses out on the *contain* run at higher recall levels. With many smaller elements in the index it finds many relevant `<collectionlink>` elements which, due to their small size add little to recall, but are wholly relevant, thus leading to high precision. If a relevant `<collectionlink>` element is retrieved, any relevant ancestor nodes are not allowed in the results list, making it hard to improve recall with other element from that article. The *element+link* run scores best on very early

Table 2. Results for the Ad Hoc Track Focused Task

Run	iP[0.00]	iP[0.01]	iP[0.05]	iP[0.10]	MAiP
element	0.5672	0.4599	0.3137	0.2339	0.0707
element+link	**0.5999**	0.4745	0.3321	0.2753	0.0850
element+pool	0.5287	0.4705	0.3547	0.2729	0.0916
element+pool+link	0.5337	0.4779	0.3624	02938	0.1048
contain	0.5371	0.4728	0.3545	0.2952	0.0956
contain+link	0.5541	**0.4949**	0.3746	0.3156	0.1117
contain+pool	0.5289	0.4774	**0.3749**	0.2974	0.1011
contain+pool+link	0.5309	0.4821	0.3734	**0.3173**	**0.1157**

precision. The link prior clearly moves relevant elements to the top of the results list and shows a consistent improvement over the base run. For the *element* run, the pool filter has a huge impact, filtering out all the <collectionlink> and many other small elements, so that after the subsequent list based overlap removal, the relevant ancestors of these small elements are retained. The pool of target elements is very small. The only elements that are mentioned as target elements in this years CAS topics are <article>, <body>, <section>, <p>, <figure>, <image> and <title>. The *contain* index has only the larger elements and <name> elements, making the pool filter much less effective, although it still has a positive effect on overall precision. The combination of link evidence and structural hints improves matters further. Although not effective at the highest ranks (iP[0.00]), it consistently improves on all three runs; base, link and pool, further down the results list.

3.2   Relevant in Context Task

For the Relevant in Context task, we use the Focused runs and cluster all elements belonging to the same article together, and order the article clusters by the highest scoring element. Table 3 shows the results for the Relevant in Context Task. Comparing the two base runs, the elements in the *contain* run match the relevant text within articles much better than those in the *element* run. Given the better early precision of the *element* run, the larger elements in the *contain* run have more relevant text. The link prior here improves the article

Table 3. Results for the Ad Hoc Track Relevant in Context Task

Run	gP[5]	gP[10]	gP[25]	gP[50]	MAgP
element	0.1805	0.1566	0.1232	0.0891	0.0770
element+link	0.1838	0.1584	0.1216	0.0860	0.0814
element+pool	0.2373	0.2037	0.1523	0.1197	0.1117
element+pool+link	0.2336	0.2048	0.1529	0.1221	0.1125
contain	0.2156	0.1882	0.1484	0.1181	0.1066
contain+link	0.2315	0.1966	0.1504	0.1174	0.1085
contain+pool	**0.2497**	0.2069	0.1576	0.1239	0.1177
contain+pool+link	0.2456	**0.2144**	**0.1584**	**0.1271**	**0.1191**

Table 4. Results for the Ad Hoc Track Best in Context Task

Run	gP[5]	gP[10]	gP[25]	gP[50]	MAgP
element	0.2089	0.2048	0.1673	0.1291	0.1194
element+link	0.2334	**0.2283**	**0.1804**	0.1348	**0.1316**
element+pool	0.2373	0.2193	0.1684	0.1323	0.1232
element+pool+link	**0.2423**	0.2218	0.1712	0.1364	0.1238
contain	0.2075	0.2060	0.1700	0.1356	0.1243
contain+link	0.2319	0.2212	0.1710	0.1356	0.1273
contain+pool	0.2304	0.2140	0.1693	0.1360	0.1283
contain+pool+link	0.2343	0.2246	0.1729	**0.1388**	0.1297

ranking of the early ranks, but after 25 articles (50 in the *contain* run) the base run is better. Recalling that the link prior showed consistent improvement in precision, it seems that it pushes the articles with less relevant text up in the ranking. The pool filtered runs show consistent improvement over the base runs, especially for the *element* runs and for the first 5 retrieved articles. By Filtering out the smaller elements, the *element* run retains much more relevant text after overlap removal. Reranking the pool filtered run using the link prior further boosts scores. In contrast to the effect of the link prior on the base runs, for the pool filtered CAS runs, after rank 5, the article ranking of both runs improves. This could be explained by the pool filtered runs having a better article ranking than the base runs, and thus more relevant articles in the local link graph.

3.3 Best in Context Task

The aim of the Best in Context task is to return a single result per article, which gives best access to the relevant elements. Table 4 shows the results for the Best in Context Task. The two base runs show similar performance. The link prior has a huge impact on the article ranking—the link prior only affects the article ranking of runs—of both base runs up to rank 10. After that, it still has a positive effect on the *element* run, but almost no effect on the *contain* run. We see a similar effect with the pool filtered CAS runs. The biggest impact is on the first 10 results. Combining the pool filter and the link prior leads to a further improvement in early precision and seems to be more effective for the *contain* run than the *element* run. To summarise, link evidence and structural hints are both effective for improving early precision for both base runs and are complementary to some extent.

4   Link Detection Experiments

In this section, we discuss our participation in the Link The Wiki (LTW) track. LTW is aimed at detecting missing links between a set of topics, and the remainder of the collection, specifically detecting links between an origin node and a destination node. Existing links in origin nodes were removed from the 90 topics, making these articles 'orphans.' The task was to detect these links again and

Table 5. Statistics of Types of Links in the 90 un-orphaned LTW articles

Type	Uniq	Total	All		Link in Article	
			1×	Max	1×	Max
`<collectionlink>`	5,786	8,868	4,275	51	5,781	15
`<unknownlink>`	1,308	1,458	1,201	14	1,271	7
`<outsidelink>`	807	851	772	5	778	5
`<imagelink>`	197	212	195	15	197	15
`<languagelink>`	79	1,147	12	66	1,147	1
`<wikipedialink>`	59	60	58	2	58	1
`<weblink>`	27	28	26	2	26	2
Total	8,263	12,624	6,513	-	9,232	-

find the correct destination node ('fosters'), thus detecting links both on element and article level.

There are several types of links in the topics. These links have been implemented in the Wikipedia collection using XLink. An overview of the occurrence of these types of links in the un-orphaned (original) topics is presented in Table 5. For example, if we regard all the links as one distribution, then the `<languagelink>` has 79 different types (appearing once), but the same types are used 1147 times, of which the single link `<languagelink lang="de">` is used as often as 66 times, which means the same language links are reused in the articles. When we look at each file separately, then a language link appears only once in a file.

For the LTW task, three type of links are used for detection: `<collectionlink>`, `<wikipedialink>`, and `<unknownlink>`. The `<collectionlink>` comprises of the bulk of the links in the orphaned articles (70.0%). When looking at all orphaned articles, there are 5,786 unique type of collection links, out of the total of 8,868. The number of collection links that only occurs once is 4,275, which is 73.9% of the different types of collection links, and 48.2% out of all collection links. The collection link to article 35524.xml is occurring most often: 51 times, but it surprisingly does not to exist in the 2007 collection that we used. When we look at the links in the files separately, then 5,781 of the 8,868 collection links appear only once (65.2%), an outlier is the collection link 10829.xml (*"Florida"*), which is occurring 15 times in the topic 150340.xml (*"Miss Universe"*). On average, there are 98.5 *outgoing* collection links per topic, of which 64.3 per topic are unique, thus occurring once.

We also found that there is a significant strong positive relationship between the length of a Wikipedia article (excluding structure) and the number of links appearing in that article (Spearman's rho = 0.85, $p < 0.01$), i.e. longer articles have more links than shorter articles. Moreover, the average length of an anchor text is 12.3 characters, only 62 (0.7%) collection links are 3 characters or shorter.

4.1   Approach

Information Retrieval methods have been employed to automatically construct hypertext on the Web [1, 2], as well for specifically discovering missing links in Wikipedia [4]. To detect whether two nodes are implicitly connected, it is necessary to search the Wikipedia pages for some text segments that both nodes share. Usually it is only one specific and extract string [1]. Our approach is mostly based on this assumption, where we defined one text segment as a single line, and a string that both nodes share is a relevant substring. A substring of a string $T = t_1 \ldots t_n$ is a string $\hat{T} = t_{i+1} \ldots t_{m+i}$, where $0 \leq i$ and $m + i \leq n$. Only relevant substrings of at least 3 characters length are considered in our approach, because anchor texts of 3 characters or less do not occur frequently, and to prevent detecting too many false positives.

We also assume that pages that link to each other are somehow related in text content. We adopt a *breadth m–depth n* technique for automatic text structuring for identifying candidate anchors and text node, i.e. a fixed number of documents accepted in response to a query and fixed number of iterative searches. So the similarity on the document level and text segment level is used as evidence. The latter is used as a precision filter. So our approach consisted of two steps:

1. First, we detect links on the article level. We focus on the global similarity by collecting a set of similar or related pages using the set of topics. We search in the collection by retrieving the top N similar documents by using the whole document (including stopwords, stemmed with Porter stemmer, no XML structure) as a query against the index of the Wikipedia collection. We use the Vector Space Model (VSM) to retrieve related documents (articles). Our vector space model is the default similarity measure in Lucene [10], i.e., for a collection D, document d and query q:

$$sim(q, d) = \sum_{t \in q} \frac{tf_{t,q} \cdot idf_t}{norm_q} \cdot \frac{tf_{t,d} \cdot idf_t}{norm_d} \cdot coord_{q,d} \cdot weight_t, \qquad (5)$$

where $tf_{t,X} = \sqrt{\text{freq}(t, X)}$; $idf_t = 1 + \log \frac{|D|}{\text{freq}(t,D)}$; $norm_q = \sqrt{\sum_{t \in q} tf_{t,q} \cdot idf_t^2}$; $norm_d = \sqrt{|d|}$; and $coord_{q,d} = \frac{|q \cap d|}{|q|}$.

2. Second, we detect links on the element level. We search on the local level with text segments. Normalized lines (lower case, removal of punctuation and trailing spaces) are matched with string processing. At the same time we parse the XML and keep track of the absolute path for each text node and calculate the starting and end position (offset) of the identified anchor text by looking up the index of the string. For all our official runs, we blindly select the first instance of a matching line, and continue with the next line so an anchor text only has one link.

INEX LTW Task focuses on structural links, which have an anchor and refers to the Best Entry Point of another page (on the element level). Our Best Entry Point for both incoming and outgoing links was the start of an article, or */article[1]/name[1]* element, because in the current Wikipedia, links often point directly to entire articles or sections of these articles as logical units.

We do not assume that links are reciprocal, so we have different approaches for detecting outgoing and incoming links, though we set a threshold of 250 for both type of links and do not allow duplicated links as requested in the LTW task specification.

Detecting Outgoing Links. This is a link from an anchor text in the topic file to the Best Entry Point of existing related articles, which in our case was always the text-node of the */article[1]/name[1]* element. There is an outgoing link for topic t, when $S_{1...n} = T_{q...r}$, where S is the title of a foster article, and T is a line in a orphan article.

Detecting Incoming Links. This type of link consists of a specified path expression (anchor) from text nodes in the target articles to the */article[1]/name[1]* node of one of the 90 topics. There is an incoming link for topic t, when $T_{1...n} = S_{q...r}$, where $T_{1...n}$ is the title of t, and S is a line in a foster article.

Links also appear locally within an article to improve navigation on that page, but this was outside the scope of the LTW track. We extract for each topic the title enclosed with the `<name>` tag with a regular expression and store that in a hash-table for substring matching. We do not apply case-folding, but we do remove any existing disambiguation information put between brackets behind the title, e.g. *"What's Love Got to Do with It (film)"* becomes the substring *"What's Love Got to Do with It."*

4.2   Link the Wiki Track Findings

For the evaluation, only article-to-article links are considered in the scores. The threshold for the number of incoming and outgoing links was each set to 250 for each topic. Table 6 shows the mean number and range of incoming and outgoing links. For all runs there were more incoming links than outgoing links. Compared to the frequencies of the original articles as depicted in Table 5, we seem to have under-generated the number of outgoing links. This is a limitation of the Vector Space Model, as links in Wikipedia do not always relate to textually related or similar documents. It also shows that as we increase the pool of candidate target pages retrieved with the VSM (top 100, 200, 250, 300, 400), the number of detected links is also increased for incoming and outgoing links. However, this does not mean necessarily that retrieval performance is also improved as Table 7 shows. We achieved best performance by setting the threshold of the result list to the top 300 ($MAP_{in} = 0.3713$). The run name400 stands for the top 400 of the hit list retrieved with the VSM, and with name-matching post-processing. It shows that while the recall improves, which has slight positive effect on the performance for the outgoing links, the precision drops and thus the fallout also increases for the incoming links.

In summary, we experimented with the Vector Space Model and substring match for detecting missing links in Wikipedia. We used entire orphaned articles as query. We showed that exact substring matching improves the performance as compared to generating plain article-to-article links. This approach worked well, especially for the early precision. Our assumption that pages that link to each

Table 6. Results Link The Wiki: Number of Outgoing and Incoming Links

Run	Outgoing					Incoming				
	Mean	(SD)	Min	Median	Max	**Mean**	(SD)	Min	Median	Max
Name100	12.80	(7.58)	2	11	35	38.40	(34.67)	0	24.5	100
Name200	16.63	(10.39)	2	14	55	62.09	(65.49)	0	33.5	200
Name250	17.83	(11.40)	2	14.5	65	72.10	(79.84)	0	35.5	250
Name300	19.08	(12.52)	2	16	74	77.77	(85.97)	0	38.5	250
Name400	22.72	(14.78)	3	19	83	82.34	(90.26)	0	37.5	250

Table 7. Results for the Link The Wiki Track

Run	Outgoing			Incoming		
	MAP	**R-Prec**	**P@5**	**MAP**	**R-Prec**	**P@5**
Article100	0.1518	0.2277	0.5711	0.2646	0.3062	0.7311
Name100	0.1533 +1.0%	0.1781	0.7489	0.2906 +9.8%	0.3134	0.8000
Article200	0.1629	0.2389	0.5711	0.3075	0.3529	0.7311
Name200	0.1739 +6.8%	0.2073	0.7356	0.3471 +12.9%	0.3835	0.8044
Article250	0.1658	0.2406	0.5711	0.3193	0.3628	0.7311
Name250	0.1783 +4.9%	0.2147	0.7267	0.3618 +13.3%	0.3998	0.8044
Article300	0.1678	0.2407	0.5711	0.3274	0.3691	0.7311
Name300	0.1825 +8.8%	0.2233	0.7178	**0.3713** +13.4%	0.4101	0.8044
Name400	**0.1836**	0.2405	0.6844	0.3117	0.3757	0.6067

other are related or similar in content may not necessarily hold, thus reducing the pool of relevant pages that can be linked. Our experiments focused on exact string matching, but we have not explored yet techniques with best matching of substrings, e.g. semantic clustering of words, which could further improve the performance.

5   Interactive Experiments

In this section, we discuss the interactive experiments of the INEX 2006 Interactive Track (which has run well into INEX 2007). For details about the track and set-up we refer to [11]. For the interactive track, we conducted an experiment where we took part in the concerted effort of Task A, in which we compare element and passage retrieval systems. We reported the result of the track based on the users responses on their searching experience for each task and comparative evaluation on the element and passage retrieval systems. The element and passage retrieval systems evaluated are developed in a java-based retrieval system built within the Daffodil framework by the track organizers.

We participated in task A with nine test persons in which seven of them completed the experiment. Two persons failed to continue the experiment due to systems down time. Each test person worked with four simulated tasks in the Wikipedia collection. Two tasks were based on the element retrieval and the other two tasks were based on the passage retrieval. The track organizer provided a multi-faceted set of 12 tasks in which the test person can choose from. The 12 tasks consist of three task types (decision making, fact finding and information gathering) which further slit into two structural kinds (hierarchical and parallel). The experiment was conducted in accordance with the track guideline.

5.1   Post Task Questionnaire

For each task, each test person filled in questionnaires before and after each tasks, and before and after the experiment, resulting in 70 completed questionnaires. The questionnaire focuses on the users' searching experience with the systems and the usefulness of system features. Table 8 shows the post task questionnaire.

Table 9 shows the responses for the post-task questionnaire. First, we look at the result for all tasks. We found that the test persons were positive regarding both systems. Next, we look at responses for the element and passage system, without considering the task types and structures. We found that the element system is rated higher in terms of the amount of time used (Q2), certainty of completing the task (Q3), easiness of task (Q4), and satisfaction (Q5). As for the experience rate (Q1) and the usefulness of presentation (Q6), the passage retrieval system is rated higher.

Furthermore, we asked the test persons to rate the usefulness of system features. The answer categories used a 5-point scale with 1=not at all useful, 3=somewhat, and 5=extremely useful. When asked what helped the test persons in their searching tasks, five persons mentioned the table of content. The usefulness of table of content was rated at 3.90. The reasons were the table of content was detailed and it gave a good overview of the document. As one person noted, "the table of content was useful to get the overview of the document

Table 8. Post-task questionnaire

Q1 How would you rate this experience?
(1=frustrating, 3=neutral, 5=pleasing)
Q2 How would you rate the amount of time available to do this task?
(1=much more needed, 3=just right, 5=a lot more than necessary)
Q3 How certain are you that you completed the task correctly?
(For Q3 until Q6, 1=not at all, 3=somewhat, 5=extremely)
Q4 How easy was it to do the task?
Q5 How satisfied are you with the information you found?
Q6 To what extent did you find the presentation format (interface) useful?

Table 9. Post-task responses on searching experience: mean scores and standard deviations (in brackets)

	Q1	Q2	Q3	Q4	Q5	Q6
All tasks	3.11 (1.45)	3.63 (1.28)	3.30 (1.32)	3.30 (0.99)	3.33 (1.21)	3.48 (0.70)
Element	2.93 (1.44)	3.64 (1.22)	3.43 (1.22)	3.36 (1.01)	3.36 (1.22)	3.43 (0.76)
Passage	3.31 (1.49)	3.62 (1.39)	3.15 (1.46)	3.23 (1.01)	3.31 (1.25)	3.54 (0.66)

and it helped me to get back to where I want (by clicking on it)." However, the table of content was not so appreciated in short documents. For example, one person noted, "the table of content is useful, but to go through document I was only scrolling, because the text was not too long." Another reason why the table content was not useful in short documents is because in passage retrieval often the table of content only consisted of one item, thus the table seemed to be useless.

Four test persons mentioned result list helped them in their searching tasks and the test persons rated the usefulness of result presentation at an average of 3.90. The result list provided the test persons with detailed information of relevant paragraphs. However, we found out that result list was not sufficient enough because in some cases it returned too many irrelevant document. As one person who noticed the problem noted, "the ranking was poor, but I could immediately reject non-relevant result based on the shown information."

Paragraph highlighting was rated at an average of 3.04. Two test persons mentioned that paragraph highlighting was useful while two other test persons mentioned it as not useful. The reason of the mix-answers was because sometimes the relevant information was not highlighted by the system. As one person noted, "I did not find the paragraph highlighting useful since I found the relevant information at the non-highlighted passages."

Links in the document were appreciated by the test persons. It is observed that sometimes the system did not return relevant documents, thus the test persons just clicked the available links and found the relevant information through the links in the document. As mentioned by one person, "the most relevant pages (the general description of castle and fortress page) were not on the result list, but I just found them by clicking the links in the document."

Related terms function was rated the least with an average of 0.8. Six test persons commented that they did not use the related terms at all. We found out that the related terms provided by the system were not useful for the tasks because they were too long and not relevant.

5.2   Post Experiment Questionnaire

After each completed task, the test persons filled in a post-experiment questionnaire on ease of use and ease of learning of element and passage systems. The answer categories used a 5-point scale with 1=not at all, 3=somewhat, and 5=extremely. With respects to ease of use, element retrieval is rated higher ($\underline{M}$= 4.29, $\underline{SD}$=0.488) then passage retrieval ($\underline{M}$=3.86, $\underline{SD}$=0.90). Also with respect to ease of learning, element retrieval is rated higher ($\underline{M}$=4.14, $\underline{SD}$= 0.378) then passage retrieval ($\underline{M}$= 3.86, $\underline{SD}$=0.69).

We can see that there is a tendency to favor element retrieval system. This also shown by the answers of the post experiment questionnaire where the test persons were more positive for the element retrieval system. In comparison between the two systems, the element retrieval system seemed to give a more complete table of content compare to the passage retrieval system, resulting a better overview to see the relations between sections. Furthermore, the result list in the passage system seemed to give a poorer result in the result list and in some cases it missed the relevant documents.

5.3   Interactive Track Findings

From the result of the experiment, we focus on the comparison of element and passage retrieval systems and the usefulness of system features. From the quantitative result, we discovered that test persons appreciated both systems positively and found only small difference between element and passage retrieval systems. Passage retrieval seemed favorable in post-task questionnaire but element retrieval was rated higher in the comparative questions. However, it is too early to conclude that element retrieval is better then passage retrieval on this experiment. Because our finding is based on a small user test that only involved seven test persons. Furthermore, the system performance was slow and we think that this might influence our result. Over the whole experiment, perhaps the most striking result is that in the beginning of the post-experiment questionnaire, two test persons did not notice the differences between element and passage systems at all. They started to notice the differences after they were presented screenshots of both systems. In addition, table of content and result list were found to be the most useful features of the system. The test persons argued that the content of table gave them a good overview for long documents and the result list provided them with detailed information about the document. The least appreciated feature of the system was the related terms feature. From the comments we found out that the related terms did not help the test persons because the terms were too long and not relevant.

6   Discussion and Conclusions

In this paper, we documented our efforts at INEX 2007 where we participated in the Ad hoc Track, the Link the Wiki Track, and the Interactive Track that continued from INEX 2006.

For the Ad Hoc Track, we investigated the effectiveness of incorporating link evidence into the model, and of a CAS filtering method exploiting the structural hints in the INEX topics. We found that link priors improve our base runs, especially in early precision, for all tasks. The CAS pool filtering method is effective for all three tasks as well, showing more consistent improvement throughout the results lists. Combining the two methods improves performance further.

For the Link the Wiki Track, we investigated the relative effectiveness of link detection based on the VSM by using an entire article as a query. First, we established article-to-article links. We continued by detecting links on the element level by filtering with the names of Wikipedia articles. We show that name-filtering the results obtained with the VSM improves the precision. We achieved best performance by setting the threshold to 300.

For the Interactive Track, we took part in the interactive experiment comparing an element retrieval system with a passage retrieval system. Our test-persons showed a weak preference for the element retrieval system over the passage retrieval system. Of course, due to its small scale the study warrant general conclusions on the usefulness of passage-based approaches in XML retrieval.

Acknowledgments. Jaap Kamps was supported by the Netherlands Organization for Scientific Research (NWO, grants # 612.066.302, 612.066.513, 639.072.601, and 640.001.501), and by the E.U.'s 6th FP for RTD (project MultiMATCH contract IST-033104). Marijn Koolen was supported by NWO under grant # 640.001.501. Khairun Nisa Fachry and Junte Zhang were supported by NWO under grant # 639.072.601.

References

[1] Agosti, M., Crestani, F., Melucci, M.: On the use of information retrieval techniques for the automatic construction of hypertext. Information Processing and Management 33, 133–144 (1997)

[2] Allan, J.: Building hypertext using information retrieval. Information Processing and Management 33, 145–159 (1997)

[3] Denoyer, L., Gallinari, P.: The Wikipedia XML Corpus. SIGIR Forum. 40, 64–69 (2006)

[4] Fissaha Adafre, S., de Rijke, M.: Discovering missing links in wikipedia. In: LinkKDD 2005: Proceedings of the 3rd international workshop on Link discovery, pp. 90–97. ACM Press, New York (2005)

[5] Hawking, D., Craswell, N.: Very large scale retrieval and web search. In: TREC: Experiment and Evaluation in Information Retrieval, ch. 9, pp. 199–231. MIT Press, Cambridge (2005)

[6] Hiemstra, D.: Using Language Models for Information Retrieval. PhD thesis, Center for Telematics and Information Technology, University of Twente (2001)

[7] ILPS. The ILPS extension of the Lucene search engine (2007),
    `http://ilps.science.uva.nl/Resources/`

[8] Kamps, J., Koolen, M., Sigurbjörnsson, B.: Filtering and clustering XML retrieval results. In: Fuhr, N., Lalmas, M., Trotman, A. (eds.) INEX 2006. LNCS, vol. 4518, pp. 121–136. Springer, Heidelberg (2007)

[9] Kraaij, W., Westerveld, T., Hiemstra, D.: The importance of prior probabilities for entry page search. In: Proceedings of the 25th Annual International ACM SIGIR Conference on Research and Development in Information Retrieval, pp. 27–34. ACM Press, New York (2002)

[10] Lucene. The Lucene search engine (2007), `http://lucene.apache.org/`

[11] Malik, S., Tombros, A., Larsen, B.: The interactive track at INEX 2006. In: Fuhr, N., Lalmas, M., Trotman, A. (eds.) INEX 2006. LNCS, vol. 4518, pp. 387–399. Springer, Heidelberg (2007)

[12] Sigurbjörnsson, B.: Focused Information Access using XML Element Retrieval. SIKS dissertation series 2006-28, University of Amsterdam (2006)

[13] Sigurbjörnsson, B., Kamps, J.: The effect of structured queries and selective indexing on XML retrieval. In: Fuhr, N., Lalmas, M., Malik, S., Kazai, G. (eds.) INEX 2005. LNCS, vol. 3977, pp. 104–118. Springer, Heidelberg (2006)

[14] Sigurbjörnsson, B., Kamps, J., de Rijke, M.: An Element-Based Approach to XML Retrieval. In: INEX 2003 Workshop Proceedings, pp. 19–26 (2004)

[15] Sigurbjörnsson, B., Kamps, J., de Rijke, M.: Mixture models, overlap, and structural hints in XML element retreival. In: Fuhr, N., Lalmas, M., Malik, S., Szlávik, Z. (eds.) INEX 2004. LNCS, vol. 3493, pp. 196–210. Springer, Heidelberg (2005)

[16] Trotman, A., Geva, S.: Passage retrieval and other XML-retrieval tasks. In: Proceedings of the SIGIR 2006 Workshop on XML Element Retrieval Methodology, pp. 43–50. University of Otago, Dunedin New Zealand (2006)

[17] Wikipedia. The free encyclopedia (2006), http://en.wikipedia.org/

GPX: Ad-Hoc Queries and Automated Link Discovery in the Wikipedia

Shlomo Geva

Faculty of IT, Queensland University of Technology
Brisbane, Australia
s.geva@qut.edu.au

Abstract. The INEX 2007 evaluation was based on the Wikipedia collection. In this paper we describe some modifications to the GPX search engine and the approach taken in the Ad-hoc and the Link-the-Wiki tracks. In earlier version of GPX scores were recursively propagated from text containing nodes, through ancestors, all the way to the document root of the XML tree. In this paper we describe a simplification whereby the score of each node is computed directly, doing away with the score propagation mechanism. Results indicate slightly improved performance. The GPX search engine was used in the Link-the-Wiki track to identify prospective incoming links to new Wikipedia pages. We also describe a simple and efficient approach to the identification of prospective outgoing links in new Wikipedia pages. We present and discuss evaluation results.

Keywords: GPX, INEX, XML, Information Retrieval, Link Discovery.

1 Introduction

In this paper we describe the submission of QUT and the GPX search engine in the Ad Hoc and Link the Wiki tracks of INEX 2007. Both the Ad Hoc track and the Link the Wiki track are described in some detail elsewhere in these proceedings and so we restrict our brief introduction to some background that is related to our specific approach. Having reported on GPX over several years now, we refer the reader to previous proceedings of INEX [2,3] for a comprehensive general overview of the Ad Hoc track and focus our attention more on the Link the Wiki task.

The Wikipedia is a free online document repository written collaboratively by wiki contributors around the world. The INEX collection is composed of about 660,000 articles and it offers many attractive features as a corpus for information retrieval tasks. The INEX Wikipedia collection has been converted from its original wiki-markup text into XML [1]. That collection is composed of a set of XML files where each file corresponds to an online article in Wikipedia. Search as well as retrieval could benefit from rich semantic information in the XML Wikipedia collection, where it exists. However, the XML semantics is not rich and relates mostly to structure. This is arguably a deficiency that hinders taking advantage of the XML technology which offers semantic annotation capacity.

N. Fuhr et al. (Eds.): INEX 2007, LNCS 4862, pp. 404–416, 2008.

The semi-structured format provided by the XML-based collection offers a useful property for the evaluation of various semi-structured retrieval techniques. In 2007 we have made a modification to GPX [3] ranking algorithm and this is discussed in the following sections. We omit further literature review of link discovery since this can be found in the Link-the-Wiki track overview paper, elsewhere in theses proceedings, but references [5-12] herein provide good coverage.

There are essentially two immediate approaches that come to mind when setting out to link the Wikipedia. One is to perform Link Analysis on the existing connectivity graph in the wikipedia, and the other is to look for semantic connections between documents. Link analysis relies on the already existing semantic links (previously generated by users) while as context analysis does not assume semantic links, but attempts to discover such from scratch. As it turns out, these approaches are very effective and both were in fact represented in the Link the Wiki track in 2007.

The remainder of this paper is organized as follows. In sections 2, 3 and 4 we report and analyse the performance of GPX in the various tasks of the Ad Hoc track. The Link the Wiki task is discussed in sections 5 and 6. We conclude in section 7.

2 The GPX Search Engine

In this section we provide a very brief description of GPX. The reader is referred to earlier papers on GPX in INEX previous proceedings [3] for a more complete description.

2.1 GPX Inverted List Representation

The GPX search engine is based on XPath inverted lists. For each term in the collection we maintain an inverted list of XPath specifications. This includes the file name, the absolute XPath identifying a specific XML element, and the term position within the element. The actual data structure is designed for efficient storage and retrieval of the inverted lists which are considerably less concise by comparison with basic text retrieval inverted lists.

The GPX search engine is using a relational database implementation (Apache Derby) to implement an inverted list data structure. It is a compromise solution which provides the convenience of a DBMS at the cost of somewhat reduced performance compared to what might otherwise be possible with an optimized file structure.

Consider the XPath:

/article[1]/bdy[1]/sec[5]/p[3]

This could be represented by two expressions, a Tag-set and an Index-set:

Tag-set: **article/bdy/sec/p**

Index-Set: 1/1/5/3

The original XPath can be reconstructed from the tag-set and the index-set. There are over 48,000 unique tag-sets, and about 500,000 unique index-sets in the collection. We assign to each tag set and each index-set a hash code and create auxiliary database tables mapping the hash-codes to the corresponding tag-set and index-set entries. These hash tables are small enough to be held in memory and so decoding is efficient.

The GPX database tables are then:

Term-Context = { **Term-ID**, File-ID, XPath-Tag-ID, XPath-IDX-ID, Position }
       Terms =   { **Term**, Term-ID }
       Files =   { **File-Name**, File-ID }
       TagSet =   { **XPath-Tag-ID**, Tag-Set }
       IndexSet =   { **XPath-IDX-ID**, Index-Set }
       XPathSize = { **XPath-ID**, Node-Size }

Given a search term the database can be efficiently accessed to obtain an inverted list containing the context of all instances where the term is used (identified by File Name, full XPath, and term position). Having retrieved a set of inverted lists, one for each term in the query, the lists are merged so as to keep count of query terms in each node and also keeping the term positions. Stop words are actually indexed, but too frequent terms are ignored by applying a run-time stop-word frequency threshold of 300,000. We also used plural/singular expansion of query terms. We have found that - on average - the use of a Porter stemmer is not adding to system performance and so it was not used.

Having collected all the nodes that contain at least one query term, the system proceeds to compute node scores. Calculation of node relevance score from its content is based on a variation of TF-IDF. We used the inverse collection frequency of terms rather than the inverse document frequency (TF-ICF). The score is then moderated by a step function of the number of unique terms contained within the node. The more unique terms the higher the score. The score is further moderated by the proximity within which the terms are found. Additionally, the scores of all article nodes that contained query terms in the *name* node were further increased. All this can be calculated with the information in the inverted lists.

2.2 Calculation of Text Nodes Score

GPX 2007 deviates significantly from earlier versions with respect to the way that ancestor node scores are calculated. For clarity we shall refer to GPX-2007 to denote the current system and GPX to denote the older system. In the earlier version GPX computed node scores on the basis of direct text content (having a text node in the DOM model) and then the scores were propagated upwards in the XML tree. GPX accumulated all children node scores for a parent and reduced the score by a decay factor (typically about 0.7) to account for reduced specificity as one moved upwards in the XML tree. In GPX 2007 the scores are computed directly from the node text content, direct, or indirect. That means that any node is scored by the text it contains regardless of whether it has a direct text node in the DOM representation – all the text in the node and its descendents is used.

Naturally, nodes closer to the root could receive a higher score on account of more query terms in descendent nodes. A common variation to TF-IDF is to normalise the score by taking into account the document size. The motivation is to account for the increased probability of finding query terms in larger documents and hence biasing the selection towards larger documents. The motivation here is similar with a slight twist. Node normalisation in the XML score calculation is motivated by the need to compensate for the reduced specificity of larger nodes. We are aiming for focused retrieval and look for nodes of "just the right size" (whatever that may be.) Node normalisation introduces a penalty in a parent node that contains large amounts of irrelevant text in descendent nodes and which do not contribute towards an increased score. However, when two nodes have a similar size but contain different amount of relevant text then the more relevant node will score higher.

But there is another twist here. We also know that nodes that are too small are unlikely to satisfy a user information need (except perhaps in factoid type QA). At least with the Wikipedia we know that the most common element selected by assessors is a paragraph (or passage). Very small passages are not common in the qrels of past experiments. Therefore, we do not want to normalise the scores of too small nodes thereby unduly increasing their score relative to otherwise similarly scoring nodes which are somewhat larger. Node scores are normalised by dividing the raw score by the node size (measured as the number terms), but all nodes with size of below 75 terms are normalised by 75.

Equation 1: Calculation of S, node size for normalisation

$$S = \begin{cases} NodeSize & if\ (NodeSize > 75) \\ 75 & if\ (NodeSize \leq 75) \end{cases} \tag{1}$$

The value of S, the node size for the purpose of normalization, is thus equal to 75 for nodes smaller than 75 terms, but taken as the actual node size for nodes with more terms.

This heuristic is convenient in the XML case because when breaking ties in node selection (focused retrieval) we prefer the ancestor to the descendant when the scores are equal. This means that we prefer parent nodes as long as the parent is larger than the descendant and below 75 terms in size. For example, this means that a very deep XML branch with no breadth will be collapsed to an ancestor of up to size 75 terms (if such exists). So in summary, node size normalisation is biasing the selection towards passages of 75 terms, both from above and from below. We experimented with other values for node size from 50 to 150 with little difference in results.

Since GPX 2007 computes node scores over much larger text segments it is necessary to take account of term proximity. The intuition is that we should award higher scores to nodes in which search terms are found in closer proximity to each other. In earlier versions of GPX this was not critical since node scores were computed at text nodes and these were typically paragraphs, titles, captions, and other such relatively small nodes. A proximity function was defined and incorporated into the score calculation.

Equation 2: Calculation of P, node terms proximity score

$$Pr = 10 \sum_{i=1}^{n} \exp\left(-\left(\frac{p_i - p_{i+1} + 1}{5}\right)^2\right) \tag{2}$$

Here terms are processed in the order in which they appear in the text node. P_i is the position of term i in the text node. Note that for immediately successive terms Pr=10. This is a Gaussian function with a maximum value of 10 and decaying exponentially with increased term distance between successive terms. The function is depicted in Figure 1, for two terms separation. Note that in practice, a table lookup is more efficient than the numerical calculation.

So finally we have the following score calculation:

Equation 3: Calculation of element relevance score from its content

$$L = \frac{\mathrm{Pr}}{S} K^{n-1} \sum_{i=1}^{n} \frac{t_i}{f_i} \tag{3}$$

Here n is the count of unique query terms contained within the element, and K is a small integer (we used $K=5$). The term K^{n-1} is a step function which scales up the score of elements having multiple distinct query terms. This heuristic of rewarding the appearance of multiple distinct terms can conversely be viewed as taking more strongly into account the absence of query terms in a document. Here it is done by rewarding elements that do contain more distinct query terms. The system is not sensitive to the value of K and a value of $k=5$ is adequate [3]. The summation is performed over all n terms that are found within the element where t_i is the frequency of the ith query term in the element and f_i is the frequency of the ith query term in the collection.

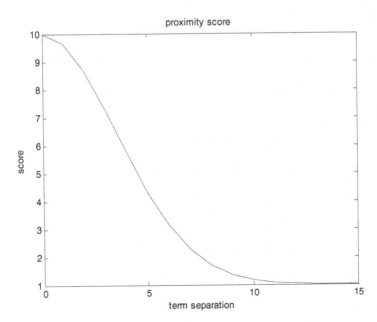

Fig. 1. Proximity score as a function of term separation

Finally, nodes that contain query terms that are preceded by a minus sign (undesirable) are eliminated.

At this point we have computed the score of all (overlapping) nodes in each article that contains query terms. The score of the <article> node itself is then added to all nodes in the article. This lifts the scores of all nodes that appear in a high scoring article. The intuition is that an article with many scoring nodes is more likely to be relevant and so all its scoring elements are ranked higher on account of more scoring nodes appearing in the same article. Without this modification, two similar nodes, one being an isolated instance of a relevant node in an article, and the other being one of many relevant nodes in an article, would receive a similar score.

As we will see below, results suggest an improved performance in GPX. The runs labeled RIC_04 and BIC_04 were produced with the 2006 GPX version (score propagation) while BIC_07 and RIC_07 were run with the GPX_07 version with direct score calculation. The GPX 07 version seems to perform better than the earlier GPX version. It does not require any magic numbers (decay constants), it treats each node as if it were a document, and it is therefore conceptually less arbitrary and more appealing.

2.3 GPX and Ad-Hoc Retrieval Tasks

The Ad-Hoc track at INEX 2007 consisted of 3 tasks – Focused, Relevant in Context, and Best in Context. These tasks are described elsewhere in this proceedings collection. We briefly describe the approach taken to each of the tasks in our best performing run.

2.3.1 Focused Retrieval
Focused Retrieval starts with the thorough results recall base. Within each article the highest scoring elements on a path are selected by keeping only elements that have a higher score than any of their descendents or ancestors. The submission consists of the remaining overlap free focused elements, sorted by descending score.

2.3.2 Relevant in Context Retrieval (RIC)
The objective of the task was to balance article retrieval and element retrieval. Whole articles are first ranked in descending order of relevance and within each article a set of non-overlapping most focused elements are grouped. We have used the focused results, which were overlap free already, but grouped the elements within articles and sorted the articles by article score.

2.3.3 Best in Context Retrieval (BIC)
We tested a straightforward approach here – we simply kept the highest scoring element in each document appearing in the focused recall base.

3 Ad Hoc Retrieval Results

The GPX system performed well and produced particularly good results in the Relevant in Context and Best in Context tasks of the Ad-hoc track.

Table 1. Comparative results for GPX vs. the best performing run at INEX 2007. RIC and BIC tasks are measured by MagP while Focused is measured by interpolated precision at 0.01 recall.

Task	Best run	Best GPX	Run Rank	System rank
RIC	0.1552	**0.1489**	6/66	2/17
BIC	0.1919	**0.1831**	2/71	2/19
Focused	0.5271	**0.4924**	15/79	8/25

Relatively good results were achieved in terms of precision at early recall levels on most of the tasks. We have submitted several variations of the GPX search engine in order to compare the performance. The GPX 2007 variation was compared with the GPX 2006 system, unchanged. The results were similar and are depicted in Table 2.

Table 2. Comparative results for GPX 2007 vs. GPX2006

Task	GPX 2006	GPX 2007	GPX 2007*
RIC	0.1298	**0.1489**	0.1369
BIC	0.1808	**0.1831**	0.1519
Focused	0.4924	**0.4828**	0.4235

Overall the performance of the GPX 2007 version (bold) is a little better than the 2006 version. The column titled GPX 2007* corresponds to GPX 2007 without the term proximity correction in equation 3. It is worth noting that the results with the correction for term proximity are considerably better than the results without this correction. The proximity correction is necessary because when computing node scores directly from content, in very large nodes (e.g. whole article) the terms may well be independent of each other. On the other hand, when dealing with short documents, the correction is not necessary. The GPX 2006 search engine computes node scores in relatively small passages – typically paragraphs. Term proximity is not particularly advantageous when documents (elements) are short. Our experiments with term proximity in previous years, with GPX 2006, provided insignificant variations in the scores when switched on or off.

4 Link the Wiki

The Link the Wiki task is described in detail elsewhere in this proceedings collection. The objective of this task was to identify a set of incoming links and a set of outgoing links for new Wikipedia pages. In practice, the topics were existing Wikipedia pages that were stripped of exiting links. The links were only at the article-to-article level.

4.1 Incoming Links

Incoming links were identified by using the GPX search engine to search for elements that were *about* the topic name element. For each topic the *name* element was used to construct a standard NEXI query: //article[about(.,*name*)]

We have used the SCAS task setting whereby the results were interpreted strictly. In this case it only means that articles nodes were returned. This was sufficient since only article-to-article links were needed. Results were ordered by article score with the more likely relevant articles returned earlier in the list. The process took an average of 8.5 seconds per topic on a standard mid-range PC. Considering that the task that is supported by the LTW process is the creation of new documents in the collection (this is the use case), 8 seconds for identifying recommended links is not significant. By comparison, the creation of a new Wikipedia topic may in some cases be measured in minutes, but it is more likely to take hours, even days. Hence the response time is quite adequate.

4.2 Outgoing Links

We have adopted a very simple approach to this task. All existing page names in the Wikipedia were loaded into an in-memory hash table. With 660,000 articles this is not an onerous task. The identification of potential links was based on a systematic search for anchor text that matches existing page names. In the first stage we have extracted the text of the topic (eliminating all markup information.) Prospective anchors for outgoing links were identified by running a window over the topic text and looking for matching page names in the collection. The window size varied from 12 words down to 1 word, and included stop words. Longer anchors were ranked higher than shorter ones, motivated by the trivial observation that the system was less likely to hit on a longer page name by accident. A naïve approach perhaps, but quite effective as it turns out. The process is purely computational and does not incur any I/O operations. The process took an average of 0.6 seconds per topic.

While it is straight forward to obtain candidate anchors by systematic comparison of substrings (of various lengths) against exiting page titles in the collection, numerous matches arise, and not all are useful. A pruning strategy is needed. We adopted the following process:

- Identify all candidate phrase-anchors of length 12 words down to 2, in that order.
- Append candidate year anchors
- Append all single term anchors

No ordering was performed other than the above. Phrases were ordered by length, followed by years, followed by single terms. Within these groups the ordering was in the sequence in which the anchors were encountered. The heuristic is simply that we are unlikely to encounter long phrases, which happen to be page names, by accident. On the other hand, single terms matches are more likely to be accidental in an encyclopedic collection and thus more risky in recommending at higher rank. Of course had we performed a deeper analysis of the anchor text context and the target document context we may have been able to resolve ambiguities, but that would have complicated the approach considerably.

5 Link the Wiki Results

The official results of the evaluation are depicted in figures 2 and 3. It should be noted that better results were subsequently obtained by the University of waterloo for

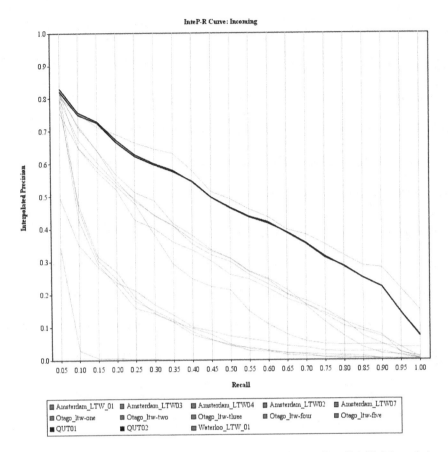

Fig. 2. Link-the-Wiki Incoming links, interpolated precision vs. Recall (official results)

Outgoing Links and we incorporate these unofficial improved results in Figure 4 for comparison.

It is evident from the official evaluation plots that the performance of our system was quite good – when viewed in terms of conventional IR. However, the Link the Wiki task is far more demanding than the conventional web search engine task. Here it is *not* sufficient to identify several good results. The user is interested in numerous anchors and links – almost exhaustive recall (within the limits of reason). All of the proposed links have to be checked by the user because it is highly undesirable to have inappropriate links. Although precision at early recall points may be at 0.8, it is even more desirable to achieve a high MAP value (i.e. high precision through a long tail of recall levels.)

Figure 4 presents the precision-recall curves for the two systems. "Anchors 90" is the Waterloo system and "Page Titles 90" is the QUT system. Both are shown for the 90 INEX topics. The link analysis approach (Waterloo) is better at almost all recall points, however, in a collection that is not already extensively linked it will not be

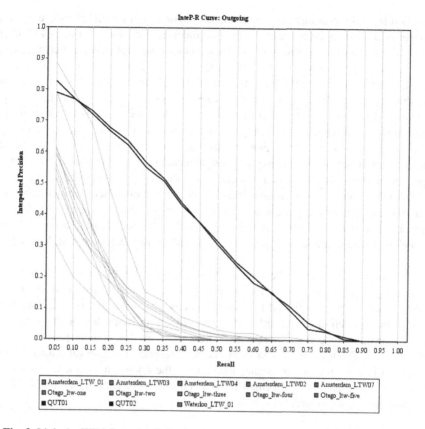

Fig. 3. Link-the-Wiki Outgoing links, interpolated precision vs. Recall (official results)

applicable. The QUT approach is independent of existing links. The Waterloo system performance in Figure 4 provides the current best performance and it is included here to provide a baseline for comparison.

To test the scalability of automated link discovery we additionally ran an extensive experiment on the collection. We randomly extracted 1% of the 660,000 documents and re-ran the experiment. The experiment was run on a PC with 2GB memory and 1.6GHz clock speed. It took 6 minutes to complete the process, processing in excess of 1,100 documents per minute. Figure 4 also presents the recall-precision curve for that run, labeled Page Titles 6600. It can be seen that performance over a very large number of topics selected at random is similar to the performance achieved over the hand picked INEX set, suggesting that 90 hand picked topics are sufficient to measure the performance of link discovery systems. Importantly in the context of the INEX evaluation, it is feasible to manually assess 90 topics whereas it would not be feasible to assess 6,600 (using the resources available to INEX). Manual assessment in future cycles of Link the Wiki would allow us to study more deeply the nature of link discovery, to identify those links returned by automatic systems that have not yet been identified by Wikipedia authors, and those automatic link links that already exist in the Wikipedia and which are not useful (e.g. *year* links are common, yet often of no use).

In order to assess the contribution of each component (phrase/year/term), we created separate submissions for each component. Figure 5 presents the recall-precision curves. Most surprisingly, the contribution of the year links is small; they are ubiquitous throughout the Wikipedia and were expected to contribute considerably to performance. However, the precision of year links at low recall levels is relatively low. Single terms contribute more than years not only at low recall levels, but over all. This is because there are many more terms that could be linked. However, it is difficult to avoid irrelevant links using only single term anchors. The phrase links achieve higher precision and recall than terms and years. Phrases (and long phrases in particular) that match page names are less likely to occur in a document without also being related to the page of the same name. Years and single terms frequently match a page name, but not in a meaningful context. The combination of phrases, years, and terms is very effective as can be seen from the combined curve labeled Phrases Years Terms in Figure 5.

It is possible to improve the ranking of single terms by estimating the likelihood that a term will be a page name. We estimate this likelihood as the ratio of the number of times that the page is linked to over the number of times that the page name appears in the collection (using either the term collection frequency or the term document frequency). Indeed the ranking of single terms in this manner provides further improvement. The top 2 curves in in Figure 5 correspond to these variations. The improvement is only marginally greater when using the document frequency in place of collection frequency. The performance of the GPX system with this correction is significantly better (see figure 4).

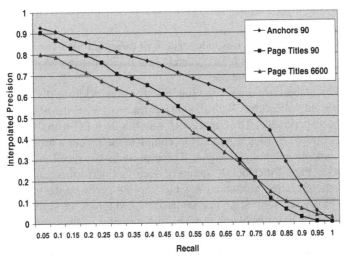

Fig. 4. Link-the-Wiki Outgoing links, interpolated precision vs. Recall

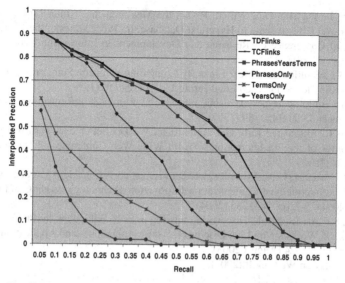

Fig. 5. Link-the-Wiki Outgoing links, interpolated precision vs. Recall

6   Conclusions

The GPX search engine was briefly described and modifications to the earlier version were described. Results indicate that performance is marginally better in terms of precision and recall, however, the approach is easier to implement. At any rate, both GPX 2006 and GPX 2007 produced highly competitive results in 2007.

Although relatively good results were produced for the Link the Wiki task, performance may be improved through analysis of the contexts surrounding the anchor texts and the corresponding target documents. This remains as a task for the next round of Link-the-Wiki evaluations in 2008.

References

1. Denoyer, L., Gallinari, P.: The Wikipedia XML Corpus. SIGIR Forum. 40(1), 64–69 (2006)
2. Comparative Evaluation of XML information Retrieval Systems 5th International Workshop of the Initiative for the Evaluation of XML Retrieval, INEX 2006, Dagstuhl Castle, Germany, December 17-20. LNCS. Springer, Heidelberg (2007) ISBN 978-3-540-73887-9
3. Geva, S.: GPX - Gardens Point XML IR at INEX 2006. In: Comparative Evaluation of XML information Retrieval Systems 5th International Workshop of the Initiative for the Evaluation of XML Retrieval, INEX 2006, Dagstuhl Castle, Germany, December 17-20. LNCS, pp. 137–150. Springer, Heidelberg (2007)
4. Robertson, S.: Understanding Inverse Document Frequency: On theoretical arguments for IDF. Journal of Documentation 60(5), 503–520 (2004)
5. Wilkinson, R., Smeaton, A.F.: Automatic Link Generation. ACM Computing Surveys 31(4) (December 1999)

6. Ellis, D., Furner-Hines, J., Willett, P.: On the Measurement of Inter-Linker Consistency and Retrieval Effectiveness in Hypertext Database. In: Proceedings of the 17th Annual International Conference on Research and Development in Information Retrieval, Dublin, Ireland, pp. 51–60 (1994)
7. Green, S.J.: Building Hypertext Links By Computing Semantic Similarity. IEEE Transactions on Knowledge and Data Engineering 11(5), 713–730 (1999)
8. Allan, J.: Building Hypertext using Information Retrieval. Information Processing and Management 33(2), 145–159 (1997)
9. Green, S.J.: Automated Link Generation: Can We Do Better than Term Repetition? In: Proceedings of the 7th International World Wide Web Conference, Brisbane, Australia, pp. 75–84 (1998)
10. Zeng, J., Bloniarz, O.A.: From Keywords to Links: an Automatic Approach. In: Proceedings of the International Conference on Information Technology: Coding and Computing (ITCC 2004), 5-7, pp. 283–286 (2004)
11. Adafre, S.F., de Rijke, M.: Discovering missing links in Wikipedia. In: Proceedings of the SIGIR 2005 Workshop on Link Discovery: Issues, Approaches and Applications, Chicago, IL, USA, pp. 21–24 (August 2005)
12. Jenkins, N.: Can We Link It (2007),
    http://en.wikipedia.org/wiki/User:Nickj/Can_We_Link_It

University of Waterloo at INEX2007: Adhoc and Link-the-Wiki Tracks

Kelly Y. Itakura and Charles L.A. Clarke

University of Waterloo, Waterloo, ON N2L3G1, Canada
{yitakura,claclark}@cs.uwaterloo.ca

Abstract. In this paper, we describe University of Waterloo's approaches to adhoc and Link-the-Wiki tracks. For the adhoc track, we submitted runs for the focused and the best-in-context tasks. We again show that Okapi BM25 works well for XML retrieval. We also analyze why our element-based best entry point result is better than our passage-based counterpart. Finally, we present our baseline algorithm for embedding incoming and outgoing links in Link-the-Wiki track.

1 Introduction

In 2007, University of Waterloo participated in adhoc and Link-the-Wiki tracks. For the adhoc track, we implemented passage retrieval and element retrieval to turn these results into submissions for the focused and the best-in-context tasks. For the focused task, we only submitted an element retrieval result that used the same algorithm as Waterloo's focused submission in INEX2004 [3]. In the best-in-context task, we submitted element results based on both element and passage retrieval. In Link-the-Wiki track, since it is the first year of its existence, we decided to submit runs using relatively simple techniques that might be suitable as a baseline for future work.

This paper is organized as follows. In Section 2, we describe our approaches to adhoc track, and in Section 3, we describe our approaches to Link-the-Wiki track. In Section 4, we describe related works on embedding links within Wikipedia. We conclude this paper with directions for future work in Section 5.

2 Ad Hoc Track

For the adhoc track, we used two retrieval schemes, element retrieval and passage retrieval to return XML elements for the focused task and best entry points for the best-in-context task.

Both element and passage retrieval work in essentially the same manner. We converted each topic into a disjunctive of query terms, removing negative query terms. We located positions of all query terms and XML tags using Wumpus [2]. We then used a version of Okapi BM25 [10] to score passages and elements. The score of an element/passage P is defined as follows.

$$s(P) \equiv \sum_{t \in Q} W_t \frac{f_{P,t}(k_1 + 1)}{f_{P,t} + k_1(1 - b + b\frac{pl_P}{avgdl})} \, , \tag{1}$$

N. Fuhr et al. (Eds.): INEX 2007, LNCS 4862, pp. 417–425, 2008.

where Q is a set of query terms, $f_{P,t}$ is the sum of term frequencies in a passage P, pl_P is a passage length of P, and $avgdl$ is an average document length in Wikipedia collection. We tuned parameters k and b using INEX2006 adhoc track focused and best-in-context tasks and the accompanying nxCG and BEPD metrics respectively. The actual parameters used for element retrieval for focused task is $k = 1.2$ and $b = 0.9$, for best-in-context task, $k = 0.8$ and $b = 0.7$. For passage retrieval in best-in-context task, we chose $k = 1.4$ and $b = 0.7$. This is interesting because when we worked on INEX2004/2005 IEEE collection [4] [6], we speculated that a large k is necessary for Okapi-based passage retrieval to work. However, it seems that this is not the case for Wikipedia corpus.

In element retrieval, we scored all of the following elements in corpus.

```
<p>, <section>, <normallist>, <article>, <body>, <td>, <numberlist>,
<tr>, <table>, <definitionlist>, <th> ,<blockquote>, <div>, <li>,
<u>.
```

For passage retrieval, we scored all possible passages. For both algorithms, we ignored elements/passages of size less than 25 word-long.

2.1 Focused Task

In the focused task, we returned the top 1500 elements obtained from element retrieval after removal of nestings. In INEX2007 official metric using interpolated precision at 0.01 recall, we ranked 3rd among different organizations.

2.2 Best-in-Context Task

For the first submission, we used element retrieval to obtain the top 1500 elements with distinct files. For the second submission, we used passage retrieval to choose the best scoring passage for each file. We then chose the top 1500 among these. We returned the XML tags listed above nearest to these 1500 passages that is closer to the beginning of the article.

The official INEX2007 results show that our element-based approach ranked 3rd among different organizations. However, we were surprised that that our passage-based approach did not work as well as our element-based approach. Our initial assumption was that since elements are passages, the highest scoring passage would give a better best entry point than the highest scoring element. After looking at the official assessments set, which we will treat as a gold standard for the purpose of our analysis, we speculate two causes for our under-achieving passage-based result. First, highest scoring passages do not tend to appear at the beginning of an article, whereas as shown in [7], the best entry points tend to appear at the beginning of an article. However, this does not explain why highest scoring elements give a better result. By examining the assessment set, we speculate that the performance of our passage-based approach is largely explained by the gap in relevant information content between the highest scoring passage and the best entry point derived from it. This is because XML elements returned as the best entry point from the top passages almost always have much lower score than the highest scoring element, which indicate that there is a lot of irrelevant material between the start of the highest scoring passage and the best entry point associated with it. Therefore, we think that the best entry point must be very close to the highest scoring passage. This leads to a preference towards either highest scoring passages or highest scoring elements over elements starting before the highest scoring passages. Raw passage results, however, do not seem to appear frequently in the relevance assessment. Moreover, we could see our passaged-based element best entry point to be *context BEPs* as in [7], and since there are many relevant passages in the gold standard, we think that the preference is more towards the highest-level element that contain all relevant passages, termed *container BEPs* [7]. The exact same phenomena also apply to results of our training set on INEX2006.

For future work, instead of returning the nearest significant XML elements that start before the highest scoring passages do, we plan to return the nearest significant XML elements that start *after* the highest scoring passages. We hope that in this way, there would be no irrelevant material between the proposed best entry point and the highest scoring passages. Additionally, we hope that by tuning Okapi parameters well the resulting best entry points would be closer to the beginning of articles. Another way to avoid an information gap is to set passages to start at element boundaries, which is a generalization of element-based best entry point that we performed well.

3   Link the Wiki Track

This year, we decided to submit a result set made from a simple algorithm to act as a baseline. Before working on incoming or outgoing links, we removed all topic files from corpus. When creating a list of anchor-destination pairs for each corpus file, we also ignored pairs that have a topic file as the destination.

3.1   Incoming Links

We decided to work at an article level for incoming links. That is, both a source and the destination are articles. For each topic title, we chose the first 250 pages using Wumpus [2] that have the topic title without an intra-corpus link from the title. We then returned a result set that consists of the first 250 pages as the source and the topic title as the destination. For example, for a topic 10005.xml, "Educational progressivism", a page 16187.xml contains the phrase, "educational progressivism" without a link from it, so we suggest a link from a page 16187.xml to a topic 10005.xml.

The official result in Fig. 1 shows that although our precision decreases as recall increases, the curve is relatively linear. We expect that if we did not simply choose the first 250 pages to return, our precision would increase overall.

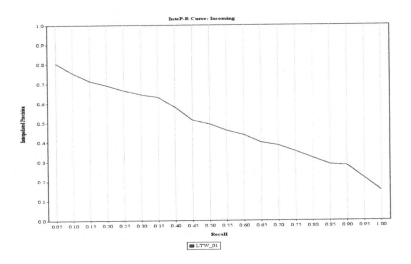

Fig. 1. Interpolated Precision and Recall for Incoming Links

3.2 Outgoing Links

To create outgoing links from topic files, we first created for each file in the corpus, a list of outgoing links specified by an anchor phrase a and the destination file d. We then selected the most frequent target d for each anchor phrase a over all titles and then computed the following ratios γ.

$$\gamma = \frac{\sharp \text{ of pages that has a link from anchor } a \text{ to a file } d}{\sharp \text{ of pages in which } a \text{ appears at least once}}$$

We set all destinations to the entire articles. We only picked those terms whose γ value is above certain threshold, in this case, 0.6.

For example, an anchor phrase, *bacteria*, appears most often with the destination file 3752.xml for 1197 times. There are 1981 number of files that contain the term *bacteria*. The value of γ for *bacteria* is then $1197/1981 = 0.604$ which is over 0.6. Similarly, there is another anchor phrase, *proteobacteria* with the most frequent destination file 24863.xml for 159 times. There are 161 number of files that contain the term *proteobacteria*, and the value of $\gamma = 159/161 = 0.988$ is also above the threshold of 0.6. Therefore, we add both *bacteria* and *proteobacteria* to our list of anchors.

Next, we found the first positions of each anchor phrase in every topic file using Wumpus [2], then linked the anchor phrases to the corresponding destinations. If an anchor phrase a is a substring of another anchor phrase b, we chose the longer anchor phrase to make a link from.

For example, suppose a position 1234 in topic files contain a term *proteobacteria*. Then we make a link to a file 24863.xml, not to a file 3752.xml.

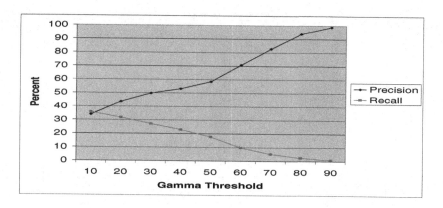

Fig. 2. Precision and Recall Plot at Various Thresholds

Table 1. Ranked v.s. Unranked Using Official Metrics

	MAP	R-Prec	P5	P10	P20	P30	P50
Official Unranked Outgoing	0.092	0.103	0.613	0.490	0.322	0.231	0.151
Unofficial Ranked Outgoing	0.607	0.628	0.849	0.816	0.75	0.698	0.614
Official Best of All Org.	0.318	0.415	0.767	0.683	0.579	n/a	0.440

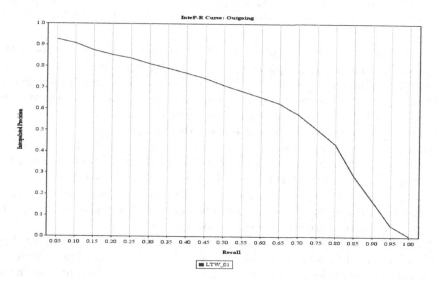

Fig. 3. Interpolated Precision and Recall for Ranked Outgoing Links

To see how we perform for various probability thresholds, we plotted a precision/recall graph for thresholds varying from 10% to 90%. We computed precision by how many outgoing links we embedded in topics appear in the original

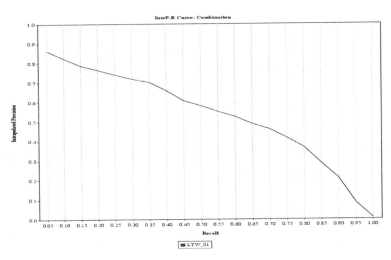

Fig. 4. Combined Interpolated Precision and Recall for Incoming and Outgoing Links

Table 2. Incoming and Combined Results Using Official Metrics

	MAP	R-Prec	P5	P10	P20	P30	P50
Official Incoming	0.465	0.512	0.662	0.653	0.603	0.57	0.516
Official Combined w/ Unranked Outgoing	0.154	0.171	0.637	0.56	0.42	0.329	0.234
Unofficial Combined w/ Ranked Outgoing	0.527	0.564	0.744	0.725	0.669	0.627	0.561

topic files in corpus. We computed recall as how many outgoing links in the original topic file in corpus appear in our embedded topic files. Figure 2 shows that precision increases as the threshold increases, and the precision is generally good. The recall decreases as thresholds increases as expected, however, the overall recall is fairly low. Therefore, it suggests that we need additional ways to identify outgoing links while still keeping the high accuracy.

Official results for outgoing links show that we achieve quite high precisions at early levels. This is because with 60% threshold, we did not return many outgoing links, and so recall is low as in Figure 2. We discovered that we did not return a ranked list of anchors as specified in the use case, but instead returned all anchors in the order of appearance. Therefore, in this paper, we decided to use a similar methodology to return a ranked list of outgoing links.

Instead of making a list of anchor phrases by ignoring anchors with γ values below a certain threshold, we decided to make a list of anchor phrases with the values of γ. We then found in topic files all occurrences of anchor phrases in the list, and returned the anchor phrases with the top 250 γ values. Figure 3 and Table 1 show that ranking by the γ values greatly increase the scores in official metrics and achieve the highest scores among all organizations participated for outgoing links. Concavity of Fig. 3 may be due to files that have less than 250 results.

Figure 4 is the final result for our ranked outgoing and incoming links combined using the official evaluation software. Table 2 shows scores of different official metrics for our incoming and combined submissions.

Anchor Density. Rather than use a fixed γ for all articles, we select γ based on the density of links it would produce. For a given article, we define the *anchor density* δ as

$$\delta = \frac{\text{\# number of anchor strings in the article}}{\text{length of the article in bytes}}.$$

Average anchor density is fairly consistent across the full range of article lengths, as illustrated in Figure 5. For each article in the INEX corpus, we allocate it to a bin according to its length, where each bin represents an 8KB range of lengths. For each bin, we computed the average anchor density for the articles it contains. At the lower end, the number of anchors increases linearly with article length. The irregularities in the upper bins reflect the small number of articles they contain, with the 58th bin (456KB to 464KB) containing no articles at all.

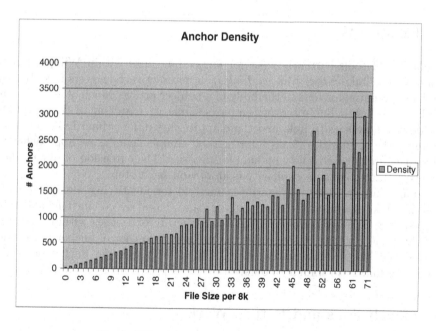

Fig. 5. File size vs. number of anchor strings

Across all articles in the INEX collection, the average density is 3.584 anchors/KB. We use this density to determine a cut-off for the number of anchors to suggest. For a given article, we rank potential links according to γ and suggest enough links to achieve this average anchor density.

Table 3. Combining link ratio and anchor density

	Precision	Recall
INEX test articles	60.52	51.94
new test articles	36.55	38.23

In order to test the effectiveness of the anchor density method, we created a second test set of 90 articles, not appearing in the INEX corpus, which were taken from a more recent version of the Wikipedia. The removed links were used as a gold standard for evaluating suggested links and precision and recall were computed as defined before.

Table 3 shows the recall and precision values achieved by combining link ratio and anchor density for both the INEX articles and new articles. For the INEX articles, performance is outside the range of what can be achieved by adjusting γ without considering anchor density. The lower performance on the new test articles may be partially due to the lag between the creation of the INEX collection in 2006 and the selection of the new test articles in early 2008. The re-computation of γ values over a more recent collection may improve performance.

4   Related Work

A number of tools currently exist for embedding links within Wikipedia. For example, the linking tool *Can we Link It* [9] interactively suggests links by recognizing potential anchors in the body of a subject article. Much of the previous work on automatic link discovery employs similarity matching of documents to decide which articles to link, and these approaches differ primarily in how they compute the similarities [5, 11]. Our method complements these approaches by suggesting links based solely on simple statistics. The extension of our method to incorporate document similarity is an obvious next step.

Adafre and de Rijke [1] describe a method for identifying missing links in existing Wikipedia articles. They cluster pages based on co-citations and then suggest new links by comparing the link structure of articles within a cluster. Mihalcea and Csomai [8] separated the problem into that of anchor phrase detection and link disambiguation. One of their techniques for anchor detection employs a link ratio similar to ours.

5   Conclusions and Future Work

We implemented a simple element retrieval technique and a more sophisticated passage retrieval technique to return result sets for adhoc focused and best-in-context tasks. We showed that our implementation of focused task along with Okapi BM25 scoring scheme works well for both IEEE collection [4] and Wikipedia collection. We speculate that the reason the passage-based best entry point retrieval did not work well is because the best entry point should start with relevant passages. Another reason is that the most relevant passage tend not to be at the beginning of an article, whereas the best entry point tend to be [7]. Therefore, we think that our passage-based retrieval may improve by returning the first element in the highest scoring passage.

We implemented a baseline algorithm for embedding incoming and outgoing links for Link-the-Wiki track. We showed that our selection of outgoing links has a high accuracy, but a raw recall. However, our ranked outgoing links performs very well against the official metrics. The use of anchor density seems to help improve the overall performance. Our result for incoming links show that the simple algorithm generally perform well overall, but need to increase precision more at an early stage.

References

1. Adafre, S.F., de Rijke, M.: Discovering missing links in Wikipedia. In: Proceedings of the 3rd International Workshop on Link Discovery, pp. 90–97 (2005)
2. Büttcher, S.: The Wumpus Search Engine (2007), http://www.wumpus-search.org
3. Clarke, C.L., Tilker, P.L.: MultiText experiments for INEX 2004. In: Fuhr, N., Lalmas, M., Malik, S., Szlávik, Z. (eds.) INEX 2004. LNCS, vol. 3493, pp. 85–87. Springer, Heidelberg (2005)
4. Clarke, C.L.A.: Controlling Overlap in Content-oriented XML retrieval. In: SIGIR 2005: Proceedings of the 28th annual international ACM SIGIR conference on Research and development in information retrieval, pp. 314–321. ACM, New York (2005)
5. Green, S.J.: Automated link generation: Can we do better than term repetition? Computer Networks 30(1-7), 75–84 (1998)
6. Itakura, K.Y., Clarke, C.L.A.: From Passages into Elements in XML Retrieval. In: SIGIR 2007 Workshop on Focused Retrieval (2007)
7. Kamps, J., Koolen, M., Lalmas, M.: Where to start reading a textual xml document? In: SIGIR 2007: Proceedings of the 30th annual international ACM SIGIR conference on Research and development in information retrieval, pp. 723–724. ACM, New York (2007)
8. Mihalcea, R., Csomai, A.: Wikify!: Linking documents to encyclopedic knowledge. In: Proceedings of the 16th ACM Conference on Information and Knowledge Management, pp. 233–242 (2007)
9. Nick, J.: Can we link it — Wikipedia, the free encyclopedia (2008) (accessed February 19, 2008)
10. Robertson, S., Walker, S., Beaulieu, M.: Okapi at trec-7: Automatic ad hoc, filtering, vlc and interactive track. In: 7th Text REtrieval Conference (1998)
11. Wilkinson, R., Smeaton, A.F.: Automatic link generation. ACM Computing Surveys 27 (1999)

Wikipedia *Ad Hoc* Passage Retrieval and Wikipedia Document Linking

Dylan Jenkinson and Andrew Trotman

Department of Computer Science
University of Otago
Dunedin
New Zealand
{djenkins,andrew}@cs.otago.ac.nz

Abstract. *Ad hoc* passage retrieval within the Wikipedia is examined in the context of INEX 2007. An analysis of the INEX 2006 assessments suggests that fixed sized window of about 300 terms is consistently seen and that this might be a good retrieval strategy. In runs submitted to INEX, potentially relevant documents were identified using BM25 (trained on INEX 2006 data). For each potentially relevant document the location of every search term was identified and the center (mean) located. A fixed sized window was then centered on this location. A method of removing outliers was examined in which all terms occurring outside one standard deviation of the center were considered outliers and the center recomputed without them. Both techniques were examined with and without stemming.

For Wikipedia linking we identified terms within the document that were over-represented and from the top few generated queries of different lengths. A BM25 ranking search engine was used to identify potentially relevant documents. Links from the source document to the potentially relevant documents (and back) were constructed (at a granularity of whole document). The best performing run used the 4 most over-represented search terms to retrieve 200 documents, and the next 4 to retrieve 50 more.

1 Introduction

The University of Otago participated in new tasks introduced to INEX in 2007. In the passage retrieval task three runs were submitted to each of the focused, relevant-in-context and best-in-contest tasks (and a fourth run was not submitted). In the Link-the-Wiki track five runs were submitted. In all cases performance was adequate (average or better).

An analysis of the 2006 INEX assessments (topics version:2006-004, assessments version:v5) shows that documents typically contain only one relevant passage, and that that passage is 301 characters in length. This leads to a potential retrieval strategy of first identifying potentially relevant documents, then from those identifying the one potentially relevant passage (of a fixed length). In essence this has reduced the passage retrieval problem to that of placing a fixed sized window on the text.

N. Fuhr et al. (Eds.): INEX 2007, LNCS 4862, pp. 426–439, 2008.

The approach we took was to identify each and every occurrence of each search term within the document. From there the mean position was computed and the window centered there. Outliers could potentially affect the placement of the window so an outlier reduction strategy was employed. All occurrences lying outside one standard deviation of the mean were eliminated and the mean recomputed. This new mean was used to place the window.

Porter stemming [6] was tested in combination with and without outlier reduction. Of interest to XML-IR is that our approach does not use document structure to identify relevant content. Kamps & Koolen [4] suggest relevant passages typically start (and end) on tag boundaries, however we leave exploitation of this to future work.

Our best passage retrieval runs when compared to element retrieval runs of other participants ranked favorably.

In the Link-the-Wiki task we again ignored the document structure and used a naive method. A score for each term in the orphaned document was computed as the ratio of length normalized document frequency to the expected frequency computed from collection statistics. Terms were ranked then queries of varying length (from 1 to 5 terms) were constructed from the top ranked terms in the list.

No attempt was made to identify anchor text or best entry points into target documents – instead linking from document to document was examined. We found that in this kind of linking query lengths of 4 terms performed best.

2 *Ad Hoc* Passage Retrieval

The INEX evaluation forum currently investigates subdocument (focused) information retrieval in structured documents, specifically XML documents. Focused retrieval has recently been defined as including element retrieval, passage retrieval and question answering [11]. In previous years INEX examined only element retrieval but in 2007 this was extended to include passage retrieval and book page retrieval. Common to all these paradigms is the requirement to return (to the user) only those parts of a document that are relevant, and not the whole document.

These focused searching paradigms are essentially identical and can be compared on an equal basis (using the same queries and metrics). If an XML element is specified using the start and end word number within a document (instead of XPath) then an XML element can be considered a passage. The same principle is true of a book page if word numbers are used instead of page numbers. A question answer within the text can also be considered a passage if it, too, is consecutive in the text.

Our interest in passage retrieval is motivated by a desire to reduce the quantity of irrelevant text in an answer presented to a user, that is, to increase focused precision. We believe that element granularity is too coarse and that users will necessarily be presented with irrelevant text along with their answers because any element large enough to fully contain a relevant answer is also likely to be sufficiently large that it contains some irrelevant text. Exactly this was examined by Kamps & Koolen [4] who report that, indeed, the smallest element that fully contains a relevant passage of text often contains some non-relevant text. The one way to increase precision is to remove the irrelevant text from the element, and one obvious way to do this is to shift to a finer granularity than element, perhaps paragraph, sentence, word, or simply passage.

2.1 INEX 2007 Tasks

There were three distinct retrieval tasks specified at INEX 2007: focused retrieval; relevant-in-context retrieval; and best-in-context retrieval. In focused retrieval the search engine must generate a ranked non-overlapping list of relevant items. This task might be used to extract relevant elements from news articles for multi-document summarization (information aggregation).

The relevant-in-context task is user-centered, and the aim is to build a search engine that presents, to a user, a relevant document with the relevant parts of that document highlighted. For evaluation purposes documents are first ranked on topical relevance then within the document the relevant parts of the document are listed.

Assuming a user can only start reading a document from a single point within a document, a search engine should, perhaps, identify that point. This is the aim of the best-in-context task, to rank documents on topical relevance and then for each document to identify the point from which a user should start reading in order to satisfy their information need.

For all three tasks both element retrieval and passage retrieval are applicable. For both it is necessary to identify relevant documents and relevant text within those documents. For element retrieval it is further necessary to identify the correct granularity of element to return to the user (for example, paragraph, sub-section, section, or document). For passage retrieval it is necessary to identify the start and end of the relevant text. It is not yet known which task is hardest, or whether structure helps in the identification of relevant text within a document. It is known that the precision of a passage retrieval system must, at worst, be at least equal to that of an element retrieval system.

2.2 Passage Retrieval

Passages might be specified in several different ways: an XML element, a start and end word position, or any granularity in-between (sentences, words, and so on). The length of a passage can be either fixed or variable. Within a document separate passages might either overlap or be disjoint.

If element retrieval and passage retrieval are to be compared on an equal basis it must be possible to specify an XML element as a passage. This necessitates a task definition that allows variable sized passages. Interactive XML-IR experiments show that users do not want overlapping results [10], necessitating a definition of disjoint passages. The INEX passage retrieval tasks, therefore, specify variable length non-overlapping passages that start and end on word boundaries. We additionally chose to ignore document structure as we are also interested in whether document structure helps with the identification of relevant material or not.

2.3 Window Size

Previous experiments suggest that fixed sized windows of between 200 and 300 words is effective [2]. To determine the optimal size for the Wikipedia collection an analysis of the INEX 2006 results was performed.

In 2006 INEX participants assessed documents using a yellow-highlighting method that identified all relevant passages within a document. For each passage the start and

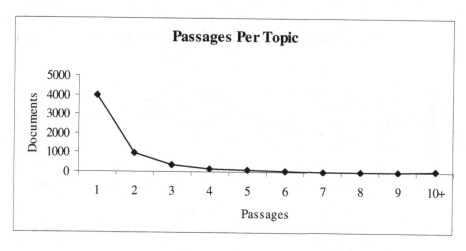

Fig.1. Number of documents containing the given number of passages

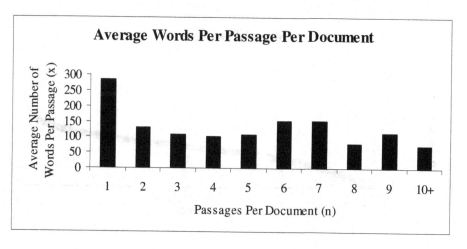

Fig. 2. Passage length varies with number of passages per document

end location are given in XPath and the length is given in characters. Best entry points are also specified.

Kamps & Koolen [4] performed a thorough analysis of the assessments and report a plethora of statistics. We reproduce some of those analyses, but present results in a different way.

Fig.1 presents the number of relevant documents in the assessment set that contain the given number of passages. The vast majority of relevant documents (70.63%) contain only one relevant document. This suggests that any passage retrieval algorithm that chooses to identify only one relevant passage per document will be correct the majority of the time. Because it is reasonable to expect only one relevant passage per document the tasks can be simplified to identifying *the relevant passage* in a

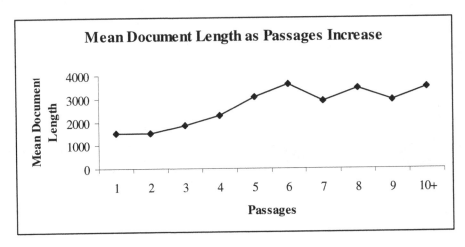

Fig. 3. Mean document length as the number of passages increases

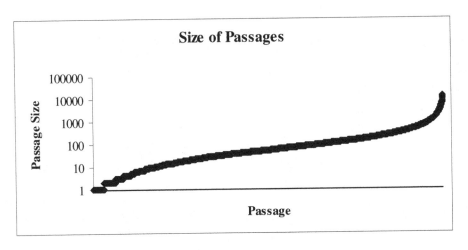

Fig. 4. Log of passage size for all relevant passages

document, not the relevant *passages* within a document. 17.27% contain 2 passages and 12.10% contain 3 or more passages.

Fig. 2 presents the mean passage length (in words) of a passage as the number of passages within a document increases. It was reasonable to expect that as the number of passages increased that the mean length of the passage would decrease as there is a natural limit on the sum of the lengths (the document length). Instead it can be seen that the average length is about constant. In a multiple-assessor experiment on the same document collection Trotman *et al.* [12] asked assessors whether they preferred to identify fixed-sized passages or variable sized passages and found that half preferred fixed sized passages of about a paragraph in length. This is consistent with the observation that passages are all about the same length – when a single passage is seen the mean is 283 words, but if more than one passage is sent then it varies between 73 and 153 words. Given this is the case then it is reasonable to expect that the length of a document is related to the number of passages it contains – this is shown to be the case in Fig. 3 where it can be seen that document length increases with the number of passages.

The mean relevant content per document is 301 words. In Fig. 4 the length of all relevant passages in all documents is presented – very few passages are long (over 1000 words) or short (under 10 words).

Given the mean length of relevant content in a document is about 300 words, and that only one passage is expected per document, it is reasonable to develop a passage retrieval algorithm that identifies one passage of 300 words. There does, however, remain the problem of identifying where, within a document, that passage should be placed.

2.4 Window Location

A heat map of the document can be built by noting the location of all search terms within the document. Areas where search terms do not occur (cold areas) are unlikely to be relevant to the user's query; conversely areas where there are many occurrences of the search terms (hot areas) are likely to be relevant.

Our hypothesis is that centering the one fixed-sized window over the middle of the dense areas will be an effective retrieval strategy. This method ignores the structure of the document, which we believe makes the comparison to element-retrieval systems of particular interest.

For each document identified as potentially relevant the XML structure is removed and the location of all occurrences of all search terms is identified. The mean of these locations is considered to be the center of relevance and so the window is centered on this point. If the window extends outside the document (before the beginning for example) then the window is truncated at the document boundary.

Problematically, in a well structured document it is reasonable to assume search terms will occur in the abstract and conclusions, but for the relevant text to occur elsewhere, in the body of the document for example. Several early or late term occurrences might shift the window towards the outliers which will in turn reduce precision. A method is needed to identify and remove outliers before the window is placed. We hypothesize that removing outliers will increase precision.

Two window placement methods were implemented: *meanselection* and *stddevselection*. With *meanselection* the center point (mean) of all occurrences of all search terms was used. With *stddevselection* the mean search term position was found and the standard-deviation computed. Then all occurrences outside one standard deviation from the mean were discarded. A new mean was then computed from the pruned list, and this was used as the passage midpoint.

2.5 Stemming

The identification of search terms within the document is essential to the performance of the window placement technique. It is reasonable to expect authors to use different morphological variants and synonyms of search terms within their documents. The inclusion of these in the algorithms is, therefore, important. We experimented with Porter's stemming algorithm [6].

2.6 Potentially Relevant Documents

The identification of relevant documents in *ad hoc* retrieval has been studied extensively by others. Several effective methods have been presented including language models [13], pivoted cosine normalization [9], and BM25 [7]. We chose BM25.

BM25 is parametric and requires scores for *k1*, *k3* and *b*. We used genetic algorithms [1] and trained on the INEX 2006 data to obtain good scores. The details are not important and we just report that the training resulted in the values 0.487, 25873, and 0.288 for *k1*, *k3* and *b* respectively.

Stemming was not used during training and was not used to identify potentially relevant documents

2.7 Best Entry Points

Kamps *et al.* [5] show a correlation between the best entry point and the start of the first relevant passage. They report 67.6% of best entry points in a single-passage document lying at the start of the passage (17.16% before and 15.24% after). For a document with two passages these numbers are substantially different. The chance that the best entry point coincides with the start of the first passage in the document is reduced to 35.33%, whilst the chance that the best entry point is before the first passage is increased to 45.21%. The chance of the best entry point coming after the first passage is about 19.46%. Fig. 5 presents our analysis. It shows, for all documents with a single relevant passage, the distance (in characters) from the start of that passage to the best entry point. The vast majority of all passages start at or very close to the best entry point. This suggests a best entry point identification strategy of "just choose the start of the first relevant passage".

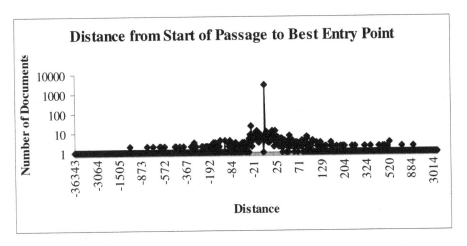

Fig. 5. Distance (in characters) of the best entry points from the start of the first passage. Negative are before the first passage.

3 *Ad Hoc* Experiments

3.1 *Ad Hoc* Runs

We conducted two experiments: the first was the effect of stemming, the second was the effect of removing outliers. This gave 4 possible combinations (runs) for each task as outlined in Table 1. However, as we were only permitted to submit 3 official runs per task and so the last run was scored informally. We expect the performance with standard-deviation and stemming to be most effective as this run will be better at identifying occurrences of search terms, while also better at removing outliers.

The same runs were submitted to each of the *ad hoc* tasks (focused, relevant-in-context, and best-in-context) and the runs differ only in name.

Table 1. Runs submitted to the INEX 2007 *ad hoc* track

Run	Focused	Relevant-in-context	Best-in-context
1	DocsNostem-PassagesStem-StdDevYes-Focused	DocsNostem-PassagesStem-StdDevYes	DocsNostem-PassagesStem-StdDevYes-BEP
2	DocsNostem-PassagesStem-StdDevNo-Focused	DocsNostem-PassagesStem-StdDevNo	DocsNostem-PassagesStem-StdDevNo-BEP
3	DocsNostem-PassagesNoStem-StdDevNo-Focused	DocsNostem-PassagesNoStem-StdDevNo	DocsNostem-PassagesNoStem-StdDevNo-BEP
4	DocsNostem-PassagesNoStem-StdDevYes-Focused	DocsNostem-PassagesNoStem-StdDevYes	DocsNostem-PassagesNoStem-StdDevYes-BEP

3.2 *Ad hoc* Results

Table 2 presents the scores and relative rank of the focused runs. The best run used stemming but not the *stddevselection* method. The relative rank of all runs is similar and the differences are small.

Of particular note is that of the 79 runs submitted to the task our runs that did not use document structure performed adequately (in the top 33%).

Table 2. Focused task results computes at 0.01 recall. ⁺values computed locally.

Run	iMAP	iMAP⁺	Rank
DocsNostem-PassagesStem-StdDevYes-Focused	0.4659	0.4609	30
DocsNostem-PassagesStem-StdDevNo-Focused	0.4716	0.4698	26
DocsNostem-PassagesNoStem-StdDevYes-Focused	-	0.4645	-
DocsNostem-PassagesNoStem-StdDevNo-Focused	0.4705	0.4688	28

In the tables in this section column 3, marked +, represents scored computed at the University of Otago using the released INEX evaluation software whereas column 2 represents the official score released on the INEX website (so the score for the fourth run is not given).

The performance of the runs submitted to the relevant-in-context task is shown in Table 3. Here there is no material difference in the score of the runs. Of 66 runs submitted to the task our top run that ignores structure performed averagely (32nd).

Table 3. Relevant-in-context results. [+]values computed locally.

Run	MAgP	MAgP[+]	Rank
DocsNostem-PassagesStem-StdDevYes	0.1028	0.1010	33
DocsNostem-PassagesStem-StdDevNo	0.1021	0.1014	34
DocsNostem-PassagesNoStem-StdDevNo	0.1033	0.1020	32
DocsNostem-PassagesNoStem-StdDevYes	-	0.1012	-

The performance with respect to the best-in-context task is shown in Table 4. Here outlier reduction was effective but stemming was not. The relative system performance of our best submitted run was 42 of 71.

Table 4. Best-in-context results. . +values computed locally.

Run	MAgP	MAgP[+]	Rank
DocsNostem-PassagesNoStem-StdDevYes-BEP	-	0.1101	-
DocsNostem-PassagesStem-StdDevYes-BEP	0.1061	0.1083	43
DocsNostem-PassagesStem-StdDevNo-BEP	0.1064	0.1066	42
DocsNostem-PassagesNoStem-StdDevNo-BEP	0.1060	0.1062	44

3.3 Discussion

We chose to ignore document structure and submitted run that, instead, simply used term locations to place a fixed sized window on the text. From the relative system performance it is reasonable to conclude that selecting a single fixed sized passage of text produces reasonable results.

The stemming experiment shows that stemming is not important for choosing the location of the window. When searching a very large document collection it is reasonable to ignore stemming because any relevant document will satisfy the user's information need. This should not be the case when looking within a single document where missing some occurrences of morphological variants of search terms has an effect on window placement and system performance – further investigation is needed

The use of the *stddevselection* method for selecting the centre point of a passage typically produced better results then the *meanselection* method. That is, there are, indeed, outliers in the document that affect window placement.

4 Link-the-Wiki

In 2007 INEX introduced a new track, Link-the-Wiki. The aim is to automatically identify hypertext links for a new documents when added to a collection [3]. The task contains two parts, the identification of out-going links to other documents in the collection and the identification of in-going links from other documents to the new document. In keeping with the focused retrieval theme, links are from passages of text (anchor text) to best entry points in a target document. In 2007, as the task is new, a reduced version of the track was run in which the task is simply document to document linking (both incoming and outgoing) [3]. Participants were also asked to supply information about the specifications of the computer used to generate the results, and the time taken to perform the generation. We used Intel Pentium 4, 1.66GHz, single core, no hyper-threading, and only 512MB memory. Our execution times were all less than 4 minutes and are presented in Table 5.

4.1 Themes

Almost all words or phrase in a document could be linked to another document (if for no other reason than to define the term). The task, therefore, is not the identification of links, but the identification of salient links. The approach we took was the identification of themes (terms) that are over-represented within the document, and the identification of documents about those themes. Our approach is based on that of Shatkay & Wilbur [8].

An over-represented term is a term that occurs more frequently in the source document than expected, that is, the document is more about that term that would be expected if the term was used *ordinarily*. The actual frequency (af) of a term within the document is computed as the term frequency (tf) over the document length (dl).

$$af = \frac{tf}{dl}$$

The expected frequency (ef) of the term is computed on the prior assumption that the term does occur within the document. Given the collection frequency (cf) and the document frequency (df), and the average length of a document (ml), this is expressed as

$$ef = \frac{cf}{df \times ml}$$

The amount by which the term is over-represented (*repval*) in the document is the ratio of the actual frequency to the expected frequency.

$$repval = \frac{af}{ef}$$

Terms that occur in a document but not the collection are assigned negative scores.

4.2 Link-the-Wiki Runs

We generated document to document linking runs using a relevance ranking search engine that used BM25 ($k1$=0.421, $k3$=242.61, b=0.498). Incoming links and outgoing links were strictly reciprocal, that is, the list of incoming links was generated from the outgoing list by reversing the direction of each link (and maintaining the relative rank order).

First the source (orphan) document was parsed and a list of all unique terms and *repval* scores was generated. Stop words were removed from the list.

Five runs were generated from the term list. In the first the single most over-represented term was used to generate a query for which we searched the collection returning the top 50 documents. The second term was then used to identify the next 50 documents, and so on until 250 documents had been identified.

In the second run the top two terms were used and 100 documents identified. 100 more for the third and fourth term, and 50 for the sixth and seventh term. In the third run triplets of terms were used to identify 150 documents each. In the fourth run quads of terms were used, and in the final run sets of 5 terms were used to identify all 250 documents. The details are outlined in Table 5.

In our experiment the total length of the result set was held constant (at 250) and the number of documents retrieved per search terms was held constant (at 50). The aim of our experiment was to identify whether or not there was a query-length effect in identifying related documents.

Table 5. Runs submitted to the Link-the-Wiki track

Run	Query length	Results per query	Time
ltw-one	1	50/50/50/50/50	134s
ltw-two	2	100/100/50	170s
ltw-three	3	150/100	161s
ltw-four	4	200/50	225s
ltw-five	5	250	124s

4.3 Results

The performance of the runs measured using mean average precision (MAP) is presented in Table 6. The relative rank order of our runs for both incoming and outgoing links was the same. The best run we submitted performed 4th of 13 submitted runs.

Fig. 6 graphs outgoing precision (and Fig. 7 incoming precision) at early points in the results list. Comparing the two, the technique we used is far better at identifying incoming links than outgoing links. When compared to runs from other participants, our best incoming precision at 5 and 10 documents ranked first.

Table 6. Link-the-Wiki results

Run	Outgoing		Incoming	
	MAP	Rank	MAP	Rank
ltw-four	0.102	4	0.339	4
tw-five	0.101	5	0.319	5
ltw-three	0.092	7	0.318	6
ltw-two	0.081	8	0.284	7
ltw-one	0.048	13	0.123	9

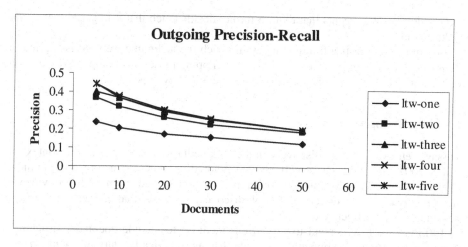

Fig. 6. Precision – Recall of outgoing links

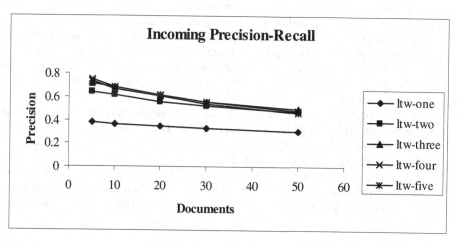

Fig. 7. Precision – Recall of incoming links

4.4 Discussion

We experimented with queries of different length and discovered that queries of 4 terms work better than either longer or shorter queries. When adding search terms to a query there comes a point at which the query becomes general resulting in the retrieval any an increasing number of irrelevant documents. This point appears to be 4 terms.

Of particular interest to us is the difference in performance of incoming and outgoing links. We constructed outgoing links from a document using a simple technique to identify terms that were over-represented. Incoming links were simply the same list inverted in direction. The technique appears capable of identifying the salient concepts within the document (such that it might be beneficial to link to), but not extracting from a document concepts that require further details (such that it might be beneficial to link from).

Our results suggests a future strategy in which the technique we used is applied to all documents to identify incoming links, and flipping those to get outgoing links for a document. This is, however, likely to be computationally expensive.

5 Conclusions

Passage retrieval and link discovery in the Wikipedia was examined in the context of INEX 2007. For both tasks methods that ignored document structure were studied. We found mixed results for both stemming and outlier reduction with no evidence that either was always effective. In link discovery we found that queries containing 4 search terms was effective.

In future work we intend to extend our methods to include document structures. Others have already shown that relevant passages typically start and end on tag boundaries, none the less we chose to ignore structure. Methods of using structure in passage length identification will be examined for passage retrieval and use for Best Entry Point identification will be used for link identification.

We intent to examine the granularity of structural markup necessary before good ranking performance can be expected. Even though we chose to ignore structure the performance of our runs was reasonable when compared to those of others. This raises the question of the value of the structural markup within a document when used for relevance ranking.

The Link-the-Wiki runs we submitted also performed adequately. Queries of various length were constructed from concept terms. The concept terms were extracted from the orphaned document by taking terms overly represented in the document. The best query length we found was 4 terms.

The technique was better at identifying incoming links than outgoing links – that is, the technique identifies the concepts of the document and not concepts that require further expansion. Future work will examine fast and efficient ways to identify outgoing links.

Acknowledgements

Funded in part by a University of Otago Research Grant.

References

[1] Holland, J.H.: Adaptation in natural and artificial systems. University of Michigan Press, Ann Arbor (1975)
[2] Huang, W., Trotman, A., O'Keefe, R.A.: Element retrieval using a passage retrieval approach. Australian Journal of Intelligent Information Processing Systems (AJIIPS) 9(2), 80–83 (2006)

[3] Huang, W.C., Trotman, A., Geva, S.: Collaborative knowledge management: Evaluation of automated link discovery in the Wikipedia. In: Proceedings of the SIGIR 2007 Workshop on Focused Retrieval, pp. 9–16 (2007)

[4] Kamps, J., Koolen, M.: On the relation between relevant passages and XML document structure. In: Proceedings of the SIGIR 2007 Workshop on Focused Retrieval, pp. 28–32 (2007)

[5] Kamps, J., Koolen, M., Lalmas, M.: Where to start reading a textual XML document? In: Proceedings of the 30th ACM SIGIR Conference on Information Retrieval (2007)

[6] Porter, M.: An algorithm for suffix stripping. Program 14(3), 130–137 (1980)

[7] Robertson, S.E., Walker, S., Beaulieu, M.M., Gatford, M., Payne, A.: Okapi at TREC-4. In: Proceedings of the 4th Text REtrieval Conference (TREC-4), pp. 73–96 (1995)

[8] Shatkay, H., Wilbur, W.J.: Finding themes in medline documents probabilistic similarity search. In: Proceedings of the Advances in Digital Libraries, pp. 183–192 (2000)

[9] Singhal, A., Buckley, C., Mitra, M.: Pivoted document length normalization. In: Proceedings of the 19th ACM SIGIR Conference on Information Retrieval, pp. 21–29 (1996)

[10] Tombros, A., Larsen, B., Malik, S.: The interactive track at INEX 2004. In: Fuhr, N., Lalmas, M., Malik, S., Szlávik, Z. (eds.) INEX 2004. LNCS, vol. 3493, pp. 410–423. Springer, Heidelberg (2005)

[11] Trotman, A., Geva, S., Kamps, J.: Proceedings of the SIGIR 2007 workshop on focused retrieval (2007)

[12] Trotman, A., Pharo, N., Jenkinson, D.: Can we at least agree on something? In: Proceedings of the SIGIR 2007 Workshop on Focused Retrieval, pp. 49–56 (2007)

[13] Zhai, C., Lafferty, J.: A study of smoothing methods for language models applied to information retrieval. Transactions on Information Systems 22(2), 179–214 (2004)

The INEX 2007 Multimedia Track

Theodora Tsikrika[1] and Thijs Westerveld[2,*]

[1] CWI, Amsterdam, The Netherlands
[2] Teezir Search Solutions, Ede, The Netherlands

Abstract. The INEX Multimedia track focuses on using the structure of XML documents to extract, relate, and combine the relevance of different multimedia fragments. This paper presents a brief overview of the track for INEX 2007, including the track's test collection, tasks, and goals. We also report the approaches of the participating groups and their main results.

1 Introduction

Structured document retrieval from XML documents allows for the retrieval of XML document fragments, i.e., XML elements or passages, that contain relevant information. The main INEX Ad Hoc task focuses on text-based XML retrieval. Although text is dominantly present in most XML document collections, other types of media can also be found in those collections. Existing research on multimedia information retrieval has already shown that it is far from trivial to determine the combined relevance of a document that contains several multimedia objects [5].

The objective of the INEX Multimedia track is to exploit the XML structure that provides a logical level at which multimedia objects are connected, in order to improve the retrieval performance of an XML-driven multimedia information retrieval system. To this end, it provides an evaluation platform for the retrieval of multimedia documents and document fragments. In addition, it creates a discussion forum where the participating groups can exchange their ideas on different aspects of the multimedia XML retrieval task.

This paper reports on the INEX 2007 Multimedia track and is organised as follows. First, we introduce the main parts of the test collection: documents, tasks, topics, and assessments (Sections 2–5). Section 6 presents the approaches employed by the different participants and Section 7 summarises their main results. Section 8 concludes the paper and provides an outlook on next year's track.

2 Wikipedia Collections and Additional Resources

In INEX 2007, the Multimedia track employed the following two Wikipedia-based collections (the same as in 2006):

---

* Part of this work was carried out when the author was at CWI, Amsterdam, The Netherlands.

N. Fuhr et al. (Eds.): INEX 2007, LNCS 4862, pp. 440–453, 2008.

Table 1. Wikipedia XML collection statistics

Total number of XML documents	659,388
Total number of images	344,642
Number of unique images	246,730
Average number of images per document	0.52
Average depth of XML structure	6.72
Average number of XML nodes per document	161.35

Wikipedia XML collection: This is a structured collection of 659,388 Wiki-text pages from the English part of Wikipedia, the free content encyclopedia (http://en.wikipedia.org), that have been converted to XML [1]. This collection has been created for the Ad Hoc track. Given, though, its multi-media nature (as indicated by its statistics listed in Table 1), it is also being used as the target collection for a multimedia task that aims at finding rel-evant XML fragments given a multimedia information need (see Section 3).

Wikipedia image XML collection: This is a collection consisting of the im-ages in the Wikipedia XML collection, together with their metadata which have been formatted in XML. These metadata usually contain a brief cap-tion or description of the image, the Wikipedia user who uploaded the image, and the copyright information. Figure 1 shows an example of such a docu-ment consisting of an image and its associated metadata. Some images from the Wikipedia XML collection have been removed due to copyright issues or parsing problems with their metadata, leaving us with a collection of 170,370

1116948: AnneFrankHouseAmsterdam.jpg

AnneFrankHouseAmsterdam.jpg

Anne Frank House - The Achterhuis - Amsterdam. Photo taken by User:RossrsRossrs mid 2002 PD-self

es:Image:AnneFrankHouseAmsterdam.jpg

Category:Building and structure images

Fig. 1. Example Wikipedia image+metadata document from the Wikipedia image XML collection

images with metadata. This collection is used as the target collection for a multimedia/image retrieval task that aims at finding images (with metadata) given a multimedia information need (see Section 3).

Although the above two Wikipedia-based collections are the main search collections, additional sources of information are also provided to help participants in the retrieval tasks. These resources are:

Image classification scores: For each image, the classification scores for the 101 different MediaMill concepts are provided by UvA [6]. The UvA classifier is trained on manually annotated TRECVID video data and the concepts are selected for the broadcast news domain.

Image features: For each image, the set of the 120D feature vectors that has been used to derive the above image classification scores is available [3]. Participants can use these feature vectors to custom-build a content-based image retrieval (CBIR) system, without having to pre-process the image collection.

These resources were also provided in 2006, together with an online CBIR system that is no longer available. The above resources are beneficial to researchers who wish to exploit visual evidence without performing image analysis.

3   Retrieval Tasks

The aim of the retrieval tasks in the Multimedia track is to retrieve relevant (multimedia) information, based on an information need with a (structured) multimedia character. To this end, a structured document retrieval approach should be able to combine the relevance of different media types into a single ranking that is presented to the user.

For INEX 2007, we define the same two tasks as last year:

MMfragments task: Find relevant XML fragments in the **Wikipedia XML collection** given a multimedia information need. These XML fragments can correspond not only to XML elements (as it was in INEX 2006), but also to passages. This is similar to the direction taken by the INEX Ad Hoc track. In addition, since MMfragments is in essence comparable to the ad hoc retrieval of XML fragments, this year it ran along the Ad Hoc tasks. As a result, the three subtasks of the Ad Hoc track (see [2] for detailed descriptions) are also defined as subtasks of the MMfragments task:

1. FOCUSED TASK asks systems to return a ranked list of elements or passages to the user.
2. RELEVANT IN CONTEXT TASK asks systems to return relevant elements or passages clustered per article to the user.
3. BEST IN CONTEXT TASK asks systems to return articles with one best entry point to the user.

The difference is that MMfragments topics ask for multimedia fragments (i.e., fragments containing at least one image) and may also contain visual hints (see Section 4).

MMimages task: Find relevant images in the **Wikipedia image XML collection** given a multimedia information need. Given an information need, a retrieval system should return a ranked list of documents(=image+metadata) from this collection. Here, the type of the target element is defined, so basically this is closer to an image retrieval (or a document retrieval) task, rather than XML element or passage retrieval. Still, the structure of (supporting) documents, together with the visual content and context of the images, could be exploited to get to the relevant images (+their metadata).

All track resources (see Section 2) can be used for both tasks, but the track encourages participating groups to also submit a baseline run that uses no sources of information except for the target collection. This way, we hope to learn how the various sources of information contribute to the retrieval results. Furthermore, we also encourage each group to submit a run that is based on only the <mmtitle> field of the topic description (see Section 4). All other submissions may use any combination of the <title>, <castitle>, <mmtitle> and <description> fields (see Section 4). The fields used need to be reported.

4   Topics

The topics used in the INEX Multimedia track are descriptions of (structured) multimedia information needs that may contain not only textual, but also structural and multimedia hints. The structural hints specify the desirable elements to return to the user and where to look for relevant information, whereas the multimedia hints allow the user to indicate that results should have images similar to a given example image or be of a given concept. These hints are expressed in the NEXI query language [8].

The original NEXI specification determines how structural hints can be expressed, but does not make any provision for the expression of multimedia hints. These have been introduced as NEXI extensions during the INEX 2005 and 2006 Multimedia tracks [9,10]:

- To indicate that results should have images similar to a given example image, an *about* clause with the keyword *src:* is used. For example, to find images of cityscapes similar to the image at `http://www.bushland.de/hksky2.jpg`, one could type:

```
//image[about(.,cityscape) and
                about(.,src:http://www.bushland.de/hksky2.jpg)]
```

In 2006, only example images from within the Wikipedia image XML collection were allowed, but this year it was required that the example images came from outside the Wikipedia collections.

– To indicate that the results should be of a given concept, an *about* clause with the keyword *concept:* is used. For example, to search for cityscapes, one could decide to use the concept "building":

`//image[about(.,cityscape) and about(.,concept:building)]`

This feature is directly related to the concept classifications that are provided as an additional source of information (see Section 2). Therefore, terms following the keyword *concept:* are obviously restricted to the 101 concepts for which classification results are provided.

It is important to realise that all structural, textual and visual filters in the query should be interpreted loosely. It is up to the retrieval systems to decide how to use, combine or even ignore this information. The relevance of a document, element or passage does not directly depend on these hints, but is determined by manual assessments.

4.1 Topic Format

The INEX Multimedia track topics are similar to the Content Only + Structure (CO+S) topics of the INEX Ad Hoc track. In INEX, "Content" refers to the textual or semantic content of a document part, and "Content-Only" to topics or queries that use no structural hints. The Ad Hoc CO+S topics include structural hints, whereas the Multimedia CO+S topics may also include visual hints.

The 2007 Multimedia CO+S topics consist of the following parts:

`<title>` The topic `<title>` simulates a user who does not know (or does not want to use) the actual structure of the XML documents in a query and who does not have (or want to use) example images or other visual hints. The query expressed in the topic `<title>` is, therefore, a Content Only (CO) query. This profile is likely to fit most users searching XML digital libraries and also corresponds to the standard web search type of keyword search.

`<castitle>` A NEXI expression with structural hints.

`<mmtitle>` A NEXI expression with structural and visual hints.

`<description>` A brief, matter of fact, description of the information need. Like a natural language description one might give to a librarian.

`<narrative>` A clear and precise description of the information need. The narrative unambiguously determines whether or not a given document or document part fulfils the given need. It is the only true and accurate interpretation of a user's needs. Precise recording of the narrative is important for scientific repeatability - there must exist, somewhere, a definitive description of what is and is not relevant to the user. To aid this, the `<narrative>` should explain not only what information is being sought, but also the context and motivation of the information need, i.e., why the information is being sought and what work-task it might help to solve.

In previous years, both structural and visual/multimedia hints were expressed in the `<castitle>` field. This year, the `<castitle>` contains only structural hints, while the `<mmtitle>` is an extension of the `<castitle>` that also incorporates the additional visual hints (if any). The introduction of a separate `<mmtitle>` is particularly useful, since it makes it easier for systems to compare runs using structural hints to those using structural+visual hints, without having to modify the query expression. In addition, Multimedia CO+S topics can now also be used in Ad Hoc tasks, since they contain fields (all, except `<mmtitle>`) that can be directly processed by an Ad Hoc system.

The fact that the MMfragments task is similar to ad hoc retrieval, not only led to the decision to run the MMfragments tasks along the Ad Hoc ones, but also to include the MMfragments topics as a subset of the Ad Hoc ones. This means that submissions for the INEX 2007 Ad Hoc track also considered the subset of topics used for the MMfragments task. This allows us to compare ad hoc XML retrieval systems submissions on the MMfragments topic subset (i.e., submissions that retrieve XML document parts by using any of the available fields except `<mmtitle>`) to multimedia XML retrieval submissions on the same topic subset (i.e., to submissions that can use any of the topic fields, together with the knowledge that a multimedia XML fragment is required as a retrieval result).

MMimages, on the other hand, runs as a separate task with a separate set of topics. Given that MMimages requires retrieval at the document level, rather than elements or passages, the queries in the `<castitle>` and `<mmtitle>` fields are restricted to: `//article[X]`, where X is a predicate using one or more *about* functions with textual and/or multimedia hints.

4.2   Topic Development

The topics in the Multimedia track are developed by the participants. Each participating group has to create 2 multimedia topics for the MMfragments task and 4 topics for MMimages. Topic creators first create a 1-2 sentence description of the information they are seeking. Then, in an exploration phase, they obtain an estimate of the amount of relevant information in the collection. For this, they can use any retrieval system, including their own system or the TopX system [7] provided through the INEX organisation. The topic creator then assesses the top 25 results and abandons the search if fewer than two or more than 20 relevant fragments are found. If between 2 and 20 fragments are found to be relevant, the topic creator should have a good idea of what query terms should be used, and the `<title>` is formulated. Using this title a new search is performed and the top 100 elements are assessed. Having judged these 100 documents, topic creators should have a clear idea of what makes a fragment relevant or not. Based on that, they could then first write the narrative and then the other parts of the topic. After each created topic, participants are asked to fill a questionnaire that gathers information about the users familiarity with the topic, the expected number of relevant fragments in the collection, the expected size of relevant fragments and the realism of the topic. The submitted topics are analysed by the INEX Multimedia organisers who check for duplicates and inconsistencies before distributing the full set of topics among the participants.

Table 2. Statistics for the INEX 2007 MM topics

	MMfragments	MMimages	All
Number of topics	19	20	39
Average number of terms in <title>	3.21	2.35	2.77
Number of topics with <mmtitle>	6	10	16
Number of topics with src:	2	7	9
Number of topics with concept:	4	6	10
Number of topics with both src: and concept:	0	3	3

Table 2 shows the distribution over tasks as well as some statistics on the topics. The MMfragments topics correspond to Ad Hoc topics 525-543. Their average number of terms in <title> (3.21) is slightly lower than the average number of terms in the remaining 80 Ad Hoc topics (3.92). This is to be expected, since users who submit multimedia topics express their requirements not only by textual, but also by visual hints. Table 2 indicates that not all topics contain visual/multimedia hints; this corresponds well with realistic scenarios, since users who express multimedia information needs do not necessarily want to employ visual hints.

5   Assessments

Since XML retrieval requires assessments at a sub-document level, a simple binary judgement at the document level is not sufficient. Still, for ease of assessment, retrieved fragments are grouped by document. Since the INEX 2007 MMfragments task was run in parallel with the Ad Hoc track, the assessments for this task were arranged by the Ad Hoc track organization as follows. Once all participants have submitted their runs, the top N fragments for each topic are pooled and grouped by document. The documents are alphabetised so that the assessors do not know how many runs retrieved fragments from a certain document or at what rank(s) the fragments were found. Assessors then look at the documents in the pool and highlight the relevant parts of each document. The assessment system stores the relevance or non-relevance of the underlying XML elements and passages.

We did not give any additional instructions to the assessors of multimedia topics, but assumed that topic creators who indicated that their topics have a clear multimedia character would only judge elements relevant if they contain at least one image. We analysed the assessed fragments to verify this. We looked at the number of <image> elements in highlighted passages and contrasted the findings for the MMfragments topics with the findings for other Ad Hoc topics, and found that indeed the fragments assessed relevant for MMfragments topics contain many more images than the relevant fragments for Ad Hoc topics. On average, a relevant passage for an Ad Hoc topic contains 0.14 images. An average relevant passage for a MMfragments topic contains 0.62 images. The box plot in Figure 2 shows the minimum, median and maximum of the average number of images per highlighted

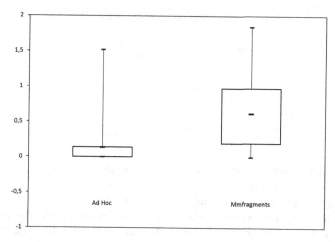

Fig. 2. Number of images per highlighted passage (minimum, median, maximum and 25th to 75th percentile)

passage over the topics in the category, the box shows the data falling between the 25th and 75th percentile. Even though not all highlighted MMfragment passages contain images, the difference with Ad Hoc topics is clear. For around 25% of the MMfragments topics the average number of images per passage is above 1, for half the topics this number is greater than 0.5.

The MMimages task is a document retrieval task. A document, i.e., an image with its metadata, is either relevant or not. For this task, we adopted TREC style document pooling of the documents and binary assessments at the document (i.e., image with metadata) level. In 2006, the pool depth was set to 500 for the MMimages task, with post-hoc analysis showing that pooling up to 200 or 300 would have given the same system ordering [10]. This led to the decision to pool this year's submissions up to rank 300, resulting in pools of between 348 and 1865 images per topic, with both mean and median around 1000 (roughly the same size as 2006).

6   Participants

Only four participants submitted runs for the INEX 2007 Multimedia track: CWI together with the University of Twente (CWI/UTwente), IRIT (IRIT), Queensland University of Technology in Australia (QUTAU), and University of Geneva (UGeneva). For the MMfragments task, three of the participants (CWI/UTwente, IRIT and QUTAU) submitted a total of 12 runs, whereas for the MMimages task, all four participants submitted a total of 13 runs.

Table 3 gives an overview of the topic fields used by the submitted runs. For MMfragments, six submissions used the topics' <title> field, and six submissions used the <castitle> field; the mmtitle field was not used by any paricipant. For MMimages, seven submissions used the topics' <title> field, and six sub-

Table 3. Topic fields used by the submitted runs

topic field	#MMfragments runs using it	#MMimages runs using it
title	6	7
castitle	6	0
mmtitle	0	6
description	0	0
narrative	0	0

missions used the `<mmtitle>` field; no submissions used the `<castitle>` field which is to be expected since this is a document retrieval task.

Table 4 gives an overview of the resources used by the submitted runs. Not all groups detailed the resources they used, but judging from the descriptions it appears most submissions only used the target Wikipedia collection of the task at hand. It seems the Wikipedia images collection and the UvA features and classification scores have not been used in the MMfragments task this year. In the MMimages task, the visual resources provided are used by IRIT and UGeneva, whereas some runs also used the main Wikipedia XML collection.

Below we briefly discuss the appproaches taken by the groups that participated in the Multimedia track at INEX 2007.

CWI/UTwente. CWI/UTwente participated in both MMfragments and MMimages tasks of the INEX 2007 Multimedia track. For MMfragments, they limited their system to return only fragments that contain at least one image that was part of the Wikipedia images XML collection. They did not use any further multimedia processing and experimented with traditional text based approaches based on the language modelling approach and different length priors. For MMimages, they represented each image either by its textual metadata in the Wikipedia image XML collection, or by its textual context when that image appears as part of a document in the (Ad Hoc) Wikipedia XML collection. Retrieval was then based on purely text-based approaches.

IRIT. IRIT participated in both the MMfragments and MMimages tasks of the INEX 2007 Multimedia track, with methods based on the context (text and structure) of images to retrieve multimedia elements. For MMimages topics, the "MMI" method proposed last year that uses 3 sources of evidence (descendant,

Table 4. Resources used by the submitted runs

resource	#MMfragments runs using it	#MMimages runs using it
wikipedia	12	4
wikipedia_IMG	0	8
UvAfeatures	0	1
UvAconcepts	0	2

sibling, and ascendant nodes) is compared to a new method "MMIConc" that uses in addition images classification scores. For the MMfragments task, the "MMF" method based on the "XFIRM Content and Structured" method and the "MMI" method were evaluated. In future work, IRIT plan to extend images context by using links.

QUTAU. No description of their approaches has been provided.

UGeneva. For their first participation at INEX MM, they submitted three runs to the MMimages task: (1) a baseline run based only on text-based retrieval, (2) an improvement of (1) with additional proper noun detection, and (3) a multi modal fusion approach using a hierarchical SVM approach.

For the simple text-based baseline run (1), the ready-to-use Matlab library TMG [11] is applied to the MMimages collection. It creates a term-document matrix filled with term frequencies of the textual input. The retrieval is done based on the Vector Space Model (VSM). In (2) the simple baseline run is improved by adding to the approach a proper noun detection based on Google result counts. This proved to be an easy and inexpensive way to reliably detect proper nouns. The multi modal fusion run (3) used all available features: textual and visual (color and texture histogram) low level features, plus the visual concepts provided by the University of Amsterdam. The approach was set up hierarchically. First a VSM-based retrieval on the extended term-document matrix was executed. Then the result list was classified into N classes with the k-NN algorithm of the TMG library. The documents of the cluster containing the most relevant documents were taken as input for a hierarchical Support Vector Machine (SVM) classification, which processes first each modality alone, before fusing all result lists in a final step.

Université de Saint-Etienne/JustSystems. These two groups did not submit any official runs for the track, but they did help with assessments for the MMimages task, and plan to use the track's data for future studies.

7   Results

This section presents the results for the submitted runs in each of the tasks.

7.1   MMfragments

Three participating groups (CWI/UTwente, IRIT and QUTAU) submitted a total of 12 MMfragments runs (5 Focused, 2 Relevant in Context and 5 Best in Context runs). Of these submissions, 6 used the topics' title field and 6 used the castitle field; the mmitle field was not used by any pariciant in the MMfragments task. Not all groups detailed the resources they used, but judging by the descriptions it appears that all submissions only used the main wikipedia collection for this task. It seems that the wikipedia images collection and the UvA features and classification scores have not been used in the MMfragments task this year.

Table 5. MMfragments Results for Focused task

MAiP	iP[0.00]	iP[0.01]	iP[0.05]	iP[0.10]	Group	Run
0.1169	0.4158	0.3389	0.2921	0.2546	utwente	article_MM
0.0910	0.3744	0.3039	0.2160	0.1713	qutau	COS_Focused
0.1218	0.2989	0.2947	0.2790	0.2382	qutau	CO_Focused
0.0991	0.2471	0.2467	0.2422	0.2294	utwente	star_loglength_MM
0.0042	0.3448	0.0595	0.0000	0.0000	utwente	star_lognormal_MM

Table 6. MMfragments Results for Relevant in Context task

MAgP	gP[5]	gP[10]	gP[25]	gP[50]	Group	Run
0.1043	0.1729	0.1763	0.1528	0.1193	qutau	CO_RelevantInContext
0.0900	0.2072	0.1787	0.1441	0.1085	qutau	COS_RelevantInContext

Table 7. MMfragments Results for Best in Context task

MAgP	gP[5]	gP[10]	gP[25]	gP[50]	Group	Run
0.1783	0.3210	0.3039	0.2558	0.2099	qutau	CO_BestInContext
0.1533	0.3671	0.3084	0.2334	0.1761	qutau	COS_BestInContext
0.0541	0.1423	0.1394	0.0784	0.0437	irit	iritmmf06V2_BIC
0.0506	0.1133	0.1319	0.1267	0.0943	irit	iritmmf06V1
0.0458	0.1164	0.1316	0.1114	0.0876	irit	iritmmf06V3_BIC

These runs have been evaluated using the standard measures as used in the Ad Hoc track [4]: interpolated Precision (iP) and Mean Average interpolated Precision (MAiP) for the Focused task and non-interpolated generalized precision at early ranks gP[r] and non-interpolated mean average generalized precision MAgP). Tables 5-7 show the results.

Since the MMfragments topics were mixed with the Ad Hoc topics we received many more submissions that were not tailored to answering information needs with a multimedia character. We evaluated these runs on the subset of 19 multimedia topics. Tables 8–10 show the results of these runs for the top 5 performing groups. Compared to the tables above, for none of the tasks the best performing run was an official multimedia submission. That shows that for this task standard text retrieval techniques are competitive. This does not necessarily lead to the conclusion

Table 8. Ad Hoc runs for the MMfragments topics for Focused task

MAiP	iP[0.00]	iP[0.01]	iP[0.05]	iP[0.10]	Group	Run
0.0649	0.5367	0.4435	0.1960	0.1393	mines	EMSE,boolean,Prox200NF,0010
0.1059	0.4494	0.4219	0.2952	0.2272	qutau	FOC_02
0.1175	0.3961	0.3856	0.3176	0.2888	justsystem	VSM_CO_02
0.1338	0.3962	0.3853	0.3199	0.2558	unigordon	Focused-LM
0.1050	0.5793	0.3715	0.2990	0.2796	maxplanck	TOPX-CAS-Focused-exp-all

Table 9. Ad Hoc runs for the MMfragments topics for Relevant in Context task

MAgP	gP[5]	gP[10]	gP[25]	gP[50]	Group	Run
0.1323	0.1838	0.2035	0.1740	0.1438	udalian	DUT_03_Relevant
0.1120	0.2129	0.2151	0.1467	0.1152	rmit	zet-okapi-RiC
0.1044	0.1729	0.1763	0.1528	0.1193	qutau	CO_RelevantInContext
0.0951	0.1282	0.1500	0.1281	0.0980	utwente	star_logLP_RinC
0.0949	0.1987	0.1919	0.1346	0.1049	unigordon	RelevantInContext-LM

Table 10. Ad Hoc runs for the MMfragments topics for Best in Context task

MAgP	gP[5]	gP[10]	gP[25]	gP[50]	Group	Run
0.2275	0.4306	0.3610	0.2725	0.2090	rmit	zet-okapi-BiC
0.1889	0.4306	0.3610	0.2857	0.2210	inria	ent-ZM-BiC
0.1879	0.2505	0.2377	0.1949	0.1660	udalian	DUT_02_Best
0.1852	0.3588	0.3324	0.2243	0.1647	justsystem	VSM_CO_14
0.1839	0.3381	0.3052	0.2244	0.1838	unigordon	BestInContext-LM

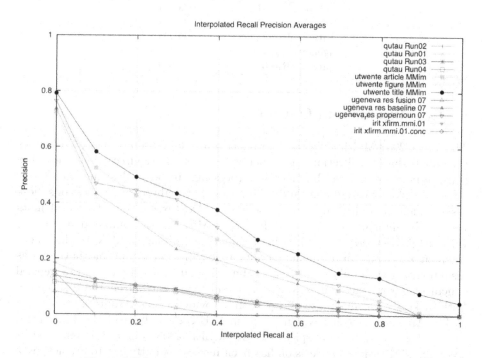

Fig. 3. MMimages: Interpolated Recall Precision Averages

that specific treatment of multimedia topics is ineffective. It may still be the case that a combination of techniques from the top performing Ad Hoc and Multimedia submissions would give better results on these topics than either alone.

7.2 MMimages

The four participating groups (CWI/UTwente, IRIT, QUTAU, and UGeneva) submitted a total of 13 MMimages runs. Figure 3 shows the interpolated recall precision graphs of these runs and Table 11 shows their mean average precision scores. Similarly to last year, the top performing runs do not use any image analysis or visual processing; they are purely text-based.

Table 11. Mean average precision (MAP) for submitted MMimages runs

group	run	MAP
utwente	title_MMim	0.2998
ugeneva	res_propernoun_07	0.2375
utwente	article_MMim	0.2240
ugeneva	res_baseline_07	0.1792
utwente	figure_MMim	0.1551
qutau	Run03	0.0482
irit	xfirm.mmi.01	0.0448
qutau	Run01	0.0447
irit	xfirm.mmi.01.conc	0.0445
qutau	Run04	0.0411
ugeneva	res_fusion_07	0.0165
qutau	Run02	0.0011

8   Conclusions and Outlook

The INEX 2007 Multimedia track provides a nice collection of related resources (Wikipedia-based collections, together with a set of resources that are either starting points for or results of visual processing) to be used in the track's two retrieval tasks: MMfragments and MMimages. The main research questions these tasks aimed at addressing are the following: Do textual and structural hints need to be interpreted differently for the MMfragments compared to the Ad Hoc tasks? How do visual hints in the query help image and XML document fragment retrieval? Since the number of participants in the multimedia track was disappointing with only four groups submitting runs, it is hard to draw general conclusions from the results. What we could see so far is that the top runs in both tasks did not make use of any of the provided visual resources.

The Multimedia track will not run in INEX 2008. Instead the MMimages task will run under the auspices of ImageCLEF 2008, where it is renamed as wikipediaMM task. This decision has been made in an attempt to attract more participants, since ImageCLEF provides a more natural habitat for such an image retrieval task. The set of related collections and resources, makes this task an interesting playground, both for groups with a background in information retrieval, and for groups with a deeper understanding of computer vision or image analysis.

Acknowledgements

Theodora Tsikrika and Thijs Westerveld (while he was at CWI, Amsterdam, The Netherlands) were supported by the European Union via the European Commission project VITALAS (contract no. 045389). The authors would also like to thank Jaap Kamps for producing the evaluation results for the Ad Hoc runs on the MMfragments topics, Saadia Malik for valuable technical support, and the reviewer for useful comments that helped us improve this paper.

References

1. Denoyer, L., Gallinari, P.: The Wikipedia XML Corpus. SIGIR Forum. 40(1), 64–69 (2006)
2. Fuhr, N., Kamps, J., Lalmas, M., Malik, S., Trotman, A.: Overview of the INEX 2007 ad hoc track. In: Fuhr, N., Lalmas, M., Trotman, A., Kamps, J. (eds.) INEX 2006, Springer, Heidelberg (2008)
3. von Gemert, J.C., Geusebroek, J.-M., Veenman, C.J., Snoek, C.G.M., Smeulders, A.W.M.: Robust scene categorization by learning image statistics in context. In: Proceedings of the 2006 Conference on Computer Vision and Pattern Recognition Workshop, Washington, DC, USA, p. 105. IEEE Computer Society, Los Alamitos (2006)
4. Kamps, J., Pehcevski, J., Kazai, G., Lalmas, M., Robertson, S.: INEX 2007 evaluation measures. In: Fuhr, N., Lalmas, M., Trotman, A., Kamps, J. (eds.) INEX 2006, Springer, Heidelberg (2008)
5. Over, P., Awad, W., Kraaij, G., Smeaton, A.F.: TRECVID 2007 Overview. In: TREC Video Retrieval Evaluation Online Proceedings (2007)
6. Snoek, C.G.M., Worring, M., van Gemert, J.C., Geusebroek, J.-M., Smeulders, A.W.M.: The challenge problem for automated detection of 101 semantic concepts in multimedia. In: Proceedings of the 14th annual ACM international conference on Multimedia, pp. 421–430. ACM Press, New York (2006)
7. Theobald, M., Schenkel, R., Weikum, G.: An efficient and versatile query engine for topx search. In: VLDB 2005: Proceedings of the 31st international conference on Very large data bases, pp. 625–636. VLDB Endowment (2005)
8. Trotman, A., Sigurbjörnsson, B.: Narrowed Extended XPath I (NEXI). In: Fuhr, N., Lalmas, M., Malik, S., Szlávik, Z. (eds.) INEX 2004. LNCS, vol. 3493, pp. 16–40. Springer, Heidelberg (2005)
9. van Zwol, R., Kazai, G., Lalmas, M.: INEX 2005 multimedia track. In: Fuhr, N., Lalmas, M., Malik, S., Kazai, G. (eds.) INEX 2005. LNCS, vol. 3977, pp. 497–510. Springer, Heidelberg (2006)
10. Westerveld, T., van Zwol, R.: The INEX 2006 multimedia track. In: Fuhr, N., Lalmas, M., Trotman, A. (eds.) INEX 2006. LNCS, vol. 4518, pp. 331–344. Springer, Heidelberg (2007)
11. Zeimpekis, D., Gallopoulos, E.: TMG: A MATLAB toolbox for generating term-document matrices from text collections. In: Kogan, J., Nicholas, C., Teboulle, M. (eds.) Grouping Multidimensional Data: Recent Advances in Clustering, pp. 187–210. Springer, Heidelberg (2006)

Author Index